T0192455

Lecture Notes in Mathematics 2232

In recent years, the role of mathematics in the life sciences has evolved a long way from the role it played in the 1970's, in the early days of "biomathematics", and is a somewhat different one now, and its perception by the mathematical community is also different. We feel it is important for the Lecture Notes in Mathematics to reflect this and thus underline the immense significance of the life sciences as a field of application and interaction for mathematics in the 21st century.

We are particularly interested in going far beyond the traditional areas in which mathematics was applied to ecology, such as population dynamics, and would like to attract publications in areas such as cell growth, protein structures, physiology, vision, shape recognition & gestalt theory, neural dynamics, genomics, perhaps also some statistical aspects (this list is non-exhaustive).

More information about this series at http://www.springer.com/series/304

Raluca Eftimie

Hyperbolic and Kinetic Models for Self-organised Biological Aggregations

A Modelling and Pattern Formation Approach

 Springer

Raluca Eftimie
Division of Mathematics
University of Dundee
Dundee, UK

ISSN 0075-8434   ISSN 1617-9692   (electronic)
Lecture Notes in Mathematics
ISSN 2524-6771   ISSN 2524-678X   (electronic)
Mathematical Biosciences Subseries
ISBN 978-3-030-02585-4   ISBN 978-3-030-02586-1   (eBook)
https://doi.org/10.1007/978-3-030-02586-1

Library of Congress Control Number: 2018963056

Mathematics Subject Classification (2010): 92-01, 92-02, 92C15, 92D50, 35C07, 35Lxx, 35Q20, 35Q92, 35R09, 35R60, 37G40, 58J55, 65Nxx

This Springer imprint is published by the registered company Springer Nature Switzerland AG
The registered company address is: Gewerbestrasse 11, 6330 Cham, Switzerland

To Georgi

Preface

Self-organised biological aggregations (i.e., aggregations that form in the absence of a leader or external factors) have caught the interest and imagination of scientists for thousands of years. For example, Pliny the Elder, in his book *The Natural History* (published around 77–79 AD) [1], discussed the migration and movement of various flocks of birds, from thrushes and blackbirds to starlings and swallows. About starlings, he noted that "it is a peculiarity of the starling to fly in troops, as it were, and then to wheel round in a globular mass like a ball, the central troop acting as a pivot for the rest". In regard to swallows, Pliny the Elder remarked that they are "the only birds that have a sinuous flight of remarkable velocity" [1]. These comments emphasise humanity's long-term fascination with the spatial and spatio-temporal patterns displayed by various animal aggregations (including flocks of birds).

Over the last 60 years or so, researchers have used mathematical models to identify the biological mechanisms that could explain the formation and structure of these animal group patterns. One of the first studies in this area was published by Breder [2], who used algebraic equations for the repulsive-attractive forces among individuals, to understand "the basic nature of influences at work in a school of fishes as well as in other less compact aggregations". In his pioneering work, Breder proposed the idea that there are attraction and repulsion forces between fish, which vary with the distance between them and those forces are likely mediated by different sensory mechanisms. Breder [2] identified vision to be the main factor involved in the attraction of fish towards each other and suggested that repulsion could be caused not only by vision but also by water movement or sound.

Currently, most of mathematical models for self-organised aggregations assume, one way or another, that the attractive-repulsive interactions are the basic mechanism behind the formation and persistence of biological aggregations. (Here, I define self-organised aggregations as being those biological aggregations that form in the absence of a leader or some external stimulus.) In general, these social interactions are nonlocal, with repulsion acting on short distances and attraction acting on large spatial distances. Thus, many mathematical models for self-organised biological aggregations are nonlocal. Nevertheless, the incorporation into the mathematical models of these social interactions alone cannot explain all complex patterns observed empirically in cell, bacterial and animal aggregations. This led researchers to consider another social interaction: alignment/polarisation. However, alignment behaviours cannot be properly described with the help of parabolic-type partial differential equations (which focus on random movements). A more natural approach for modelling alignment behaviours sees the use of hyperbolic and kinetic transport models.

The purpose of this monograph is to introduce this research area, of mathematical approaches for the investigation of spatial and spatio-temporal patterns displayed by self-organised biological aggregations, to students and researchers not familiar with the topic. To this end, I consider a step-by-step approach to describe various 1D and 2D local hyperbolic and kinetic models (where interactions depend only on the local density of neighbours), as well as nonlocal models (where individuals can perceive, via different sensory mechanisms, their conspecifics positioned further away). I discuss the patterns obtained numerically with these models, as well as other patterns that have been shown to exist or not with the help of analytical methods. For completeness, I also give a brief overview of the most common analytical approaches used to investigate the dynamics of hyperbolic and kinetic models. In addition, I discuss briefly a variety of numerical schemes developed to approximate the solutions of different hyperbolic and kinetic models (mainly related to problems in physics, but which can be considered also in biological contexts).

The complexity and variety of these hyperbolic and kinetic models makes it difficult to include here all types of models existent in the literature (and all analytical and numerical approaches developed to investigate these models). Moreover, because the investigation of collective aggregations and movement is currently one of the most active research areas in mathematical biology, more and more models are developed every month. Since one needs to stop somewhere, I tried to focus on models that either introduced a new idea in terms of modelling self-organised aggregations or used particular analytical and numerical techniques to investigate the formation of patterns. However, there is a feeling that many modelling/analytical/numerical aspects should have been presented in more detail.

Ultimately, I hope that this monograph will offer a first overview into the use of kinetic and hyperbolic models to reproduce and investigate stationary and moving biological aggregations. Moreover, I hope that researchers interested in analytical and numerical approaches for hyperbolic and kinetic models (that have been applied so far mainly to problems in physics and engineering) will become aware of the complexity of phenomena in biology and the numerous open analytical and numerical problems associated with the models for self-organised biological aggregations.

This monograph is the result of multiple research collaborations (over the past 15 years) on various topics related to pattern formation in ecological and biological systems and discussions with colleagues and mentors. I am particularly grateful to my PhD supervisors, Prof. Mark A. Lewis and Prof. Gerda de Vries, as well as to Prof. Frithjof Lutscher (with whom I collaborated at the beginning of my PhD), who introduced me to the use of hyperbolic systems to describe 1D movement in biological/ecological aggregations and guided my first steps in the analytical and numerical investigation of these mathematical models. Prof. Mark Lewis was very supportive to extend a review article on hyperbolic and kinetic models for self-organised biological aggregations (published in 2012 in the *Journal of Mathematical Biology*) into a book. I am also very grateful to Prof. Pietro-Luciano Buono who introduced me to the fascinating field of equivariant bifurcation theory: without our collaboration on classifying the various spatial and spatio-temporal patterns exhibited by 1D nonlocal hyperbolic systems, this monograph would not have come to light. Prof. Thomas Hillen opened my eyes to the theory of hyperbolic conservation laws (through the postgraduate courses he taught at the University of Alberta while I was a PhD student). I must also thank Prof. Jose Carrillo de la Plata and Prof. Nicola Bellomo who introduced me to their research on higher dimensional kinetic equations (which led to our current collaborations). Many more colleagues and collaborators, among which I mention Prof. Razvan Fetecau and Prof. Kees Weijer, have influenced over the past years my research on pattern formation in biological systems, which was the starting point of this monograph.

Finally, I would like to thank the editorial staff at Springer, in particular Dr. Eva Hiripi, for their approachability and help with this book.

Dundee, UK   Raluca Eftimie
October 2018

References

1. P. the Elder, Birds which take their departure from us, and whither they go; The thrush, the blackbird, and the starling – Birds which lose their feathers during their retirement – The turtle-dove and the ring-dove – The flight of starlings and swallows, in *The Natural History* (Taylor and Francis, London, 1855). Translated by John Bostock, M.D., F.R.S., and H.T. Riley, Esq. B.A.
2. C.M. Breder, Ecology **35**(3), 361 (1954)

Contents

Chapter 1
Introduction

1.1 Modelling Self-organised Aggregation and Movement

Aggregation and traffic-like movement are two of the most common collective behaviours observed in animal and human communities, as well as in some cell populations [1–3]. These two behaviours can lead to the formation of a large variety of complex spatial and spatiotemporal group patterns in various organisms. Milling schools of fish [4], stationary aggregations formed by resting animals, zigzagging flocks of birds [5], pedestrian traffic jams [6], bi-directional lanes in ants traffic [7], or rippling waves in Myxobacteria swarms [8] are only a few of the observed patterns. The interest in these aggregative and movement behaviours was triggered by the desire and necessity to understand and control the resulting group patterns. For example, understanding the formation and movement of schools of fish and swarms of insects is important to control insect outbreaks [9], or to establish fishing strategies [10]. Understanding pedestrian movement is useful for improving the architectural design of buildings, train stations, stadiums, or airport terminal to increase their efficiency and safety[11–14]. In cell biology, understanding the collective movement of cells is important in morphogenesis (tissue patterning) and cancer [15, 16].

There are two types of aggregative and movement behaviours displayed by organisms: aggregation and movement in response to external factors (e.g., external chemicals), and aggregation and movement as a result of interactions with conspecifics (self-organised behaviours). In this study we focus only on the self-organised behaviours, and review some classes of mathematical models derived to investigate them. For more general reviews of mathematical models describing biological pattern formation (including patterns in response to chemotaxis) we refer the reader to [17–20].

The mathematical models derived to investigate the self-aggregation patterns observed in nature aim to propose plausible hypotheses regarding the biological

© Springer Nature Switzerland AG 2018
R. Eftimie, *Hyperbolic and Kinetic Models for Self-organised Biological Aggregations*, Lecture Notes in Mathematics 2232,
https://doi.org/10.1007/978-3-030-02586-1_1

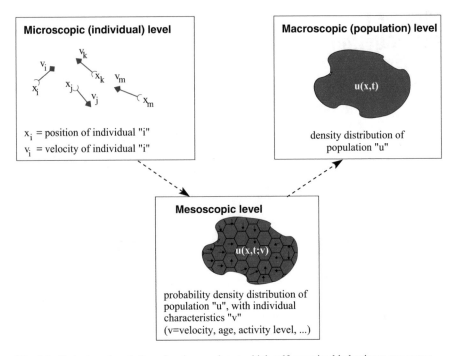

Fig. 1.1 Caricature description of various scales at which self-organised behaviours can occur

mechanisms at the microscale-level (e.g., speed of animals/cells, turning rate of animals/cells, presence/absence of attractive-repulsive social interactions, molecular pathways inside cells that control movement and cell-cell interactions) that can explain the observed group-level behaviours (e.g., the shape, size, and movement of animal or cellular aggregation). In Fig. 1.1 we exemplify the relations between the different scales over which dynamics can occur during self-organised behaviours in animal and cell communities. Note that for collective cell dynamics, various studies interpret differently the micro-scale level. For example, in [21] the authors see the micro-scale level as the level of individual cells, which move as discrete entities with specific positions and velocities (as in Fig. 1.1), while in [22] the authors see the microscale level as the level of molecular processes that take place either inside the cell or on the cell surface (e.g., receptor binding, activation of specific signalling pathways inside cells). To clarify this issue, throughout this study we will refer to the cell-level versus molecular-level dynamics (and note that in some studies the cell-level can be connected to the molecular-level, thus giving rise to mesoscale models).

Based on these different scales of cell/animal dynamics, the mathematical models for self-organised biological aggregations can be classified as follows: *individual-based models* (for microscale-level dynamics), *kinetic models* (for mesoscale-level dynamics) and *macroscopic parabolic and hyperbolic models* (for macroscale-level dynamics):

- *Individual-based models* (or agent-based models): models describing the behaviour of a number of entities (individuals/cells) with specific characteristics (e.g., spatial position, velocity) that are tracked through time. These individual characteristics can change through time in response to some well-defined rules.
- *Kinetic models:* continuum models that describe the evolution of a probability distribution of individuals/cells in a population, in a phase space generally described by the position, time and velocity (orientation) of individuals. In addition to individuals/cells velocity (or orientation), other specific characteristics can be incorporated into these models: from individuals' age and size [23], to the activation level of cells [24], or the concentration of molecules inside cells or on cells' surface [25].
- *Macroscopic parabolic and hyperbolic models:* continuum models that describe the spatial distribution of the whole population density (where all individual characteristics are averaged).

The transitions from microscopic to kinetic and further to macroscopic models involve different technical approximations, which come with analytical or computational drawbacks. In particular, the structure of these different types of models leads to their investigation either from a computational perspective, or from an analytical perspective (with relatively few models being investigated both analytically and numerically):

- *Computational approaches:* the focus is on numerical simulations to reproduce and compare the simulated solution patterns with the available data (to ascertaining whether the assumptions incorporated into the models are enough to explain the experimentally-observed aggregation patterns). This approach is commonly found in individual-based models for aggregations (i.e., models that contain a set of decision rules that govern the movement of individuals, and in particular their spatial position, speed and turning behaviours) [26–33]. In spite of the complex group patterns displayed by the individual-based models (e.g., swarms, tori, polarized groups [26]), the lack of analytical techniques to investigate them causes difficulties in understanding some of these patterns [34]. In this case, numerical simulations are sometimes the only option to investigate the behaviour of the models. We note that this approach is also very common in some nonlinear continuum models for collective behaviours in cells [35, 36], where the complexity of equations makes it difficult to investigate analytically the resulting patterns.
- *Analytical approaches:* the focus is on using existence and uniqueness results to show that the models can exhibit particular types of solutions, such as stationary steady-state aggregations [37]. Homogenisation results are also used to try to connect the dynamics at the micro-scale and macro-scale levels [38]. Such analytical approaches are very common for continuum models (i.e., models that describe the evolution of the density of individuals), and in particular for kinetic models [39–44]. Since the numerical simulations for many of these

kinetic models are too time consuming, the analytical results can offer a quicker understanding of some dynamics of the models.

Throughout this monograph we will discuss the contribution of both approaches to our understanding of pattern formation in biological self-organised aggregations. In particular, we will emphasise the necessity of combining these two approaches to obtain a complete description of the patterns that can be exhibited by the models. Since the majority of individual-based models are investigated mainly numerically (due to a lack of analytical techniques to investigate the existence and structure of patterns obtained numerically, which might impede the discovery of some new possible patterns), we decided to focus this monograph only on kinetic and macroscopic models. However, before we discuss these models, we will briefly present some individual-based models that are the starting point of kinetic and hyperbolic models for aggregation and movement.

1.1.1 Individual-Based Models

There is a vast literature of individual-based models that describe self-organised biological aggregations; see for example [21, 26, 29, 45–59], and the references therein. Generally, these models are suitable to investigate the dynamics of small and medium populations. Very large populations, such as tumour aggregations formed of $10^9 - 10^{12}$ cells, are still difficult to be simulated numerically. The behaviour of these large populations is usually modelled with the help of continuum models. To connect the micro- and macro-scales, many studies start from individual-based models and, in the limit of large numbers, derive equivalent kinetic and then macroscopic models. However, the original individual-based models and the limiting continuum models might not always agree in terms of the patterns they exhibit [51] (since the macroscopic variables are obtained by averaging microscopic variables that might play important roles in specific dynamics). One such example is the existence of double milling patterns (see Fig. 1.2b) displayed by individual-based models [51, 60], and the lack of these patterns in the corresponding continuum models [44]. The validity of continuum models will be discussed in more detail in the following sub-section.

Next, we present some individual-based models that are the starting point of some of the continuum kinetic models that we will investigate in the following chapters. We start with a few individual-based models for homogeneous populations, and conclude with a model for a heterogeneous population.

1. **Vicsek Model.** Vicsek et al. [28] introduced the following model for the movement of N self-propelled particles with constant velocity. Each particle $i = 1..N$ aligns with its neighbours up to a random noise:

$$\mathbf{x}_i(t + \Delta t) = \mathbf{x}_i(t) + \mathbf{v}_i(t)\Delta t, \tag{1.1a}$$

$$\mathbf{v}_i(t + \Delta t) = v_o \begin{pmatrix} \cos(\theta_i(t)) \\ \sin(\theta_i(t)) \end{pmatrix}, \tag{1.1b}$$

$$\theta_i(t + \Delta t) = \langle \theta_i(t) \rangle_r + \Delta\theta. \tag{1.1c}$$

Here, $\mathbf{x}_i(t) \in \mathbb{R}^2$ denotes the position vector of the i-th particle at time $t \in \mathbb{R}^+$, $\mathbf{v}_i(t) \in \mathbb{R}^2$ is the velocity vector of the i-th particle at time t, and $\theta_i(t) \in [-\pi, \pi]$ is the angle defining the direction of the velocity. (Note that throughout this monograph we use the notation $\mathbb{R}^+ = \{x \in \mathbb{R} | x \geq 0\}$.) Moreover, $\langle \theta_i(t) \rangle_r$ denotes the average direction of the velocities of particles positioned within a circle of radius r around the reference particle $\mathbf{x}_i(t)$ (and $\langle \theta_i(t) \rangle_r = \arctan[\langle \sin(\theta_i(t)) \rangle_r / \langle \cos(\theta_i(t)) \rangle_r]$). Also, $v_o \in \mathbb{R}$ is the magnitude of particles velocities. Finally, the term $\Delta\theta$ denotes a random noise (i.e., a random number chosen with a uniform probability from an interval $[-\eta/2, \eta/2])$ [28].

The authors showed that, when varying the noise level η or the number of particles N, this model exhibits transitions from a disordered behaviour (characterised by zero averaged velocity) to an ordered behaviour (characterised by nonzero average velocity), through spontaneous symmetry breaking of the rotational symmetry [28]. To characterise the overall behaviour of the collective of particles, the authors introduced an *order parameter* [28]:

$$\phi = \frac{1}{Nv_o} \left| \sum_{i=1}^{N} \mathbf{v}_i \right|. \tag{1.2}$$

Here ϕ is called an *order parameter*, since it characterises the level of order (or disorder) in a multi-particles system. Changes in this order parameter, as a result of changes in various external model parameters (e.g., magnitude of noise), can give rise to *phase transitions*, where the particles undergo a change in their collective behaviour [61]. As discussed in [61], phase transitions can be:

1. *discontinuous* (or of *the first order*), when the order parameter that measures the group structure undergoes a jump between two different values;
2. *continuous* (or of *the second order*), when the order parameter changes quickly its value, but does not undergo a discontinuous jump.

Model (1.1) was shown to exhibit second order phase transitions [28, 61].

We conclude this discussion on phase transitions, by noting that in [62] the authors mentioned the analogy between phase transitions and bifurcations, emphasizing that "a phase transition is nothing other than a bifurcation in the underlying microscopic dynamics". We will return to this aspect in Chap. 6, in the context of subcritical and supercritical bifurcations for models described by partial differential equations.

2. **Cucker-Smale Model.** Cucker and Smale [53] focused on flocks of birds and considered Vicsek's model (1.1) as a starting point for modelling changes in birds' positions and velocities. However, these authors assumed that each particle (i.e., bird in a flock) adjusts its velocity by adding to it a weighted average of the differences between its velocity and the velocities of its neighbours:

$$\mathbf{x}_i(t + \Delta t) = \mathbf{x}_i(t) + \mathbf{v}_i(t)\Delta t, \tag{1.3a}$$

$$\mathbf{v}_i(t + \Delta t) = \mathbf{v}_i(t) + \frac{\gamma}{N} \sum_{j=1}^{N} a_{ij} \left(\mathbf{v}_j(t) - \mathbf{v}_i(t) \right). \tag{1.3b}$$

As before, $\mathbf{x}_i, \mathbf{v}_i \in \mathbb{R}^n$ (with $n = 3$ in [53]) for each particle $i = 1, \ldots, N$. The weights a_{ij} are assumed to depend on the distance between particles:

$$a_{ij} := \frac{1}{\left(1 + |\mathbf{x}_i - \mathbf{x}_j|^2\right)^{\beta}}. \tag{1.4}$$

Parameter N gives the total number of particles in the system, while parameter γ gives the strength of the interactions. Parameter β describes the rate of decay of the influence between particles, as they move away from each other. The authors then found conditions on the parameter β that ensured that all particles in the aggregation will converge to a common velocity [53]. In particular, for

$$\Gamma(\mathbf{x}(t)) = \frac{1}{2} \sum_{i \neq j} |\mathbf{x}_i - \mathbf{x}_j|^2, \quad \Lambda(\mathbf{v}(t)) = \frac{1}{2} \sum_{i \neq j} |\mathbf{v}_i - \mathbf{v}_j|^2, \tag{1.5}$$

the authors show that there exists a constant B_0 such that $\Gamma(\mathbf{x}(t)) \leq B_0$, for all $t \in \mathbb{R}, t > 0$. Also, $\Lambda(\mathbf{v}(t)) \to 0$ as $t \to \infty$, and $\mathbf{x}_i - \mathbf{x}_j \to \hat{\mathbf{x}}_{ij}$ (where $\hat{\mathbf{x}}_{ij}$ is a limit vector). When $\beta < 1/2$ the convergence to a flock with a common velocity $\hat{v}$ is always guaranteed. However, for $\beta > 1/2$, the authors showed that convergence happens only for some initial conditions for birds positions $(\mathbf{x}_i(0))$ and velocities $(\mathbf{v}_i(0))$.

We note that while the Vicsek model (1.1) includes noise, the original Cucker-Smale model (1.3) does not include any noise. (However, uniform noise could be easily added to the velocity equation, as shown in [63].) Moreover, while the particles in Vicsek's model interact within well-defined spatial ranges (the authors looked at the velocities of the neighbouring particles within a circle of radius r around a reference particle i), the dynamics of the Cucker-Smale model occurs over a very long spatial range (with β controlling the decay of the interactions over this range). As shown in [53], one could slightly modify the Cucker-Smale model in the spirit of Vicsek's model by assuming that the velocity is a function of $a_{ij}(|\mathbf{x}_i - \mathbf{x}_j|^2)$ which acts only on finite distances:

$$\mathbf{v}_i(t + \Delta t) = \mathbf{v_i}(t) + \frac{\gamma}{N} \sum_{j=1}^{N} a_{ij}(|\mathbf{x}_i - \mathbf{x}_j|^2)\left(\mathbf{v}_j(t) - \mathbf{v}_i(t)\right), \qquad (1.6)$$

with

$$a_{ij}(|\mathbf{x}_i - \mathbf{x}_j|^2) = \begin{cases} 1, & \text{if } |\mathbf{x}_i - \mathbf{x}_j|^2 \leq r^2, \\ 0, & \text{otherwise}, \end{cases} \qquad (1.7)$$

where r the interaction distance between individuals. However, in this case the change in the angular velocity is abrupt, which is in contrast with the continuous change in the velocity for the Cucker-Smale model.

The weights a_{ij} are often referred to as "communication rates". However, these rates, which are usually distance-dependent, are different from the communication mechanisms we will refer to in Chap. 5. Haskovec [64] introduced an alternative derivation of these rates in terms of the topological interactions between individuals:

$$a_{ij} = \frac{K}{\left(1 + g_{ij}^2\right)^{\beta}}, \quad \text{with } g_{ij} = \sum_{k=1}^{N} \chi\left(\frac{\mathbf{x}_i - \mathbf{x}_k}{|\mathbf{x}_i - \mathbf{x}_j|}\right), \qquad (1.8)$$

with χ the characteristic function of the open unit ball in $\mathbb{R}^n$. Here, the author assumed that the important quantity is not the metric distance between two particles i and j, but rather how many intermediate particles k separate them.

The discussion on the appropriateness of using metric-distance models (in which individuals interact with all neighbours within a certain distance) versus topological-distance models (in which individuals interact only with a fixed number of neighbours) has been fuelled in the past years by various empirical evidence which, in the end, seems to support both types of model assumptions. Topological interactions can explain the dynamics of starling flocks [33], while metric-distance interactions (given by the interaction zones) can explain the dynamics of fish shoals [30] or flocking surf scoters [31]. We will return to this discussion in Chap. 5, in the context of communication mechanisms.

3. **D'Orsogna Model**. D'Orsogna et al. [60] and Chuang et al. [51] developed an individual-based model which assumed that the velocities of particles depend on an attractive-repulsive potential:

$$\frac{d\mathbf{x}_i}{dt} = \mathbf{v}_i, \qquad (1.9a)$$

$$m_i \frac{d\mathbf{v}_i}{dt} = (\alpha - \beta|\mathbf{v}_i|^2)\mathbf{v}_i - \frac{1}{N}\sum_{i \neq j} \nabla U(|\mathbf{x}_i - \mathbf{x}_j|), \quad 1 \leq i \leq N. \qquad (1.9b)$$

with

$$U(|\mathbf{x}_i - \mathbf{x}_j|) = C_r e^{-|\mathbf{x}_i - \mathbf{x}_j|/l_r} - C_a e^{-|\mathbf{x}_i - \mathbf{x}_j|/l_a}. \tag{1.10}$$

As before, $\mathbf{x}_i \in \mathbb{R}^n$ and $\mathbf{v}_i \in \mathbb{R}^n$ describe the position and velocity of particle i at time $t \in \mathbb{R}^+$, $m_i \in \mathbb{R}^+$ is the mass of this particle, $\alpha > 0$ describes the self-propulsion strength, while $\beta > 0$ is the friction. In the expression of the potential function U, C_r and C_a represent the magnitudes of the repulsive and attractive interactions, while l_r and l_a represent the potential ranges for the repulsive and attractive social interactions. In [60], the authors focus on the *H-stability* property of the system, which ensures that the system preserves inter-particle space and does not collapse as $N \to \infty$. (Note that a system of N particles interacting via a potential U is called *H-stable* if there is a constant $B > 0$ such that $U \geq -NB$. If this property does not hold, then the system is called *catastrophic*, since the aggregations collapse to their core.) The authors then go on and investigate the formation of H-stable and catastrophic patterns, such as clumps (i.e., stationary pulses), rings (i.e., vortices or mills; single mills and double mills), coherent flocks (i.e., travelling pulses) or rigid-body rotation states (particles with fixed positions that rotate around the swarm centre), in the parameter space determined by the ratios l_r/l_a and C_r/C_a. For a visual description of these patterns see Fig. 1.2, while for their definitions see Table 1.2.

Haskovec [64] adapted the model (1.9) to incorporate topological-metric interactions in the equation for the velocity:

$$m_i \frac{d\mathbf{v}_i}{dt} = (\alpha - \beta|\mathbf{v}_i|^2)\mathbf{v}_i - \frac{1}{N} \sum_{i \neq j} \nabla U_r(|\mathbf{x}_i - \mathbf{x}_j|)$$

$$- \frac{1}{N} \sum_{i \neq j} a_{ij} \frac{\mathbf{x}_i - \mathbf{x}_j}{|\mathbf{x}_i - \mathbf{x}_j|}, \quad 1 \leq i \leq N. \tag{1.11}$$

where U_r is the repulsive component of the potential U, and the attractive force depends on the relative separation a_{ij} of the i-th and j-th individuals/particles. The introduction of this particular attractive interaction has been shown to lead to chaotic movement [64].

To conclude the discussion of these models with attractive-repulsive velocities, we note that phase transitions between different types of patterns have been observed also in model (1.9), as the ratios l_r/l_a and C_r/C_a are varied. Moreover, in this class of models the transitions between certain patterns (e.g., from single to double mills, or from rigid-body rotations to single mills) can be either gradual or abrupt processes (see Fig. 8(b) in [51]), which is in contrast with the phase transitions observed in Vicsek's model [51].

4. **Generalised Langevin Model.** Chavanis and Sire [65] started with a generalised Langevin (individual-based) model, and further used it to derive a class of general kinetic and hydrodynamic/hyperbolic models for chemotactic aggregations.

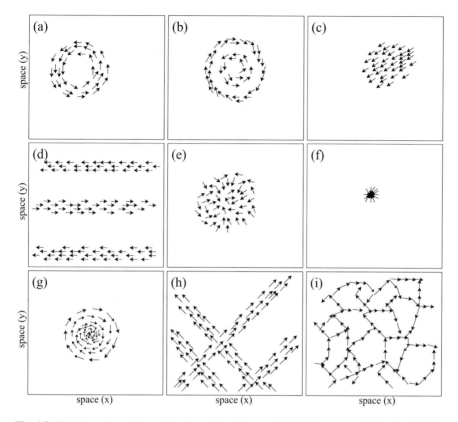

Fig. 1.2 Caricature description of various 2D patterns exhibited by one-population individual-based models: (**a**) Single rings (mills); (**b**) Double rings (mills) formed of clockwise and counterclockwise rotating rings; (**c**) Travelling groups (or pulses); (**d**) Lane patterns; (**e**) Swarm patterns; (**f**) Collapsed swarm patterns; (**g**) Rigid body rotation; (**h**) Ripples; (**i**) Network pattern (this is an example of splitting-merging behaviours in 2D)

Since we will discuss these kinetic and hydrodynamic models in Chap. 6 in the context of stochastic models for collective movement, here we briefly introduce the generalised Langevin model for the collective movement of N bacteria in response to a chemical $c(\mathbf{x}, t)$:

$$\frac{d\mathbf{x}_i}{dt} = \mathbf{v}_i, \quad i = 1, \ldots N, \tag{1.12a}$$

$$\frac{d\mathbf{v}_i}{dt} = -\xi \mathbf{v}_i + \nabla c + \sqrt{2D} R_i(t), \tag{1.12b}$$

$$\frac{\partial c}{\partial t} = -kc + D_c \Delta c + h \sum_{i=1}^{N} \delta(\mathbf{x} - \mathbf{x}_i(t)). \tag{1.12c}$$

As before, $\mathbf{x}_i, \mathbf{v}_i \in \mathbb{R}^n$, $\mathbf{x} \in \mathbb{R}^n$ and $t \in \mathbb{R}^+$. The first term in Eq. (1.12b) is
a friction force (with ξ the friction coefficient of bacteria moving on the fixed
substrate). The second term in (1.12b) is a chemotactic attraction force due to the
chemical c, while the third term in (1.12b) is a random force (where $R_i(t)$ is a
white noise and D is a diffusion coefficient). Finally, the terms in Eq. (1.12c)
describe the degradation of the chemical c at a rate k, the diffusion of this
chemical with diffusion coefficient D_c, and the production of the chemical at
a rate h by the cells themselves (i.e., at every spatial position of these cells).
Numerical simulations with this model have shown the formation of a network
pattern with filamentary structure; see also Fig. 1.2.

To conclude the discussion on these models, we note that the attractive effect
of the chemical could be replaced by an attractive-repulsive potential U [66]:

$$\frac{d\mathbf{x}_i}{dt} = \mathbf{v}_i, \quad i = 1, \ldots N, \tag{1.13a}$$

$$\frac{d\mathbf{v}_i}{dt} = -\xi \mathbf{v}_i - \frac{1}{m} \nabla U(\mathbf{x}_1, \ldots \mathbf{x}_N) + \sqrt{2D} R_i(t), \tag{1.13b}$$

with

$$U(\mathbf{x}_1, \ldots, \mathbf{x}_N) = m^2 \sum_{i<j} U_{LR}(|\mathbf{x}_i - \mathbf{x}_j|) + m^2 \sum_{i<j} U_{SR}(|\mathbf{x}_i - \mathbf{x}_j|)$$

$$+ m \sum_i U_{ext}(\mathbf{x}_i), \tag{1.14}$$

where m is the mass of particles/cells/bacteria..., U_{LR} is a long-range binary
potential, U_{SR} a short-range binary potential, and U_{ext} an external potential [66].

5. **Zmurchok–de Vries Model.** A final individual-based model that we review
   here in the context of homogeneous populations was recently introduced by
   Zmurchok and de Vries [67] to compare pattern formation with a 1D nonlocal
   two-speed kinetic/hyperbolic models previously introduced in [68, 69] (which
   will be discussed in detail in Chap. 5). The changes in the position ($x_i \in \mathbb{R}$)
   and direction (v_i) of particles/cells/individuals are described by the following
   equations in 1D:

$$x_i(t + \Delta t) = x_i(t) + \Gamma_i^\pm(y_i^\pm) v_i(t + \Delta t) \Delta t, \quad i = 1, \ldots, N, \tag{1.15a}$$

$$v_i(t + \Delta t) = \begin{cases} -v_i(t), & \text{if } \lambda_i^\pm \Delta t \geq X, \\ v_i(t), & \text{otherwise,} \end{cases} \tag{1.15b}$$

with $x_i \in \mathbb{R}$, $t \in \mathbb{R}^+$, X a uniformly distributed random variable
on $[0, 1]$ (which is updated at every time step), $\lambda_i^\pm \Delta t$ the probability
of turning during the time step Δt, and $\Gamma_i(y_i^\pm)$ is the travel speed.
Here, $\lambda_i^\pm = \lambda_1 + \lambda_2 f(y_{al,i}^\pm)$, $\Gamma_i^\pm(y_i^\pm) = \gamma g(y_{a,i}^\pm - y_{r,i}^\pm)$, with $y_{r,a,al;i}^\pm$

describing the repulsive(r)/attractive(a)/alignment(al) social interactions for each particle/cell/individual i within the respective interaction zones $Z_{r;i}^{L,R}$, $Z_{a;i}^{L,R}$ and $Z_{al;i}^{L,R}$ (see also Fig. 1.3). Note that the $\pm$ signs describe the direction of the reference individual, while the superscripts L, R describe the interaction zone to the *Left* or to the *Right* of this individual i. These social interaction forces are defined the same way as the interaction forces for the hyperbolic models in [68, 69]. For example, the attractive force experienced by the ith

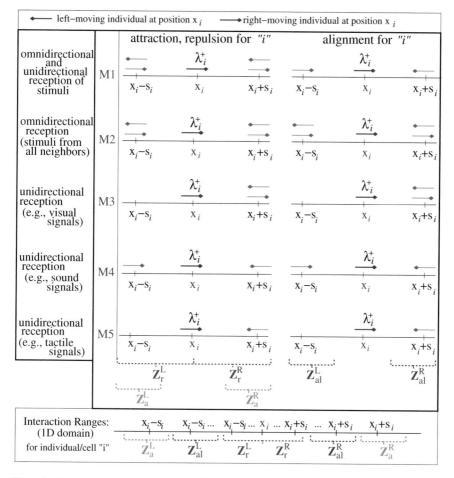

Fig. 1.3 Caricature description of five different directional 1D communication mechanisms introduced in [69], which were adapted in [67] to describe the perception of neighbours by a reference (left-moving/right-moving) individual positioned at x_i. The turning rates $\lambda_i^{\pm}$ depend on the perception of neighbours within the repulsion zone ($Z_r^{L,R}$), alignment zone ($Z_{al}^{L,R}$), or attraction zone ($Z_a^{L,R}$)

particle/cell/individual is

$$y_{a;i}^{\pm} = \pm q_a \sum_{j \in Z_{a;i}^R} K_a(|x_i - x_j|) \mp q_a \sum_{j \in Z_{a;i}^L} K_a(|x_i - x_j|), \qquad (1.16)$$

with $K_a(\cdot)$ an interaction kernel describing the width of the attractive interaction zone (e.g., a Gaussian kernel in [67], or any of the kernels summarised in Fig. 5.2). Similar interaction forces are derived for repulsion and alignment. Following the approach in [68, 69] (for hyperbolic/kinetic models), the authors in [67] have defined different types of nonlocal communication mechanisms, which influence the way social interactions terms $y_i^{\pm}$ affect the turning rates $\lambda_i^{\pm}$; see Fig. 1.3. A summary of the different y_i terms, corresponding to the communication mechanisms in Fig. 1.3 is shown in Table 1.1. (We will discuss the importance of inter-individual communication on collective behaviours in Sect. 1.2.)

In terms of pattern formation, Zmurchok and de Vries [67] have shown that these nonlocal individual-based models can reproduce the majority of the patterns

Table 1.1 The nonlocal terms used to describe the social interactions in [67]

Model	Social interactions: repulsion ($y_{r,i}^{\pm}$), attraction ($y_{a,i}^{\pm}$), alignment ($y_{al,i}^{\pm}$)
M1	$y_{r,a;i}^{\pm} = \pm q_{r,a} \sum_{j \in Z_{r,a;i}^R} K_{r,a}(d_{ij}) \mp q_{r,a} \sum_{j \in Z_{r,a;i}^L} K_{r,a}(d_{ij})$
	$y_{al;i}^{\pm} = \pm q_{al} \sum_{j \in Z_{al;i}^R, \, v_j < 0} K_{al}(d_{ij}) \mp q_{al} \sum_{j \in Z_{al;i}^L, \, v_j > 0} K_{al}(d_{ij})$
M2	$y_{r,a;i}^{\pm} = \pm q_{r,a} \sum_{j \in Z_{r,a;i}^R} K_{r,a}(d_{ij}) \mp q_{r,a} \sum_{j \in Z_{r,a;i}^L} K_{r,a}(d_{ij})$
	$y_{al;i}^{\pm} = \pm q_{al} \sum_{j \in Z_{al;i}^R, \, v_j < 0} K_{al}(d_{ij}) \pm q_{al} \sum_{j \in Z_{al;i}^L, \, v_j < 0} K_{al}(d_{ij})$
	$\mp q_{al} \sum_{j \in Z_{al;i}^R, \, v_j > 0} K_{al}(d_{ij}) \mp q_{al} \sum_{j \in Z_{al;i}^L} K_{al}(d_{ij})$
M3	$y_{r,a;i}^{+} = q_{r,a} \sum_{j \in Z_{r,a;i}^R} K_{r,a}(d_{ij}), \quad y_{r,a;i}^{-} = q_{r,a} \sum_{j \in Z_{r,a;i}^L} K_{r,a}(d_{ij})$
	$y_{al;i}^{+} = q_{al} \sum_{j \in Z_{al;i}^R, \, v_j < 0} K_{al}(d_{ij}) - q_{al} \sum_{j \in Z_{al;i}^R, \, v_j > 0} K_{al}(d_{ij}),$
	$y_{al;i}^{-} = q_{al} \sum_{j \in Z_{al;i}^L, \, v_j > 0} K_{al}(d_{ij}) - q_{al} \sum_{j \in Z_{al;i}^L, \, v_j < 0} K_{al}(d_{ij}),$
M4	$y_{r,a;i}^{\pm} = \pm q_{r,a} \sum_{j \in Z_{r,a;i}^R, \, v_j < 0} K_{r,a}(d_{ij}) \mp q_{r,a} \sum_{j \in Z_{r,a;i}^L, \, v_j > 0} K_{r,a}(d_{ij})$
	$y_{al;i}^{\pm} = \pm q_{al} \sum_{j \in Z_{al;i}^R, \, v_j < 0} K_{al}(d_{ij}) \mp q_{al} \sum_{j \in Z_{al;i}^L, \, v_j > 0} K_{al}(d_{ij})$
M5	$y_{r,a;i}^{+} = q_{r,a} \sum_{j \in Z_{r,a;i}^R, \, v_j < 0} K_{r,a}(d_{ij}), \quad y_{r,a;i}^{-} = q_{r,a;i} \sum_{j \in Z_{al;i}^L, \, v_j > 0} K_{r,a}(d_{ij})$
	$y_{al;i}^{+} = q_{al} \sum_{j \in Z_{al;i}^R, \, v_j < 0} K_{al}(d_{ij}), \quad y_{al;i}^{-} = q_{al} \sum_{j \in Z_{al;i}^L, \, v_j > 0} K_{al}(d_{ij})$

These terms are the translations of the diagrams from Fig. 1.3 into mathematical equations, after summing up the information received from all neighbours ($s_i \in (0, \infty)$). For each of the five models, the equations for $y_{r,i}^{\pm}$ and $y_{a,i}^{\pm}$ are almost identical (since the effect of the repulsive and attractive interactions is similar, but leads to opposite outcomes). Here q_a, q_r, and q_{al} describe the strength of the attraction, repulsion, and alignment interactions between the individual at x_i at its neighbours at x_j, where $d_{ij} = |x_i - x_j|$. Kernels $K_a(d_{ij})$, $K_r(d_{ij})$ and $K_{al}(d_{ij})$ give the width of the attractive ($Z_a^{R,L}$), repulsive ($Z_r^{R,L}$) and alignment ($Z_{al}^{R,L}$) interaction zones. Since the direction of the neighbours j is important for social interactions via communication, the sums distinguish between individuals with $v_j > 0$ and $v_j < 0$

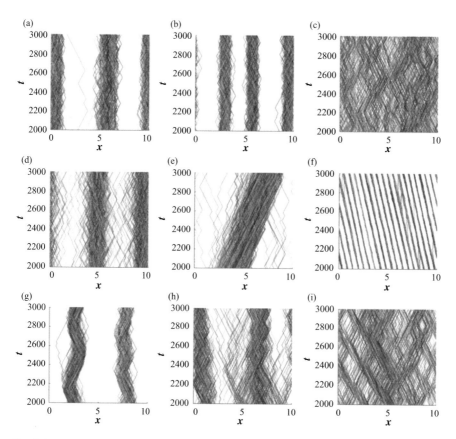

Fig. 1.4 Examples of spatio-temporal patterns exhibited by the 1D individual-based models (1.15). Here $N = 500$ individuals that randomly scattered over a 1D domain $[0, L]$, with $L = 10$ and periodic boundary conditions. Patterns reprinted from [67], under the Creative Commons Attribution License. (**a**) Stationary pulse 1. (**b**) Stationary pulse 2. (**c**) Ripples. (**d**) Feathers. (**e**) Travelling pulse. (**f**) Travelling train. (**g**) Zigzag pulse. (**h**) Breathers. (**i**) Travelling breather. Copyright ©2018 Zmurchok, de Vries

obtained in [69] with a hyperbolic model incorporating similar communication mechanisms: from stationary pulses, to travelling pulses, travelling trains, ripples, breathers, travelling breathers, feathers, zigzags, and splitting and merging patterns. These patterns are summarised in Fig. 1.4. In Chap. 5 we will introduce a kinetic/hyperbolic 1D model that is the mesoscale version of this microscale model, and show how numerical simulations combined with bifurcation and symmetry theory can help us understand the formation of some of the patterns, as well as the transitions between certain patterns. Here, we emphasise that no investigation of the phase transitions displayed by this individual-based model has been performed yet.

Even if the previous individual-based models focused on homogeneous populations, in nature the cell/bacterial/animal populations are not homogeneous.
Therefore, some recent studies have developed individual-based models that investigate the interactions between different populations [70–72]. In the following we
briefly discuss such an individual-based model, since the patterns generated by this
model (see also Fig. 1.5) are generic for the interactions between two populations

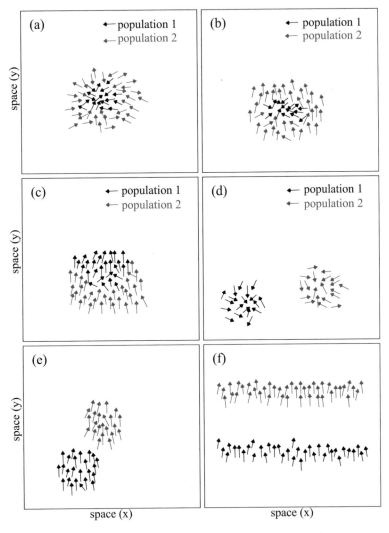

Fig. 1.5 Caricature description of the patterns that could be generated by two-population
individual-based models. (**a**) stationary aggregations with complete engulfment; (**b**) moving
aggregations (as given by the direction of the arrows) with complete engulfment; (**c**) moving aggregations with partial engulfment; (**d**) stationary segregated aggregations; (**e**) moving segregating
aggregations; (**f**) segregated line formation

(as displayed by both discrete and continuum models; see also the discussion in Sect. 6.10).

- **Pineda & Eftimie Model.** Pineda and Eftimie [72] introduced an individual-based model to describe the self-organised dynamics of two cell populations which interact not only through spatially-dependent repulsive social forces, but also through a chemical that can be secreted by the cells themselves or can be secreted externally (which leads to an externally imposed stable chemical gradient). Thus, assuming that a population of N cells is formed of two subpopulations N_A and N_B of cells with different speeds (v_i) and different chemotactic sensitivities (χ_i), then the equations for the evolution of the two populations (characterised by position $\mathbf{x}_i = (x_i, y_i) \in \mathbb{R}^2$ and velocity $\mathbf{v}_i \in \mathbb{R}^2$) are as follows:

$$\mathbf{x}_i(t + \Delta t) = \mathbf{x}_i(t) + \mathbf{v}_i(t)\Delta t, \quad i = i_A + i_B, \quad i_{A,B} = 1..N_{A,B}, \quad (1.17)$$

with

$$\mathbf{v}_i(t) = v_i \mathbf{e}_i(t) = v_i \begin{pmatrix} \cos \theta_i(t) \\ \sin \theta_i(t) \end{pmatrix}, \quad \theta_i \in [-\pi, \pi], \quad (1.18)$$

and

$$\theta_i(t + \Delta t) = \arg \left\{ \chi_i \frac{\nabla c_T(t)}{||\nabla c_T(t)||} + \sum_{j=1, i \neq j}^{N} m_{ij} \mathbf{F}_{ij}(t) + \alpha_i \boldsymbol{\sigma}_i(t) \right\} \quad (1.19)$$

It was assumed in [72] that the concentration c_T of a chemical signal is secreted by two sources: a local chemical signal c_I secreted by the cells themselves (at every position $\mathbf{x}_i$ of cells), and an externally imposed chemical gradient c_E (changing only along the y direction). Therefore, $\nabla c_T = \nabla c_I + \nabla c_E$, with

$$\nabla c_I(\mathbf{x}_i) = -p \sum_{j=1, j \neq i}^{N} K_I(||\mathbf{x}_i - \mathbf{x}_j||/l)\mathbf{x}_{ij},$$

$$\nabla c_E(y) = -c_m \sqrt{\frac{\mu_E}{D_E}} \frac{e^{-\sqrt{\mu_E D_E} y} + e^{\sqrt{\mu_E D_E} y}}{e^{-\sqrt{\mu_E D_E} y} + e^{\sqrt{\mu_E D_E} y}} \mathbf{y}. \quad (1.20)$$

In the equation for the gradient of c_I, K_I is a modified Bessel function of the second kind, $l = \sqrt{D_I \mu_I}$ (with D_I= diffusion rate of chemical c_I, μ_I= degradation rate of chemical c_I), p measures the response of the cells to the chemical gradient, and $\mathbf{x}_{ij}$ is a unitary vector directed from j to i. In the equation for the gradient of c_E (obtained after solving a simple diffusion-reaction equation for the degradation of a chemical), c_m gives the constant concentration of the chemical c_E at the upper boundary of the domain (in the y-direction), D_E is the diffusion rate of the chemical c_E, μ_E is the degradation rate of this chemical, and **y** is a unitary vector pointing in the y-direction.

The second term in Eq. (1.19) describes the repulsive interactions between cells i and j: $\mathbf{F}_{ij} = f_{ij}\mathbf{x}_{ij}$, with

$$f_{ij} = \begin{cases} 0, & \text{if } x_{ij}(t) > 2r_0, \\ 1 - \frac{x_{ij}}{2r_0}, & \text{if } 0 \leq x_{ij}(t) \leq 2r_0, \end{cases} \tag{1.21}$$

where $x_{ij}(t) = ||\mathbf{x}_i(t) - \mathbf{x}_j(t)||$ is the distance between two cells, and $2r_0$ is the interaction distance between two cells. The parameter m_{ij} (from Eq. (1.19)) controls the magnitude of the repulsive cell-cell interactions.

Finally, the last term in Eq. (1.19) describes the noise affecting cells orientation: α_i is the noise intensity, while $\boldsymbol{\sigma}_i = (\cos \xi_i(t), \sin \xi_i(t))$ is a random uniformly oriented unitary vector.

Numerical simulations with this individual-based model (where individual cell dynamics was coupled with continuous chemical gradients) have shown: (1) moving aggregations with complete engulfment of one population (with stronger chemotactic sensitivity) by the other population (as in Fig. 1.5b); (2) moving aggregations with partial engulfment of one population by the other population (as in Fig. 1.5c); (3) segregated moving cell aggregations (as in Fig. 1.5e); (4) lane formation, with the two lanes—corresponding to the two populations—moving towards the external chemical gradient (as in Fig. 1.5f). We emphasise that while all aggregation patterns in [72] are moving because cells are attracted to the external chemical gradient c_E, it is possible to obtain also stationary aggregations (as those summarised in Fig. 1.5) for different parameter values. We will return to these patterns in Chap. 6, in the context of patterns generated by two-population kinetic models.

The models in [65] and [72] could be seen as connecting micro-scale dynamics (as represented by the movement of individual cells/bacteria) to macro-scale dynamics (as represented by the time-evolution of the diffusive chemicals). Another approach that connects the micro- and macro-scale models is based on the derivation of microscopic models on a lattice with cells of length Δx. The models (sometimes called "master equations") describe the probability $p(x)$ of a particle/pedestrian, initially positioned in cell $x \in \mathbb{R}$, to move right (to $x + \Delta x$) or left (to $x - \Delta x$). (In 2D the particles can move also up and down.) In the limit $\Delta x \to 0$, this microscopic model becomes a macroscopic model for the density $u(x)$ of particles/pedestrians at x (where x is now a continuous space variable). This approach has been considered for the derivation of some local and nonlocal hyperbolic systems with constant speed [73–75], or for the derivation of local hyperbolic models with density-dependent speed [76]. We will discuss these models in more detail in Chaps. 4 and 5.

Remark 1.1 While many individual-based models report results that are visually similar to experimentally-observed collective migration (e.g., see [59, 77]), a quantitative comparison between these numerical results and experimental data is a very tedious process. In this case, one needs to perform a large number of simulations to obtain appropriate statistics (suitable for comparison with data statistics). For this reason, the majority of individual-based models in the literature ignore the quantitative comparison step.

1.1.2 Kinetic and Macroscopic Models

The kinetic (mesoscale) models describe phenomena occurring at intermediate scales, where one can investigate the behaviour of the whole population as it is influenced by individual characteristics (e.g., individual's velocity, age, size, etc.). The macroscale models can be derived from the kinetic models by taking an average over the characteristic variable. This way, the description of the statistics for the characteristic variable (e.g., velocity) in the kinetic models is reduced in continuous models to only a small number of its moments [78]. For both mesoscale and macroscale models, the spatial scale of interest is at least one order of magnitude larger than the diameter of a cell (or the size of an individual/particle/etc.).

Initially, the macroscale models for animal movement and aggregation were mainly of parabolic type (see [73, 79–86], and the references therein). These models are suitable when one is interested in measuring the population as a whole (i.e., mean square displacements, mean population drift) [87]. However, since the parabolic models unrealistically assume an infinite speed of propagation (i.e., even if the initial data has compact support, perturbations in this data can be perceived everywhere in the domain [88]), scientists started to focus more and more on hyperbolic and kinetic models [69, 73, 87, 89–94]. The use of these hyperbolic and kinetic models is helped by the recent advances in experimental techniques, which allow scientists to track the movement of individual particles/cells/animals (i.e., individual speeds and turning angle distribution) [95, 96]. When this kind of information exists, the hyperbolic and kinetic models become the natural choice of continuum models [87]. Another reason for using these two types of models is the possibility of incorporating into variables detailed descriptions of various biological functions and activities (e.g., different animal communication mechanisms [69], or different degrees of cell differentiation or cell activation—information that is characteristic to lower molecular scales [24]). Moreover, directed movement that leads to alignment/polarisation is easier incorporated into these transport models.

In principle, the use of kinetic and hyperbolic models (as well as parabolic models) requires having a large population of interacting organisms [44, 97, 98]. As emphasised in [97], mesoscale/macroscale models for biological movement and pattern formation usually incorporate the hidden assumption that statistical correlations between cells/particles can be neglected, which means that the predictions of these models might fail in those spatial regions where cell/particle numbers are small and there are strong long-range correlations in cell/particle movement. This calls for caution when applying such models to smaller populations of bacteria, cells, or animals, since it could lead to different group dynamics (as observed by Chuang et al. [51] in a swarming model which shows two different group patterns exhibited by the continuum model and the individual-based model when the number of particles is small—characterising an H-stable regime). However, these mesoscale/macroscale models are still applied occasionally to investigate the dynamics of relatively small populations (e.g., bacteria [65], cells [99]). One of the reasons for this is that they can provide some qualitative and quantitative relationships between model parameters. Thus, they can give a better insight into the underlying biological and physical principles governing the investigated problem, compared to many of the individuals-based models. To ensure that the continuum models offer a good depiction of the dynamics displayed by relatively small populations, one can derive them from individual-based models [43, 51, 65]. This approach requires, however, to compare the two models in terms of their dynamics (and to check for inconsistencies, and discuss parameter ranges of model validity) [51].

We need to emphasise that while all these multi-scale studies emphasise that the continuum models are obtained from the discrete models as the number of particles/cells N becomes very large (and thus the continuum model should be a valid approximation of the discrete model for large N), it is still unclear what one means by "small" or "large" populations of particles/cells. In [98] the author suggests that interacting bacteria or cells can be found in communities less than a few thousand particles, which implies that statistical fluctuations will be important in these communities and thus the continuum limit might not be appropriate (although one way of overcoming this problem is to use stochastic kinetic equations [98]). Moreover, in [97] the author suggests that macroscopic models should not be used to describe communities formed of a few hundreds of cells (or dynamics at the edges of cell colonies). However, in [100] a continuum version of a discrete model for self-organised movement of particles showed similar coherent moving flock solutions when the discrete model was formed of $N = 200$ particles, and similar densities of rotating vortices when the discrete model was formed of $N = 400$ particles. An even more extreme situation is presented in [99], where the solutions of an individual-based model for epithelial cell monolayers agree very well with the solutions of the corresponding continuum limit model, even for $N = 20..32$ cells.

The hyperbolic and kinetic models (as well as the parabolic ones) can be local, when immediate neighbours or local environmental effects are important [90, 101, 102], or nonlocal, when distant individuals or nonlocal environmental effects play an important role [69, 103–105]. The nonlocal models focus on the effects of three social interactions (repulsion, attraction, and alignment) on the movement of organisms. These social interactions can influence the individuals turning behaviour (e.g., organisms turn to align with their neighbours, or to move away from them [68, 69, 103, 104], or turn to align with some external feature such as a pheromone trail [106]), and/or influence their speeding behaviour (i.e., organisms speed-up or slow-down to move away from neighbours or towards them; [104, 107]). In contrast, for local models the turning and speeding behaviours are influenced only by local conspecifics [73, 101, 102, 108, 109].

Another aspect considered by these hyperbolic and kinetic models for self-organised aggregations is the possibility of having demographic effects, in addition to movement in response to local or nonlocal conspecifics [73, 91, 110–112]. These demographic effects usually occur on a timescale faster than or equal to the movement timescale. In general, they are modelled by quadratic (logistic) terms [73, 113] or cubic terms [111, 114], although competition and prey-predation effects could also be easily incorporated.

The majority of these models were derived to investigate cell and animal populations (see, for example, [69, 90, 102, 103, 115, 116]). However, a few models have also been used to describe pedestrian self-organised movement [94]. Note that these particular models were initially introduced in the context of car traffic [12, 117, 118], and later were used to describe pedestrian traffic [118, 119]. In regards to the mathematical models for pedestrian movement, it is worth mentioning that some of these models have been developed in parallel in applied mathematics/physics/engineering (see the reviews in [120–122]) and in social sciences (sociology and spatial geography) [123]. Moreover, recent studies have emphasised the necessity of considering also social psychology when investigating the collective dynamics of pedestrians [124].

In the last decade, there has been an explosion of applications of kinetic models to social modelling and opinion formation, as well as to economic modelling and financial markets [125–127]. This was mainly the result of an increase in the awareness that collective behaviours can lead to unexpected outcomes, very hard to predict. And the kinetic models are the most obvious choice when investigating how the dynamics of social and economic systems are influenced by individual behaviours (especially since the trading between agents in an economy has similarities with particles collisions as described by ideal gas dynamics).

The kinetic models for self-organised aggregations are generally based on the Boltzmann equation, which describes the evolution of particles/cells/individuals as point masses that interact via pair collisions, or on the Vlasov equation, which describes the interactions between particles as mean field actions of the field particles over the test particle [128]. The research into the kinetic equations that are used to describe biological aggregations is particularly rich, different mathematical schools contributing to it. Some of the research groups focusing on these types of models are in France [38, 41, 129–131], Italy [24, 91, 119, 126, 132–134], Germany/Swizerland/Austria [12, 37, 39], Spain [43, 44] and North America [42, 87, 135–139]. Historically, these groups focused on slightly different aspects of these aggregations. For example, the German school focused on traffic models (as part of early studies to improve highway traffic) [12, 140, 141]. The French school focused more on theoretical aspects related to kinetic models for biological aggregations: existence of solutions, limiting behaviour and homogenisation [38, 41, 131]. The Italian school, on the other hand focused mainly on applications of these models to medicine and social sciences: immunology and cell-cell interactions [91, 142, 143], or human behaviours [119, 125, 126]. However, due to the high mobility of researchers between different countries, this classification based on national schools is not appropriate anymore (e.g., research groups in the USA stared focusing also on traffic models [144] and models for biological aggregations [145]). In consequence, throughout this monograph we will ignore this classification, and discuss the various types of kinetic models existent in the literature based on their approach: a more analytical approach or a more numerical approach. This strategy will allow us to identify the analytical and numerical challenges posed by the kinetic and continuum hyperbolic models derived to investigate biological aggregations.

1.2   The Importance of Communication for Self-organised Biological Behaviours

The main difference between modelling interacting particles in a physical context (as modelled by the majority of kinetic models developed in the last few decades; see [132, 146–148]) and modelling interacting animals/cells in a biological context is that these animals/cells communicate directly with each other, and actively modify each other's behaviours. In fact, the survival of cells and animals depends on receiving and processing different information from the environment and their neighbours (regarding food availability, potential dangers, etc.). Living organisms emit/perceive information to/from conspecifics via different communication mechanisms: cell signalling pathways in cells, or visual, auditory, chemical and tactile interactions in animals (see Fig. 1.6). Therefore, through communication (which can be local or non-local, depending on the spatial location of communicating neighbours) animals/cells continuously adapt their own behaviours to the behaviours of surrounding individuals/cells. This leads to coordinated emergent behaviours at the level of the whole communities, which sometimes can be difficult to predict. (We emphasise here that physical particles can also interact with each other through gravitational, electric, magnetic forces, as well as through fluid around them, but they don't actively adapt their behaviours in response to neighbouring particles.)

Since self-organised behaviours are the result of how animals and cells communicate with each other, one of the particular aspects that we will focus on in this study is the incorporation of inter-individual communication mechanisms and inter-cell signalling pathways into the mathematical models for movement and aggregation.

The first mathematical models for self-organised collective dynamics in biology assumed that animal communication was omnidirectional, with individuals perceiving all neighbours around them, or only some neighbours within specific distances (i.e., within the repulsive, attractive or alignment interaction ranges); see Fig. 1.7a. Then, to make the models more realistic, some researchers proposed the existence of a blind-zone behind individuals, in which neighbours are undetectable [26]; see Fig. 1.7b. Eftimie et al. [69] took this approach one step further, and incorporated various communication mechanisms in a one-dimensional model for self-organised dynamics. These communication mechanisms are combinations of uni-directional and multi-directional perception and emission of information from/to neighbours positioned at different distances (see also Chap. 5). As we have seen in Sect. 1.1.1, this approach that focuses on multiple communication mechanisms was recently transposed also to individual-based models [67]. A different type of communication

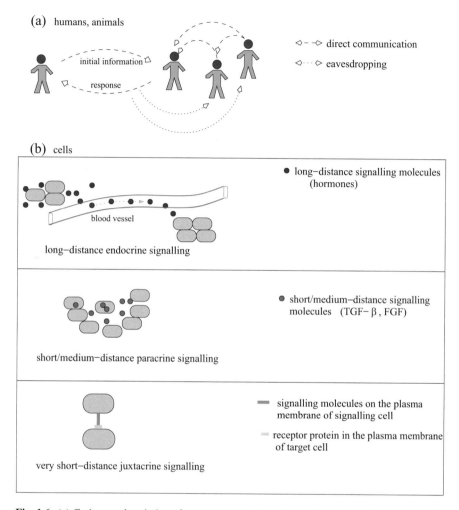

Fig. 1.6 (**a**) Caricature description of communication-based interactions among humans/animals. Individuals can communicate directly with their neighbours, but they can also eavesdrop on their neighbours' communication. (**b**) Caricature description of cell-cell communication via signalling molecules produced by the cells. These molecules can act on the cells (juxtacrine signalling), in the neighbourhood of the cell (paracrine signalling), or can travel long distances via blood vessels (endocrine signalling)

incorporated into the mathematical models occurs via chemotactic signals, where individuals in the community secrete chemicals that are perceived by (and modify the movement behaviour of) their neighbours. One of the most common examples is offered by the trail pheromone secreted by ants, which acts both as a recruitment and a orientation signal [7]. We should also mention that individuals in large communities could use different communication mechanisms at the same time (e.g., visual vs. auditory signals), depending on their physiological and psychological con-

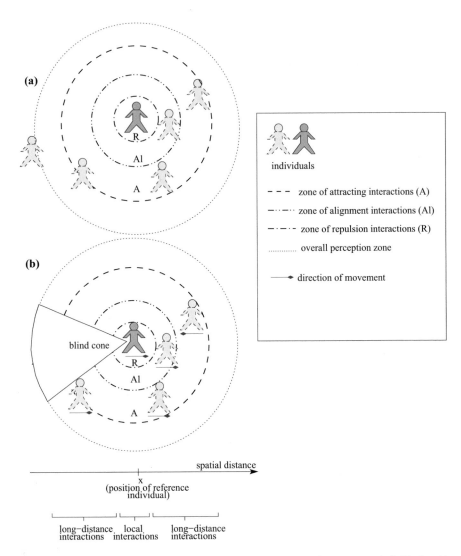

Fig. 1.7 Caricature description of communication-based interactions among individuals. (**a**) Omnidirectional perception of all neighbours. (**b**) Directional perception of neighbours, under the assumption that behind each individual (relative to its movement direction) there is a blind cone. The basic interactions could take place within the repulsion range (where individuals try to avoid each others), within the alignment/polarisation zone (where individuals try to align with each other), or within the attraction range (where individuals are attracted towards each other)

straints. This heterogeneity in inter-individual communication adds another layer of complexity in understanding the behavioural patterns of the whole community; see also Fig. 1.8.

Therefore, to understand the mechanisms involved in the formation and movement of self-organised aggregations, it is not enough to model changes in animals'

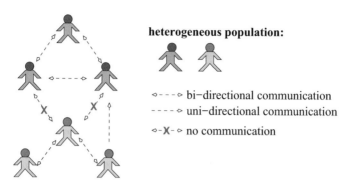

Fig. 1.8 Caricature description of a heterogeneous population formed of individuals that communicate with each others in various ways. Some individuals can avoid communication with a different sub-population. Moreover, communication can be uni-directional or omni-directional

velocity or turning rates (as many of current models in the literature do). One has to add another level of detail by studying how inter-individual communication (which can be local or nonlocal; see Fig. 1.7) affects these changes in velocity and turning behaviour.

1.3 Overview

In this monograph we will not describe in detail the fundamental analytical and numerical results obtained for these hyperbolic and kinetic models (e.g., existence results, hydrodynamic limit, parabolic limit, bifurcation results, numerical scheme used, etc.). Although we will review all these results in Chaps. 2, 7 and 8, our focus will be on the modelling of various biological aggregations and on the resulting spatial and spatio-temporal patterns. Classical existence results for hyperbolic models can be found in [118, 149–151], while classical analytical results for kinetic Boltzmann equations can be found in [148, 152–155]. LeVeque [156] reviewed the numerical methods for hyperbolic equations and systems, while numerical methods for kinetic equations can be found, for example, in [153, 157–160]. Bifurcation theory techniques for PDEs can be found, for example, in [161–164].

Some hyperbolic and kinetic models have been previously reviewed in the context of chemosensitive movement [93, 115, 116], tumour dynamics [165], collective behaviours in socio-economic sciences [125] or more generally, in the context of correlated and uncorrelated random walks [166] and local reaction transport systems [109]. A detailed investigation of car traffic models on road networks was considered in [167]. Traffic and crowd models were recently reviewed from a modelling perspective [168–170] and from an analytical perspective [118, 170]. Transport and kinetic models have been discussed in the context of their applications to car traffic [171], as well as their application to gases, semiconductors, photons, and some biological systems [128]. A more thorough discussion on the use of kinetic models to describe various biological phenomena—with a particular focus on cancer immunology—can be found in [172], while a discussion of kinetic models with applications to socio-economic systems can be found in [125, 126]. Computational approaches for kinetic models are discussed, for example, in [133, 134].

While all these reviews cover different aspects of kinetic and hyperbolic models (mainly related to modelling and analytical aspects, but also some numerical results), there is no study yet to combine all these aspects in a coherent manner in the context of pattern formation in biological phenomena. Moreover, there are no studies to focus on nonlocal models (the majority of studies mentioned above focus exclusively on local dynamics). In general, the biological systems are more complex than the physical systems, and thus the kinetic models that are applied to biological phenomena require a more careful interpretation of the assumptions. Moreover, the non-locality of biological phenomena adds another layer of complexity to the modelling aspect, as well as to the investigation of these models in terms of both numerical and analytical approaches. We emphasise that while there are studies on the analytical approaches for pattern formation (via bifurcation and symmetry theory; see [163, 164, 173]), there are no studies that focus on the application of these approaches to the investigation of pattern formation in *nonlocal hyperbolic and kinetic* models.

The purpose of this study is threefold:

1. to motivate the use of local and nonlocal hyperbolic and kinetic models for biological aggregations (and how directional communication can be incorporated into these models), since they allow for finite (and directional) propagation of information—in contrast to the more commonly used parabolic models;
2. to summarise the patterns exhibited by these complex models (which were derived to describe the complexity of self-organised biological behaviours, including nonlocal behaviours);
3. to present briefly the analytical approaches (i.e., stability results, bifurcation and symmetry results) and numerical approaches (i.e., various numerical schemes) that have been used (or could be used) to identify and classify the patterns exhibited by these models.

To achieve these goals, we use a structured approach. We start by presenting the simplest advection-reaction equations used to describe biological behaviours (as well as car traffic and pedestrian traffic behaviours). Then, we add complexity in steps by considering systems of local and nonlocal hyperbolic equations that have constant or density-dependent speeds and turning rates. We examine these aspects in the context of one-dimensional and multi-dimensional spatial models. We conclude by presenting some basic notions of stability and (equivariant) bifurcation theory, as well as presenting briefly some numerical approaches for the numerical approximation of these hyperbolic and kinetic models. This step-by-step approach allows us to elaborate on the benefits of incorporating detailed social interactions or population dynamics into the models for self-organised movement. We examine these benefits from the perspective of understanding various types of group patterns. A summary and description of the spatial and spatio-temporal patterns discussed throughout this review is shown in Table 1.2. In addition, to allow the reader to have a visual understanding of these patterns from the very beginning (especially since some of the studies presented in this monograph show analytically the existence of particular types of solutions, without any numerical simulations), in Figs. 1.9 and 1.10 we present some caricature descriptions in the $x - t$ plane of the 1D patterns listed in Table 1.2 (While many of these patterns are also observed in Fig. 1.4, we decided to present a caricature description of all these 1D patterns to emphasise their spatio-temporal structure—which is not always clear in Fig. 1.4 due to the stochasticity of the model and the simulations). In Fig. 1.11 we present some caricature descriptions in the $x - y$ plane of the 2D patterns listed in Table 1.2.

Table 1.2 Finite-amplitude and blow-up heterogeneous patterns exhibited by hyperbolic and kinetic models describing aggregation and movement of organisms in 1D

Finite amplitude solutions	Description	Figures	References
Stationary pulses	Spatially nonhomogeneous steady states u that are motionless in time (i.e., $\frac{\partial u}{\partial t} = 0$, but $\frac{\partial u}{\partial x} \neq 0$). Usually $u(\pm\infty) = 0$.	Figs. 1.9a, 1.4a, b, 5.7a, 6.2a, 5.26a(i), b(i), 5.13b, 5.12d, 5.9b	[40, 67, 69, 90, 100, 102, 104, 174]
Travelling pulses	Spatially non-homogeneous solutions u that have a fixed shape and move at a constant speed c: $u(x, t) = U(z)$, $z = x - ct$, and $U(\pm\infty) = 0$. With this definition, we can understand the stationary pulses as being travelling pulses that move with zero speed $(c = 0)$.	Figs. 1.9b, 1.4e, 5.7b, 6.2b, 6.4b, 5.26a(iii), b(iii), 5.20b, 5.19a–c, 5.13a, c, c', 5.12d, 5.11c, 5.9a	[67, 69, 101, 104]
Travelling fronts	Spatially non-homogeneous solutions u that have a fixed shape (connecting two steady states u_1 and u_2) and move at a constant speed c: $u(x, t) = U(z)$, $z = x - ct$. As $z \to \pm\infty$, the strictly monotone profile $U(z)$ connects the two steady states: $U(-\infty) = u_1$ and $U(+\infty) = u_2$.	Figs. 1.9c, 3.6b	[73, 110–113]
Travelling trains	Periodic solutions of the form $u(x, t) = U(z)$, $z = x - ct$, with U a periodic function of z.	Figs. 1.9d, 1.4f, 5.7c	[67, 69]
Ripples (standing waves)	Left-moving and right-moving travelling waves that pass through each other.	Figs. 1.9e, 1.2h, 1.4c, 5.7d, 5.12d, 5.11c	[67, 69, 90]

(continued)

Table 1.2 (continued)

Finite amplitude solutions	Description	Figures	References
Modulated standing waves	Space- and time-modulated left-moving and right-moving travelling waves that pass through each other.	Figs. 1.9f, 5.12d, 5.11c	[175]
Feathers	A particular type of stationary pulse that periodically loses density at the edges of the group.	Figs. 1.9g, 1.4d, 5.7g, 5.13a"	[67, 69]
Travelling feathers	A type of moving aggregation that periodically loses density from the back (relative to its moving direction)	Figs. 5.7k, 1.10h	[107]
Breather	Pulses that periodically expand and contract. This leads to a periodic change in the amplitude of solutions. Note that the center of mass of these pulses is always stationary.	Figs. 1.9h, 1.4h, 5.7h	[67, 69]
Travelling breathers (or modulated rotating waves)	Breather-like solutions that travel through the domain. In this case, the centre of mass of the pulse is travelling at a certain speed.	Figs. 1.9i, 1.4i, 5.7i, 5.20c, 5.12d, 5.11c	[69]
Zigzag pulses	Travelling pulses that periodically change direction. (These are 1D patterns. The 2D equivalent is represented by vortices.)	Figs. 1.10a, 5.7e, 5.26a(ii), b(ii), 5.17, 5.13b', b", c"	[67, 69]
Semi-zigzag pulses (or stop-and-go waves)	Aggregation (pulse) patterns characterised by movement in one direction (i.e., "go"), alternated by rest (i.e., "stop").	Figs. 1.10b, 5.7f	[69]
Travelling zigzags	Zigzag pulses that travel through the domain, as one run (left/right) is longer than the other run (right/left)	Figs. 5.7l, 1.10c	[107, 176]

(continued)

Table 1.2 (continued)

Finite amplitude solutions	Description	Figures	References
Amplitude blow-up (finite time)	The density $u(x, t)$ concentrates in one single point x, and becomes infinite as time t approaches a critical value $t = T < \infty$: $\lim_{t \to T} \|u(x, t)\|_\infty = \infty$. Usually, the point x is fixed, and the pattern is a "stationary blow-up pattern". However, there are cases where the aggregation moves through space as it approaches the singularity (located at (x^*, T)): $\lim_{t \to T, x \to x^*} \|u(x, t)\|_\infty = \infty$. These patterns are referred to as "moving blow-up patterns". In models describing the dynamics of left-moving (u^-) and right-moving (u^+) organisms, the stationary and moving blow-ups could happen at different spatial points: $x^* = x_1^*$ (for u^+) and $x^* = x_2^*$ (for u^-) [104]. Some models also exhibit two-point blow-ups: $\lim_{t \to T} \|u(x_1^*, t)\|_\infty = \lim_{t \to T} \|u(x_2^*, t)\|_\infty = \infty$ [177]. In 2D, it is possible to have blow-up patterns along lines [178].	Fig. 1.10c	[102, 104, 177–180]
Gradient blow-up (shocks)	The density u is bounded, but its gradient $\nabla_x u$ becomes infinite at a time point $t = T < \infty$. Note that, in some cases, the blow-up of the gradient can cause an infinite growth in the amplitude of solutions. These gradient blow-up solutions lead to the formation of shocks.	Figs. 1.10d, 3.6a, 2.4c	[12, 94]
Rarefaction waves	A type of wave that expands with time (in contrast with the shock waves that are compressed with time).	Figs. 1.10a, 2.4a	
Chaotic patterns	Patterns that are aperiodic in time.	Figs. 1.10e, 5.13, 5.23	[67, 181]
Translated homogeneous solutions	Spatially homogeneous solutions in 1D or 2D that are aligned in a preferred direction	Figs. 1.11c, 6.2c	[40]

(continued)

Table 1.2 (continued)

Finite amplitude solutions	Description	Figures	References
Lanes	Patterns (mainly in 2D) characterised by the formation of pulses that are spread in one direction over the entire domain. These groups can be formed of motionless individuals, or of individuals moving in opposite directions.	Figs. 1.11a, 6.4c	[51, 94]
Networks	Patterns (in 2D) characterised by the splitting and merging of lane-like aggregations (e.g., ant trail networks)	Figs. 1.2i, 1.11f	
Single mills	A 2D pattern where individuals display a circular-type of motion. All individuals move in the same direction (either clockwise or counterclockwise).	Figs. 6.2, 1.11b	[44, 51]
Double mills	A 2D pattern where individuals display a circular-type of motion. Some individuals move in clockwise direction, while others move in counter-clockwise direction.	Fig. 6.2	[44]
Rigid-body rotation	A 2D pattern where individuals display a circular-type of motion around the swarm centre. Unlike the single/double mills where individuals move freely within the rotating aggregation, here the individuals are at relatively fixed positions inside the rotating aggregation.	Figs. 1.2g, 6.2c	[51, 182]

Here, $u(x, t)$ describes the population density at (x, t). Some of these patterns (e.g., the zigzags) can only be observed in 1D, while other patterns (e.g., single and double mills) can only be observed in 2D. However, other patterns can be observed in both 1D and 2D (e.g. the stationary and moving pulses)

We need to stress that here we do not attempt to review all relevant hyperbolic and kinetic models. Given the intense research in this area, such an attempt would be an impossible task. Rather, we try to present some approaches taken to model and investigate group patterns that arise in self-organised cell, animal and human communities. Moreover, our focus here is on recent research on nonlocal hyperbolic and kinetic models (in contrast to previous publications in the literature that focus mainly on local models). To help our exposition, we also present briefly the analytical and numerical approaches commonly used to investigate these kinetic

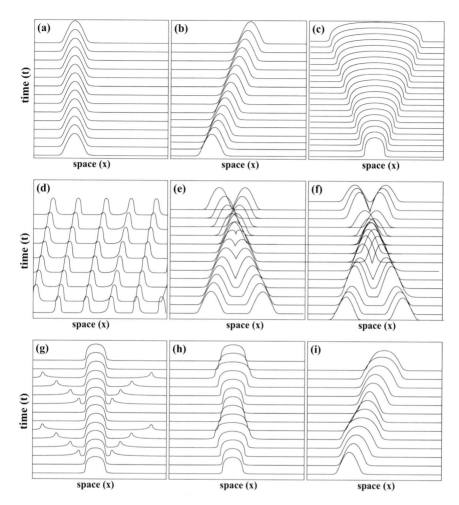

Fig. 1.9 Caricature description of the 1D patterns listed in Table 1.2. (**a**) Stationary pulses; (**b**) Travelling pulses; (**c**) Travelling fronts (left-moving and right-moving, with respect to the middle of the domain); (**d**) Travelling trains; (**e**) Ripples (or standing waves); (**f**) Modulated standing waves (or modulated ripples); (**g**) Feathers; (**h**) Breathers; (**i**) Travelling breathers

and hyperbolic models in terms of pattern formation (although these approaches are not always applied to every study).

The monograph is organised as follows. We start in Chap. 2 by introducing the terminology behind the hyperbolic equations for conservation laws. In Chap. 3 we discuss various one-equation hyperbolic models that describe population movement and traffic flows. In Chap. 4, we introduce systems of hyperbolic equations (in one spatial dimension). Here, the focus is on local hyperbolic systems. Then, in Chap. 5, we extend the discussion to nonlocal hyperbolic models which describe behaviours generated by long-distance interactions with neighbours. In Chap. 6, we discuss

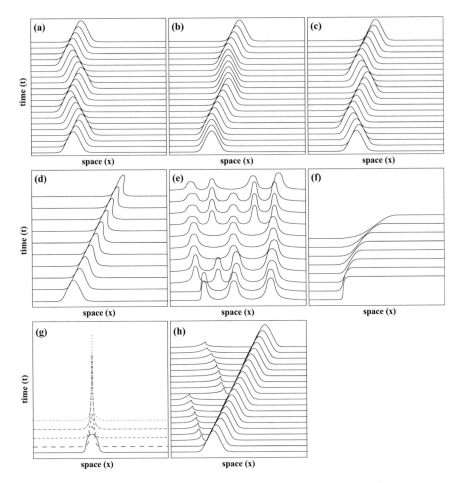

Fig. 1.10 Caricature description of the 1D patterns listed in Table 1.2 (continued from Fig. 1.9). (**a**) Zigzag pulses; (**b**) Semi-zigzag pulses/Stop-and-Go waves; (**c**) Travelling zigzags; (**d**) Gradient blow-up (shocks); (**e**) Chaotic patterns; (**f**) Rarefaction waves; (**g**) Amplitude blow-up (finite time); (**h**) Travelling feathers

generalisations of these models to higher spatial dimensions using the kinetic theory of transport processes. In Chap. 7 we present various numerical approaches used to simulate the solutions of hyperbolic and kinetic models. In Chap. 8 we present some basic techniques of bifurcation theory used to investigate some of the patterns displayed by these models. We conclude in Chap. 9 with a general discussion on the mathematical and biological relevance of the results presented in this monograph. While three chapters (i.e., Chaps. 2, 7 and 8) are dedicated exclusively to the analytical and numerical approaches used to investigate these models, each of the other chapters that review and summarise various hyperbolic/kinetic models in 1D and 2D also contains one or multiple sub-sections *"Analytical approaches*

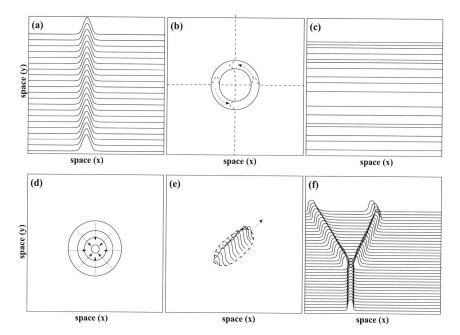

Fig. 1.11 Caricature description of the 2D patterns listed in Table 1.2. (**a**) Lanes; (**b**) Mills (vortices); (**c**) Translated homogeneous solutions; (**d**) Stationary 2D pulses (aggregations); (**e**) Travelling 2D pulses (aggregations); (**f**) Network patterns

for the investigation of patterns" that detail some analytical results (arbitrarily chosen to cover the variety of the approaches that can be used to investigate these kinetic/macroscopic models). Since transport (car/pedestrian) models in 1D or 2D, and kinetic 2D Boltzmann-type models have been reviewed in more detail in many other studies published in the literature [3, 24, 148, 152, 153, 157, 183–189], here we pay particular attention to nonlocal 1D kinetic/hyperbolic models, which have been investigated more intensely in terms of pattern formation over the last few years, but which are not as well known as the previously mentioned models.

The monograph can be used for higher undergraduate and graduate courses in mathematical modelling and pattern formation in self-organised biological aggregations, as the reviews of various models and studies presented in each of Chaps. 3–6 are complemented by detailed discussions of specific mathematical techniques applied to this class of hyperbolic/kinetic models. The combined modelling, analytical and numerical perspectives of hyperbolic and kinetic models introduced in this monograph, allow for tailoring the material to different courses focused either on modelling & numerical approaches, or on modelling & analytical approaches. Moreover, the monograph can be used for developing new research directions, by focusing on the open problems highlighted throughout the book. Finally, the material presented here could also be used for reading seminars for higher undergraduate or graduate students.

To follow the topics discussed in this review, the reader should have some basic mathematical knowledge of partial differential equations, functional analysis and numerical approaches to partial differential equations. However, no prior knowledge of modelling is assumed, and therefore, some aspects related to modelling and basic investigations of patterns are presented in more detail.

References

1. J.K. Parrish, L.E. Keshet, Science **284**, 99 (1999)
2. T. Deisboeck, M. Berens, A. Kansal, S. Torquato, Cell Prolif. **34**, 115 (2001)
3. D. Chowdhury, A. Schadschneider, N. Katsuhiro, Phys. Life Rev. **2**(4), 318 (2005)
4. K. Tunstrøm, Y. Katz, C. Ioannou, C. Huepe, M. Lutz, PLoS Comput. Biol. **9**, e1002915 (2013)
5. H. Weimerskirch, F. Bonadonna, F. Bailleul, G. Mabille, G. Dell'Omo, H.P. Lipp, Science **295**, 1259 (2002)
6. M. Muramatsu, T. Nagatani, Phys. A Stat. Mech. Appl. **275**(1–2), 281 (2000)
7. V. Fourcassié, A. Dussutour, J.L. Deneubourg, J. Exp. Biol. **213**, 2357 (2010)
8. O.A. Igoshin, R. Welch, D. Kaiser, G. Oster, Proc. Natl. Acad. Sci. USA **101**, 4256 (2004)
9. S.J. Simpson, A.R. McCaffery, B.F. Hägele, Biol. Rev. **74**, 461 (1999)
10. J.K. Parrish, Environ. Biol. Fish. **55**, 157 (1999)
11. J. Pauls, Fire Technol. **20**, 27 (1984)
12. D. Helbing, Rev. Mod. Phys. **73**, 1067 (2001)
13. D. Helbing, L. Buzna, A. Johansson, T. Wener, Transp. Sci. **39**(1), 1 (2005)
14. N. Shiwakoti, M. Sarvi, Transp. Res. C Emerg. Technol. **37**, 260 (2013)
15. P. Friedl, Y. Hegerfeldt, M. Tusch, Int. J. Dev. Biol. **48**, 441 (2004)
16. P. Røth, Ann. Rev. Cell Dev. Biol. **25**, 407 (2009)
17. P. Maini, H. Othmer (eds.), *Mathematical Models for Biological Pattern Formation* (Springer, Berlin, 2001)
18. L. Edelstein-Keshet, *Mathematical Models in Biology* (SIAM, Philadelphia, 2005)
19. J.D. Murray, *Mathematical Biology* (Springer, Berlin, 1989)
20. V. Capasso, M. Gromov, A. Hareol-Bellan, N. Morozova, L.L. Pritchard (eds.), *Pattern Formation in Morphogenesis* (Springer, Berlin, 2013)
21. C. Guven, E. Rericha, E. Ott, W. Losert, PLoS Comput. Biol. **9**(5), e1003041 (2013)
22. D. Cohen, L. Martignetti, S. Robine, E. Barillot, A. Zinovyev, L. Calzone, PLoS Comput. Biol. **11**(11), e1004571 (2015)
23. S. Chapman, M. Plank, A. James, B. Base, ANZIAM J. **49**(2), 151 (2007)
24. N. Bellomo, M. Delitala, Phys. Life Rev. **5**, 183 (2008)
25. C. Engwer, T. Hillen, M. Knappitsch, C. Surulescu, J. Math. Biol. **71**, 551 (2015)
26. I.D. Couzin, J. Krause, R. James, G. Ruxton, N.R. Franks, J. Theor. Biol. **218**, 1 (2002)
27. K. Warburton, J. Lazarus, J. Theor. Biol. **150**, 473 (1991)
28. T. Vicsek, A. Czirók, E. Ben-Jacob, I. Cohen, O. Shochet, Phys. Rev. Lett. **75**(6), 1226 (1995)
29. A. Huth, C. Wissel, J. Theor. Biol. **156**, 365 (1992)
30. J. Tien, S. levin, D. Rubenstein, Evol. Ecol. **6**, 555 (2004)
31. R. Lukeman, Y.X. Li, L. Edelstein-Keshet, Proc. Natl. Acad. Sci. **107**(28), 12576 (2010)
32. U. Börner, A. Deutsch, M. Bär, Phys. Biol. **3**, 138 (2006)
33. M. Ballerini, N. Cabibbo, R. Candelier, A. Cavagna, E. Cisbani, I. Giardina, V. Lecomte, A. Orlandi, G. Parisi, A. Procaccini, M. Viale, V. Zdravkovic, Proc. Natl. Acad. Sci. USA **105**(5), 1232 (2008)
34. R. Eftimie, G. de Vries, M. Lewis, J. Math. Biol. **59**, 37 (2009)
35. D. Trucu, P. Lin, M. Chaplain, Y. Wang, Multiscale Model Simul. **11**(1), 309 (2013)

36. P. Domschke, D. Trucu, A. Gerisch, M. Chaplain, J. Theor. Biol. **361**, 41 (2014)
37. K. Fellner, G. Raoul, Math. Models Methods Appl. Sci. **20**, 2267 (2010)
38. P. Degond, S. Motsch, Math. Models Methods Appl. Sci. **20**, 1459 (2008)
39. K. Fellner, G. Raoul, Math. Comput. Model. **53**, 1436 (2011)
40. R. Fetecau, Math. Models Methods Appl. Sci. **21**(7), 1539 (2011)
41. P. Degond, A. Frouvelle, J.G. Liu, J. Nonlinear Sci. **23**, 427 (2013)
42. J. von Brecht, D. Uminsky, T. Kolokolnikov, A. Bertozzi, Math. Models Methods Appl. Sci. **22**(1), 1140002 (2012)
43. J. Carrillo, M. Fornasier, J. Rosado, G. Toscani, SIAM J. Math. Anal. **42**, 218 (2010)
44. J. Carrillo, M. D'Orsogna, V. Panferov, Kinet. Relat. Models **2**, 363 (2009)
45. H. Reuter, B. Breckling, Ecol. Model. **75–76**, 147 (1994)
46. C.W. Reynolds, Comput. Graph. **21**, 25 (1987)
47. R. Vabø, L. Nøttestad, Fish. Oceanogr. **6**, 155 (1997)
48. U. Börner, A. Deutsch, H. Reichenbach, M. Bär, Phys. Rev. Lett. **89**, 078101 (2002)
49. J. Buhl, D.J.T. Sumpter, I.D. Couzin, J.J. Hale, E. Despland, E.R. Miller, S.J. Simpson, Science **312**, 1402 (2006)
50. H. Chaté, F. Ginelli, G. Grégoire, Phys. Rev. Lett. **99**, 229601 (2007)
51. Y.L. Chuang, M. D'Orsogna, D. Marthaler, A. Bertozzi, L. Chayes, Phys. D **232**, 33 (2007)
52. I.D. Couzin, J. Krause, Adv. Study Behav. **32**, 1 (2003)
53. F. Cucker, S. Smale, IEEE Trans. Autom. Control **52**(5), 852 (2007)
54. A. Czirók, A.L. Barabási, T. Vicsek, Phys. Rev. Lett. **82**(1), 209 (1999)
55. V. Gazi, K.M. Passino, *Proceedings of the American Control Conference*, Anchorage, AK (2002), pp. 8–10
56. G. Grégoire, H. Chaté, Phys. Rev. Lett. **92**(2), 025702 (2004)
57. S. Gueron, S.A. Levin, D.I. Rubenstein, J. Theor. Biol. **182**, 85 (1996)
58. C.K. Hemelrijk, H. Kunz, Behav. Ecol. **16**(1), 178 (2004)
59. D. Stichel, A. Middleton, B. Müller, U. Klingmüller, K. Breuhahn, F. Matthäus, NPJ Syst. Biol. Appl. **3**, 5 (2017)
60. M. D'Orsogna, Y. Chuang, A. Bertozzi, L. Chayes, Phys. Rev. Lett. **96**(10), 104302 (2006)
61. T. Vicsek, A. Zafeiris, Phys. Rep. **517**(3–4), 71 (2010)
62. A. Lesne, M. Laguës (eds.), *Scale Invariance: From Phase Transitions to Turbulence* (Springer, Berlin, 2012)
63. S.Y. Ha, K. Lee, D. Levy, Commun. Math. Sci. **7**(2), 453 (2009)
64. J. Haskovec, Phys. D **261**, 42 (2013)
65. P.H. Chavanis, C. Sire, Phys. A Stat. Mech. Appl. **384**, 199 (2007)
66. P.H. Chavanis, Phys. A Stat. Mech. Appl. **390**(9), 1546 (2011)
67. C. Zmurchok, G. de Vries, PLoS One **13**(6), e0198550 (2018)
68. R. Eftimie, G. de Vries, M.A. Lewis, F. Lutscher, Bull. Math. Biol. **69**(5), 1537 (2007)
69. R. Eftimie, G. de Vries, M.A. Lewis, Proc. Natl. Acad. Sci. USA **104**(17), 6974 (2007)
70. S. Moon, B. Nabet, N. Leonard, S. Levin, I. Kevrekidis, J. Theor. Biol. **246**, 100 (2007)
71. D. Garcia, L. Brunnet, S.D. Monte, PLoS Comput. Biol. **10**(2), e1003482 (2014)
72. M. Pineda, R. Eftimie, Phys. Biol. **14**, 066003 (2017)
73. E.E. Holmes, Am. Nat. **142**, 779 (1993)
74. T. Hillen, J. Math. Biol. **35**, 49 (1996)
75. R. Eftimie, J. Math. Biol. **65**(1), 35 (2012)
76. A. Chertock, A. Kurganov, A. Polizzi, I. Timofeyev, Math. Models Methods Appl. Sci. **81**, 1947 (2003)
77. C. Carmona-Fontaine, E. Theveneau, A. Tzekou, M. Tada, M. Woods, K. Page, M. Parsons, J. Lambris, R. Mayor, Dev. Cell **21**, 1026 (2011)
78. P. Degond, C. Appert-Rolland, M. Moussaïd, J. Pettré, G. Theraulaz, J. Stat. Phys. **152**, 1033 (2013)
79. J. Skellam, Biometrika **38**(1–2), 196 (1951)
80. E. Keller, L. Segel, J. Theor. Biol. **26**, 399 (1970)
81. M. Burger, V. Capasso, D. Morale, Nonlinear Anal. Real World Appl. **8**, 939 (2007)

82. L. Edelstein-Keshet, J. Watmough, D. Grünbaum, J. Math. Biol. **36**(6), 515 (1998)
83. A. Mogilner, L. Edelstein-Keshet, J. Math. Biol. **38**, 534 (1999)
84. A. Mogilner, L. Edelstein-Keshet, J. Math. Biol. **33**, 619 (1995)
85. M.A. Lewis, Theor. Popul. Biol. **45**, 277 (1994)
86. D. Grünbaum, J. Math. Biol. **38**, 169 (1999)
87. H.G. Othmer, S.R. Dunbar, W. Alt, J. Math. Biol. **26**, 263 (1988)
88. N. Bellomo, B. Lods, R. Revelli, L. Ridolfi, *Generalised Collocation Methods: Solutions to Nonlinear Problems* (Birkhäuser, Boston, 2008)
89. S. Goldstein, Quart. J. Mech. Appl. Math. **4**, 129 (1951)
90. F. Lutscher, A. Stevens, J. Nonlinear Sci. **12**, 619 (2002)
91. N. Bellomo, A. Bellouquid, J. Nieto, J. Soler, Math. Mod. Meth. Appl. Sci. **17**, 1675 (2007)
92. F. Filbet, P. Laurencot, B. Perthame, J. Math. Biol. **50**(2), 189 (2005)
93. T. Hillen, Math. Models Methods Appl. Sci. **12**(7), 1 (2002)
94. D. Helbing, Complex Syst. **6**, 391 (1992)
95. T. Yang, J.S. Park, Y. Choi, W. Choi, T.W. Ko, K. Lee, PLoS One **6**(6), e20255 (2011)
96. J. Killeen, H. Thurfjell, S. Ciuti, D. Paton, M. Musiani, M. Boyce, Mov Ecol. **2**(1), 15 (2014)
97. R. Grima, Curr. Top. Dev. Biol. **81**, 435 (2008)
98. P.H. Chavanis, Commun. Nonlinear. Sci. Numer. Simul. **15**, 60 (2010)
99. J. Fozard, H. Byrne, O. Jensen, J. King, Math. Med. Biol. **27**, 39 (2010)
100. H. Levine, W.J. Rappel, I. Cohen, Phys. Rev. E **63**, 017101 (2000)
101. F. Lutscher, Eur. J. Appl. Math. **14**, 291 (2003)
102. T. Hillen, A. Stevens, Nonlinear Anal. Real World Appl. **1**, 409 (2000)
103. B. Pfistner, in *Biological Motion*, ed. by W. Alt, G. Hoffmann. Lecture Notes on Biomathematics, vol. 89 (Springer, Berlin, 1990), pp. 556–563
104. R. Fetecau, R. Eftimie, J. Math. Biol. **61**(4), 545 (2010)
105. M. Lécureux-Mercier, ESAIM: Proc. **38**, 409 (2012)
106. E. Boissard, P. Degond, S. Motsch, J. Math. Biol. **66**(6), 1267 (2013)
107. R. Eftimie, Modelling group formation and activity patterns in self-organising communities of organisms. Ph.D. Thesis, University of Alberta, 2008
108. R. Mickens, SIAM Rev. **30**(4), 629 (1988)
109. K. Hadeler, Reaction transport systems in biological modelling, in *Mathematics Inspired by Biology*. Lecture Notes in Mathematics (Springer, Berlin, 1999), pp. 95–150
110. K. Hadeler, Nonlinear propagation in reaction transport systems, in *Differential Equations with Applications to Biology* (Fields Institute Communications, American Mathematical Society, Providence, 1998), pp. 251–257
111. K. Hadeler, Math. Comput. Model. **31**(4–5), 75 (2000)
112. H. Schwetlick, Ann. Inst. Henri Poincare **17**(4), 523 (2000)
113. K. Lika, T. Hallam, J. Math. Biol. **38**, 346 (1999)
114. K. Hadeler, Proc. Edinburgh Math. Soc. **31**, 89 (1988)
115. T. Hillen, K. Hadeler, Hyperbolic systems and transport equations in mathematical biology, in *Analysis and Numerics for Conservation Laws* (Springer, Berlin, 2005), pp. 257–279
116. B. Perthame, Appl. Math. **49**(6), 539 (2004)
117. M. Lighthill, G. Whitham, Proc. R. Soc. Lond. Ser. A **229**(1178), 317 (1955)
118. M. Rosini, *Macroscopic Models for Vehicular Flows and Crowd Dynamics: Theory and Applications* (Springer, Berlin, 2013)
119. F. Venuti, L. Bruno, N. Bellomo, Math. Comput. Model. **45**(3–4), 252 (2007)
120. D. Helbing, P. Monar, I. Farkas, K. Bolay, Environ. Plann. B Plann. Des. **28**, 361 (2001)
121. D. Helbing, A. Johansson, Encycl. J. Syst. Sci. Complex **16**, 6476 (2010)
122. D. Helbing, I. Farkás, P. Molnár, T. Vicsek, in *Pedestrian and Evacuation Dynamics*, ed. by M. Schreckenberg, S. Sharma (Springer, Berlin, 2002), pp. 21–58
123. P. Torrens, Ann. Assoc. Am. Geograph. **102**(1), 35 (2012)
124. A. Sieben, J. Schumann, A. Seyfried, PLoS One **12**(6), e0177328 (2017)
125. G. Naldi, L. Pareschi, G. Toscani (eds.), *Mathematical Modelling of Collective Behaviour in Socio-Economic and Life Sciences* (Birkhäuser, Basel, 2010)

126. G. Marsan, N. Bellomo, A. Tosin, *Complex Systems and Society. Modelling and Simulation* (Springer, Berlin, 2103)
127. B. Chakrabarti, A. Chakraborti, S. Chakravarty, A. Chatterjee, *Econophysics of Income and Wealth Distributions* (Cambridge University Press, Cambridge, 2013)
128. C. Cercignani, E. Gabetta (eds.), *Transport Phenomena and Kinetic Theory. Applications to Gases, Semiconductors, Photons, and Biological Systems* (Birkhäuser, Boston, 2007)
129. A. Frouvelle, Math. Models Methods Appl. Sci. **22**, 1250011 (2012)
130. C. Appert-Rolland, P. Degond, S. Motch, Netw. Heterog. Media **6**(3), 351 (2011)
131. A. Frouvelle, J.G. Liu, SIAM. J. Math. Anal. **44**(2), 791 (2012)
132. M. Colangeli, *From Kinetic Models to Hydrodynamics. Some Novel Results* (Springer, Berlin, 2013)
133. L. Pareschi, G. Russo, G. Toscani, *Modelling and Numerics of Kinetic Dissipative Systems* (Nova Science Publishers, New York, 2006)
134. L. Pareschi, G. Toscani, *Interacting Multiagent Systems: Kinetic Equations and Monte Carlo Methods* (Oxford University Press, Oxford, 2014)
135. H.G. Othmer, T. Hillen, SIAM J. Appl. Math. **62**, 1222 (2002)
136. T. Hillen, H.G. Othmer, SIAM J. Appl. Math. **61**, 751 (2000)
137. T. Hillen, Canad. Appl. Math. Quart. (CAMQ) **18**(1), 1 (2010)
138. A. Bertozzi, T. Laurent, F. Leger, Math. Models Methods Appl. Sci. **22**(1), 183 (2012)
139. Y. Huang, A. Bertozzi, Discret. Continuous Dyn. Syst. Ser. B **17**, 1309 (2012)
140. D. Helbing, A. Hennecke, V. Shvetsov, M. Treiber, Math. Comput. Model. **35**(5–6), 517 (2002)
141. D. Helbing, M. Treiber, A. Kesting, M. Schönhof, Eur. Phys. J. B **69**(4), 583 (2009)
142. N. Bellomo, N. Li, P. Maini, Math. Models Methods Appl. Sci. **18**(4), 593 (2008)
143. N. Bellomo, G. Forni, Curr. Top. Dev. Biol. **81**, 485 (2008)
144. J. Laval, C. Daganzo, Transp. Res. B Methodol. **40**(3), 251 (2006)
145. S. Motsch, D. Peurichard, J. Math. Biol. **76**, 205 (2018)
146. P. Degond, L. Pareschi, G. Russo (eds.), *Modelling and Computational Methods for Kinetic Equations* (Springer Science + Business Media, New York, 2004)
147. V. Vedenyain, A. Sinitsyn, E. Dulov, *Kinetic Boltzmann, Vlasov and Related Equations* (Elsevier, Amsterdam, 2011)
148. L. Arlotti, N. Bellomo, E. de Angelis, M. Lachowicz (eds.), *Generalized Kinetic Models in Applied Sciences* (World Scientific, Singapore, 2003)
149. A. Bressan, *Lecture Notes on Functional Analysis. With Applications to Linear Partial Differential Equations* (American Mathematical Society, Providence, 2013)
150. A. Bressan, *Hyperbolic Systems of Conservation Laws. The One-Dimensional Cauchy Problem* (Oxford University Press, Oxford, 2000)
151. A. Bressan, D. Serre, M. Williams, K. Zumbrun, *Hyperbolic Systems of Balance Laws* (Springer, Berlin, 2007)
152. C. Cercignani, *The Boltzmann Equation and Its Applications* (Springer, Berlin, 1987)
153. N. Bellomo (ed.), *Lecture Notes on the Mathematical Theory of Boltzmann Equation* (World Scientific, Singapore, 1995)
154. Y. Sone, *Kinetic Theory and Fluid Dynamics* (Birkhäuser, Boston, 2002)
155. V. Aristov, *Direct Methods for Solving the Boltzmann Equation and Study of Nonequilibrium Flows* (Springer, Berlin, 2001)
156. R. LeVeque, *Numerical Methods for Conservation Laws* (Birkhäuser, Basel, 1992)
157. P. Degond, L. Pareschi, G. Russo (eds.), *Modelling and Computational Methods for Kinetic Equations* (Birkhäuser, Boston, 2004)
158. F. Filbet, T. Rey, SIAM J. Sci. Comput. **37**(3), A1218 (2015)
159. S. Cordier, T. Goudon, M. Gutnic, E. Sonnendrücker (eds.), *Numerical Methods for Hyperbolic and Kinetic Problems* (European Mathematical Society, Zürich, 2005)
160. C.W. Shu, in *High-Order Methods for Computational Physics*, ed. by T. Barth, H. Deconinck, vol. 9 (Springer, Berlin, 1999), pp. 439–582

161. M. Haragus, G. Iooss, *Local Bifurcations, Center Manifolds, and Normal Forms in Infinite-Dimensional Dynamical Systems* (Springer, Berlin, 2010)
162. M. Golubitsky, I. Stewart, D.G. Schaeffer, *Singularities and Groups in Bifurcation Theory. Volume II* (Springer, New York, 1988)
163. M. Golubitsky, I. Stewart, *The Symmetry Perspective: From Equilibrium to Chaos in Phase Space and Physical Space* (Birkhäuser, Basel, 2002)
164. R. Hoyle, *Pattern Formation. An Introduction to Methods* (Cambridge University Press, Cambridge, 2006)
165. N. Bellomo, E.D. Angelis, L. Preziosi, J. Theor. Med. **5**(2), 111 (2003)
166. E. Codling, M. Plank, S. Benhamou, J. R. Soc. Interface **5**(25), 813 (2008)
167. B. Piccoli, M. Garavello, *Traffic Flow on Networks* (American Institute of Mathematical Sciences, Springfield, 2006)
168. N. Bellomo, C. Dogbé, SIAM Rev. **53**, 409 (2011)
169. B. Kerner, *Introduction to Modern Traffic Flow Theory and Control* (Springer, Berlin, 2009)
170. M. Treiber, A. Kesting, *Traffic Flow Dynamics* (Springer, Berlin, 2013)
171. I. Prigogine, R. Herman, *Kinetic Theory of Vehicular Traffic* (Elsevier, New York, 1971)
172. A. Bellouquid, M. Delitala, *Mathematical Modelling of Complex Biological Systems. A Kinetic Theory Approach* (Birkhäuser, Boston, 2006)
173. Y. Kuznetsov, *Elements of Applied Bifurcation Theory*, 2nd edn. (Springer, Berlin, 2000)
174. E. Geigant, K. Ladizhansky, A. Mogilner, SIAM J. Appl. Math. **59**(3), 787 (1998)
175. P.L. Buono, R. Eftimie, Math. Models Methods Appl. Sci. **24**(2), 327–357 (2014)
176. R. Eftimie, Math. Model Nat. Phenom. **8**(6), 5 (2013)
177. A. Leverentz, C. Topaz, A. Bernoff, SIAM J. Appl. Dyn. Syst. **8**(3), 880 (2009)
178. N. Vauchelet, Kinet. Relat. Models **3**(3), 501 (2010)
179. H. Hasimoto, Proc. Jpn. Acad. Ser. A Math. Sci. **50**, 623 (1974)
180. T. Hillen, H. Levine, Z. Angew. Math. Phys. **54**, 1 (2003)
181. R. Eftimie, J. Theor. Biol. **337**, 42 (2013)
182. C.M. Topaz, A.L. Bertozzi, SIAM J. Appl. Math. **65**, 152 (2004)
183. A. Bressan, *Hyperbolic Conservation Laws: An Illustrated Tutorial* (Springer, Berlin, 2013), pp. 157–245
184. R. Colombo, M. Garavello, M. Lécureux-Mercier, Math. Models Methods Appl. Sci. **22**(4), 1150023 (2012)
185. F. Golse, in *Handbook of Differential Equations. Evolutionary Equations*, ed. by C. Dafermos, E. Feireisl, vol. 2 (Elsevier B.V., Amsterdam, 2005)
186. S.Y. Ha, E. Tadmor, Kinet. Rel. Models **1**(3), 415 (2008)
187. S. Harris, *An Introduction to the Theory of the Boltzmann Equation* (Courier Corporation, Chelmsford, 2012)
188. B. Lapeyre, E. Pardoux, R. Sentis, *Introduction to Monte-Carlo Methods for Transport and Diffusion Equations* (Oxford University Press, Oxford, 2003)
189. A. Schadschneider, D. Chowdhury, K. Nishinari, *Stochastic Transport in Complex Systems. From Molecules to Vehicles* (Elsevier, Amsterdam, 2011)

Chapter 2
A Short Introduction to One-Dimensional Conservation Laws

2.1 Introduction

The simplest hyperbolic models derived to investigate the movement of animal or human populations are the advection equations. The three main advantages of these models (in contrast to the individual-based models described in the previous chapter) are:

- the models can be easily calibrated and validated;
- the existent analytical techniques allow for the theoretical investigation of different types of solutions exhibited by these models;
- the existent numerical schemes allow for accurate numerical approximations of the solutions exhibited by these models.

The analytical investigation into the existence and uniqueness of different types of solutions exhibited by one-equation hyperbolic models or by hyperbolic systems of equations has been the subject of many theoretical studies; see, for example, [1–11]. Therefore, in this chapter we do not aim to give a comprehensive review of the theory behind the hyperbolic equations and systems of equations; for such a review we refer the reader to [4, 5, 7, 12, 13]. Rather, we aim to briefly review some general aspects necessary to understand the various types of solutions (and patterns) exhibited by the models discussed throughout this monograph. Although this brief review focuses only on 1D local models, it is necessary to have it here before we discuss more complex local and nonlocal models, for which the theory is more complex (and not always fully developed).

The general advection equations that we will discuss in Chap. 3 are

$$\frac{\partial u}{\partial t} + \frac{\partial H(u)}{\partial x} = R(u), \tag{2.1a}$$

$$u(x, 0) = u_0(x). \tag{2.1b}$$

© Springer Nature Switzerland AG 2018
R. Eftimie, *Hyperbolic and Kinetic Models for Self-organised Biological Aggregations*, Lecture Notes in Mathematics 2232,
https://doi.org/10.1007/978-3-030-02586-1_2

In the context of biological behaviours, $u = u(x, t) \in \mathbb{R}$ describes the density of a population at position x and time t (with $x, t \in \mathbb{R}$ and $t \geq 0$), $H(u) : \mathbb{R} \to \mathbb{R}$ is a nonlinear function describing the advective movement (the "flux"), and $R(u) :$ $\mathbb{R} \to \mathbb{R}$ is a reaction term describing population dynamics as a result of birth and death processes, or a term describing the rate at which population enters or leaves the domain. Equation (2.1) can be derived using a "flow" approach [12] or, as we will see in Chap. 4.2, a correlated random walk approach [14]. The "flow" approach assumes that the rate of change of the density u in any fixed subdomain $D \subset \mathbb{R}$ is given by the total population density entering or leaving the domain (i.e., flow $H(u)$) through the boundary ∂D, and the birth and death ($R(u)$) of individuals inside the subdomain :

$$\frac{d}{dt} \int_D u(x, t) dx = - \int_{\partial D} H(u(x, t)) \cdot \mu d\sigma(x) + \int_D R(u(x, t)) dx. \qquad (2.2)$$

Here, μ is the outward normal to the boundary ∂D, and $d\sigma(x)$ is the surface measure. Using Gauss's divergence theorem, this equation can be written as

$$\frac{d}{dt} \int_D u(x, t) dx = - \int_D div(H) dx + \int_D R(u(x, t)) dx. \qquad (2.3)$$

Since this equation holds for any (infinitesimal) sub-domain $D \in \mathbb{R}$, we can ignore the integral and recover equation (2.1).

In general, Eq. (2.1) can be written in the quasilinear form

$$\frac{\partial u}{\partial t} + h(u) \frac{\partial u}{\partial x} = R(u). \qquad (2.4)$$

Here, $h(u) = H'(u)$ is the advection velocity. For smooth solutions, Eqs. (2.1) and (2.4) are equivalent. However, for discontinuous solutions the transport term will be the product of a discontinuous velocity $h(u)$ and a distributional derivative $\partial u / \partial x$. Therefore, while the quasilinear equation (2.4) is meaningful for continuous functions, Eq. (2.1) is meaningful also for discontinuous solutions interpreted in a distributional sense [4].

In regard to the application of these models to the movement of biological populations, we remark that the simplest quasilinear models (2.4) consider a constant advection velocity $h(u) = c$ and $R(u) = 0$. These models can be used to study the movement of a population that has an initial density profile $U(x)$. The solution at time t is described by $u(x, t) = U(x - ct)$ (which represents a *travelling wave* solution, as it will be discussed later in this chapter). Figure 2.1 shows the spatial movement, with velocity c, of such a population.

Throughout the rest of this chapter we will assume that $R(u) = 0$. In this case, we note that the population density u is conserved, as

$$\frac{d}{dt} \int_D u(x, t) dx = H(u(\partial D_{in}, t)) - H(u(\partial D_{out}, t)) = inflow - outflow,$$
$$(2.5)$$

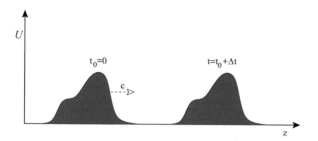

Fig. 2.1 Schematic description of the transport (with velocity c) of an initial density profile $U(z)$

where ∂D_{in} and ∂D_{out} are the inflow and outflow regions on the domain boundary (in 1D these regions reduce to specific points, e.g., $\partial D_{in} = a$ and $\partial D_{out} = b$, for $D = [a, b]$). For this reason, we refer to Eq. (2.1) as a *conservation law*.

The solutions of the hyperbolic equations (2.1) can be found using the method of characteristics [12]. Note that the characteristics of a hyperbolic PDE are the paths along which certain variables are conserved. For $R(u) = 0$, the solutions are constant along the characteristic lines $(x(t), t)$, with

$$\frac{\partial x}{\partial t} = H'(u(x(t), t)), \quad x(0) = x_0. \tag{2.6}$$

The rate of change along these characteristics is

$$\frac{du(x(t), t)}{dt} = \frac{\partial u}{\partial t} + \frac{\partial u}{\partial x}\frac{dx}{dt} = \frac{\partial u}{\partial t} + \frac{\partial H(u)}{\partial x} = 0. \tag{2.7}$$

If the characteristics speed $H'(u)$ is constant (i.e., $H(u)$ is a linear function of u), the characteristics are parallel lines. If the characteristics speed is density-dependent (i.e., $H(u)$ is a nonlinear function of u), the characteristic lines might intersect, and the first intersection time T is given by

$$T = -\frac{1}{\inf_x u_0'(x)H''(u_0(x))}, \quad \text{with} \ \inf_x u_0'(x)H''(u_0(x)) < 0. \tag{2.8}$$

Classical solutions exit only for $t \leq T$. For $t > T$, solutions defined globally in time can exist within a space of discontinuous solutions, where equation $u_t + H(u)_x = 0$ is interpreted in a distributional sense [4]. Therefore, in this case one has to ignore the classical (continuous) interpretation of the solutions and focus only on *weak solutions*. We will discuss this aspect in more detail in the next section, in the context of systems of equations. In Chap. 7 we will discuss some numerical schemes that can be used to approximate these classical/weak solutions (for single equations or systems of equations).

2.2   Fundamental Results for Systems of Conservation Laws

Since many of the models that will be discussed in Chaps. 3 and 4 are actually described by two or more hyperbolic equations (e.g., for traffic models the equations focus on changes in the density and speed of cars/pedestrians, while for populations in 1D the equations focus on changes in the left- and right-moving individuals/cells), in the following we present briefly the general theory behind conservation laws for the more general case of hyperbolic systems (which can be easily reduced to the scalar conservation laws discussed above). Our approach here follows closely the exposition in [4, 11].

Let us then consider the following general system of n conservation laws:

$$\frac{\partial u_1}{\partial t} + \frac{\partial H_1(u_1, \ldots u_n)}{\partial x} = 0, \tag{2.9a}$$

$$\ldots$$

$$\frac{\partial u_n}{\partial t} + \frac{\partial H_n(u_1, \ldots, u_n)}{\partial x} = 0. \tag{2.9b}$$

If we assume a smooth solution, and denote by $\mathbf{u} = (u_1, \ldots, u_n) \in \mathbb{R}^n$, $\mathbf{H} = (H_1, \ldots, H_n) : \mathbb{R}^n \to \mathbb{R}^n$, and by $J(\mathbf{u}) \in \mathbb{R}^n \times \mathbb{R}^n$ the Jacobian matrix associated with this system,

$$J(\mathbf{u}) := \nabla \mathbf{H}(\mathbf{u}) = \begin{pmatrix} \frac{\partial H_1}{\partial u_1} & \cdots & \frac{\partial H_1}{\partial u_n} \\ & \cdots & \\ \frac{\partial H_n}{\partial u_1} & \cdots & \frac{\partial H_n}{\partial u_n} \end{pmatrix},$$

then we can re-write the non-linear hyperbolic system in the quasi-linear form

$$\frac{\partial \mathbf{u}}{\partial t} + J(\mathbf{u}) \frac{\partial \mathbf{u}}{\partial x} = 0. \tag{2.10}$$

Definition 2.1 We say that a function $\mathbf{u}(x, t) \in C^1$ is a *classical solution* for (2.10) if and only if it satisfies system (2.10).

As discussed in the previous section, the discontinuous solutions of (2.9) need to be interpreted in a distributional sense. In this case, we work with *weak solutions*:

Definition 2.2 We say that a function $\mathbf{u}(x, t) \in \mathbb{L}^1_{loc}$ (locally integrable) is a *weak solution* for (2.9) if and only if it satisfies the following equation

$$\int \int \left[\mathbf{u} \frac{\partial \phi}{\partial t} + \mathbf{H}(\mathbf{u}) \frac{\partial \phi}{\partial x} \right] dx \, dt = 0, \tag{2.11}$$

for every $\phi \in C^1_c$ (i.e., continuously differentiable with compact support).

Note that these weak solutions are considered in the context of Sobolev spaces, on which we define the weak derivatives [5]. For a summary of spaces used throughout this study for hyperbolic and kinetic models (including the spaces discussed above, C^1, $\mathbb{L}^1_{loc}$, C^1_c), please see Table 2.1.

Table 2.1 Summary of spaces (over an arbitrary domain Ω) on which we discuss the solutions of the hyperbolic/kinetic models reviewed in this monograph

Space	Description
$\mathbb{L}^p(\Omega)$	Space of all Lebesgue measurable functions $f : \Omega \to \mathbb{R}$ that satisfy $\int_\Omega \|f(x)\|^p dx < \infty$.
$\mathbb{L}^1(\Omega)$	Space of all integrable functions on Ω, obtained from $\mathbb{L}^p$ for the special case $p = 1$.
$\mathbb{L}^1_{loc}(\Omega)$	Space of locally integrable functions $f : \Omega \to \mathbb{R}$ (with $\Omega \subseteq \mathbb{R}$ open set).
$\mathbb{L}^2(\Omega)$	Space of square integrable functions, obtained from $\mathbb{L}^p$ for the special case $p = 2$.
	$\mathbb{L}^2$ is the only Hilbert space among the $\mathbb{L}^p$ spaces (for a definition of a Hilbert space see below in this table). In the complex case, the inner product on $\mathbb{L}^2$ is $\langle f, g \rangle = \int_\Omega f(x)\overline{g(x)}dx$.
$\mathbb{L}^\infty(\Omega)$	Space of essentially bounded measurable functions on Ω, obtained from $\mathbb{L}^p$ in the limit case $p \to \infty$.
$C(\Omega)$	Space of all continuous real-valued functions $f : \Omega \to \mathbb{R}$.
$C^1(\Omega)$	Space of continuously differentiable functions.
$C^1_c(\Omega)$	Space of continuously differentiable functions with compact support.
$C^\infty_c(\Omega)$	Space of continuous functions with compact support, and which have continuous derivatives of any order.
$W^{k,p}(\Omega)$ (Sobolev space)	Sobolev space of all locally integrable functions $f : \Omega \to \mathbb{R}$ such that for every multi-index α with $\|\alpha\| \leq k$, the weak derivative $D^\alpha(f) \in \mathbb{L}^p(\Omega)$.
	The weak α-th derivative of $f \in \mathbb{L}^1_{loc}(\Omega)$, is a function $g \in \mathbb{L}^1_{loc}$ denoted as $g = D^\alpha f$ that satisfies $\int f D^\alpha \phi dx = (-1)^{\|\alpha\|} \int g\phi dx$ for all test functions $\phi \in C^\infty_c(\Omega)$.
Hilbert space H	A Hilbert space is a vector space H with an inner product $\langle \cdot, \cdot \rangle : H \times H \to \mathbf{K}$ (with $\mathbf{K} = \mathbb{R}$ or $\mathbb{C}$), which is complete with respect to the norm $\|\|x\|\| := \sqrt{\langle x, x \rangle}$.
$H^k(\Omega) := W^{k,2}(\Omega)$	Hilbert-Sobolev space obtained from the Sobolev space in the special case of $p = 2$.
	The Hilbert-Sobolev space has the inner product $\langle f_1, f_2 \rangle := \sum_{\|\alpha\| \leq k} \int_\Omega D^\alpha f_1 D^\alpha f_2 dx$.
Banach space X	A vector space X over $\mathbb{R}$ or over $\mathbb{C}$, which is equipped with a norm and which is complete with respect to that norm.
	$\mathbb{L}^p$ spaces (including $\mathbb{L}^\infty$) are examples of Banach spaces. Any Hilbert space is an example of a Banach space (but the converse is not true, as not every norm comes from an inner product).

For these definitions, we use [5]

Consider now the initial data

$$\mathbf{u}(x, 0) = \bar{\mathbf{u}}(x), \quad \bar{\mathbf{u}} \in \mathbb{L}^1_{loc}. \tag{2.12}$$

Definition 2.3 The *Cauchy problem* for a 1D system of conservation laws is given by the system together with the initial data:

$$\frac{\partial \mathbf{u}}{\partial t} + \frac{\partial \mathbf{H}(\mathbf{u})}{\partial x} = 0, \quad \text{with } \mathbf{u}(x, 0) = \bar{\mathbf{u}}(x). \tag{2.13}$$

Definition 2.4 A function $\mathbf{u}(x, t)$ (with $x \in \mathbb{R}$, $t \in [0, T]$) is a *weak solution of the Cauchy problem* (2.13) if $\mathbf{u}$ is continuous as a function from $[0, T]$ to $\mathbb{L}^1_{loc}$, and the restriction $\mathbf{u}_{\mathbb{R} \times (0,T)}$ is a distributional solution of (2.9) [6].

Definition 2.5 We say that system (2.10) is *hyperbolic* if for every $\mathbf{u} \in \mathbb{R}^n$, the Jacobian matrix $J(\mathbf{u})$ has n real eigenvalues. The system is *strictly hyperbolic* if the real eigenvalues are all distinct:

$$\lambda_1(\mathbf{u}) < \ldots < \lambda_n(\mathbf{u}). \tag{2.14}$$

If a system is strictly hyperbolic, then we can find dual bases of left eigenvectors (i.e., row vectors $(l_1(\mathbf{u}), \ldots, l_n(\mathbf{u}))$) and right eigenvectors (i.e., column vectors $(r_1(\mathbf{u}), \ldots, r_n(\mathbf{u}))$), which satisfy

$$J(\mathbf{u})r_i(\mathbf{u}) = \lambda_i(\mathbf{u})r_i(\mathbf{u}) \quad \text{and} \quad l_i(\mathbf{u})J(\mathbf{u}) = \lambda_i(\mathbf{u})l_i(\mathbf{u}), \quad \text{for } i = 1, \ldots, n. \tag{2.15}$$

For convenience, the eigenvectors are chosen to be normalised [15] such that

$$|r_i| = 1, \quad l_i \cdot r_j = \begin{cases} 1 \text{ if } i = j, \\ 0 \text{ if } i \neq j. \end{cases} \tag{2.16}$$

For each pair of eigenvalues and corresponding right eigenvectors $(\lambda_i(\mathbf{u}), r_i(\mathbf{u}))$, there is an associated characteristic field.

Definition 2.6 The i-th characteristic field is called *linearly degenerate* if

$$\nabla \lambda_i(\mathbf{u}) \cdot r_i(\mathbf{u}) = 0, \text{ for all } \mathbf{u} \in \mathbb{R}^n. \tag{2.17}$$

The i-th characteristic field is called *genuinely nonlinear* if

$$\nabla \lambda_i(\mathbf{u}) \cdot r_i(\mathbf{u}) \neq 0, \text{ for all } \mathbf{u} \in \mathbb{R}^n. \tag{2.18}$$

Here, $\nabla \lambda_i(\mathbf{u}) \cdot r_i(\mathbf{u})$ is the directional derivative of λ_i in the direction of eigenvectors r_i. In the linearly degenerate case, the eigenvalue λ_i is constant along each curve corresponding to r_i. Note that linear systems can generate only linearly degenerate fields. Nonlinear systems can generate both linearly degenerate and genuinely nonlinear fields.

The assumption that the i-th characteristic field is either *linearly degenerate* or *genuinely nonlinear* is necessary for the global existence results of solutions of (2.9). With the help of this assumption, Glimm [16] showed the existence of global weak solutions for the general hyperbolic system (2.9) with initial data having small bounded variation (BV).

Definition 2.7 Consider a function $\mathbf{u} : D \times [0, \infty) \to \mathbb{R}^n$. The *total variation* of $\mathbf{u}$ is defined as

$$\text{Tot.Var.}(\mathbf{u}) = \sup_N \{ \sum_{j=1}^N |\mathbf{u}(x_j, t) - \mathbf{u}(x_{j-1}, t)| \}, \tag{2.19}$$

where the points $x_j \in D \subseteq \mathbb{R}$ are such that $x_0 < x_1 < \ldots < x_N$. If the right-hand side of (2.19) is bounded we say that function $\mathbf{u}$ has *bounded variation*.

Returning to global solutions we note that to construct them, there are two main approaches (we will mention later another approach based on vanishing viscosity approximations [17]): (a) the Glimm scheme [16, 18]; (b) the wave-front tracking [19, 20]. Both approaches lead to a sequence of approximate solutions, whose convergence is shown using a compactness argument based on uniform bounds for the total variation of the solution (with the assumption that the total variation of the initial data is sufficiently small). Before we discuss briefly these two approaches, we need to define first the Riemann problem, which is the building block for these approaches.

Definition 2.8 The *Riemann problem* of a conservation law is the initial value problem with piecewise constant initial data:

$$\frac{\partial \mathbf{u}}{\partial t} + \frac{\partial \mathbf{H}(\mathbf{u})}{\partial x} = 0, \tag{2.20a}$$

$$\mathbf{u}(x, 0) = \begin{cases} \mathbf{u_l}, & x \leq 0, \\ \mathbf{u_r}, & x > 0. \end{cases} \tag{2.20b}$$

We note that solutions to more general Cauchy problems can be constructed by piecing together different solutions of Riemann problems. For simplicity, one can assume that the discontinuity occurs at $x = 0$ (see Fig. 2.2).

Finally, we remark that the solution of the Riemann problem is self-similar: if $\mathbf{u}(x, t)$ is a solution, then the rescaled function $\mathbf{u}(cx, ct)$ is also a solution, for any constant $c > 0$. Therefore, one can consider solutions of the form $\mathbf{u}(x, t) = \mathbf{U}(x/t)$. Moreover, the solutions of a Riemann problem are invariant under the rescaling of the independent variables [15]: if $\mathbf{u}(x, t)$ is a solution of $\partial \mathbf{u}/\partial t + \partial \mathbf{H}(\mathbf{u})/\partial x = 0$, then for any $\theta > 0$, $\mathbf{u}^\theta(x, t) := \mathbf{u}(\theta x, \theta t)$ is also a solution.

Fig. 2.2 Piecewise initial
condition for the Riemann
problem (2.20)

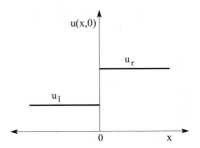

We now return to the two approaches mentioned above that can be used to construct global weak solutions [4, 6, 15]:

- *Glimm scheme.* Consider a fixed grid with mesh sizes Δt and Δx. Consider initial data $(\mathbf{u}(x, 0))$ that is assumed to be a piecewise constant function with jumps at the grid points. The corresponding Riemann problem (i.e., the initial value problem with piecewise constant initial data; see also Eqs. (2.20)) is solved by constructing solutions up to some time Δt sufficiently small such that the waves that are generated by different Riemann problems do not interact. Then, solution $\mathbf{u}(\cdot, \Delta t)$ is approximated by a piecewise constant function with jumps at the grid points. The procedure is repeated by solving the new Riemann problems at each of these jump points, thus propagating the solution up to $2\Delta t$, $3\Delta t$, etc.
- *Wave-front tracking.* As before, start with initial data approximated by a piecewise constant function, and solve the Riemann problem approximately, within the class of piecewise functions. At the first time when two waves interact, the new Riemann problem is approximately solved by a new piecewise constant function, until the next time when the waves interact. And the process is repeated. Note that while Glimm's scheme requires a priori specification of the points where the Riemann problem is to be solved, the wave-front tracking technique does not require prior specification of the location of jumps.

Having constructed a sequence of approximate solutions u_v (using either of the two approaches discussed above), then one needs to extract a subsequence converging in $\mathbb{L}^1_{loc}$ to a weak solution $\mathbf{u}(x, t)$ of the Cauchy problem, and using Helly's compactness (or selection) theorem the existence of weak solutions can be proven. For a detailed discussion of these two approaches (including a proof of Helly compactness theorem), we refer the reader to [4, 6].

As mentioned above, a different approach to show the existence of solutions uses the vanishing viscosity approximations [15, 17], where the solutions of the conservation law are obtained as limits for $\epsilon \to 0$ from solutions of the parabolic equations

$$\frac{\partial \mathbf{u}^\epsilon}{\partial t} + J(\mathbf{u}^\epsilon)\frac{\partial \mathbf{u}^\epsilon}{\partial x} = \epsilon \frac{\partial^2 \mathbf{u}^x}{\partial x^2}, \quad \text{with} \quad \epsilon \ll 1. \tag{2.21}$$

The complexity of this approach arises from the possible discontinuity of function $\mathbf{u}$, which makes it difficult to prove the convergence $\mathbf{u}^\epsilon \to \mathbf{u}$ near the discontinuity points. We note that some of the earliest proofs of existence of solutions using the vanishing viscosity method (for the scalar case) were given by Oleinik [21] in 1D, and Kružkov [22] in higher dimensions. We will return to the study by Kružkov[22] in Sect. 2.2.2, in the context of entropy conditions for the uniqueness of weak solutions. Another approach that focuses on BV solutions via vanishing viscosity was considered in [17]. Since the purpose of this study is not to discuss the solutions of the parabolic equations, we refer the reader to [15] for a detailed exposition on the vanishing viscosity approach. However, before we conclude this section, it is worth mentioning here that the vanishing viscosity approach does not consider Riemann problems, but rather focuses on viscous travelling profiles (i.e., travelling waves for the viscous equation (2.21)) that are solutions of the form $\mathbf{u}^\epsilon(x, t) = \mathbf{U}^\epsilon(x - \lambda t)$, $\lambda \in \mathbb{R}$, and lie on a suitable centre manifold [15].

In the following subsections we will discuss briefly the different types of solutions exhibited by linear and nonlinear hyperbolic systems, including travelling front solutions.

2.2.1   Travelling Waves, Rarefaction Waves, Shocks and Contact Discontinuities

Consider the following initial value problem for a linear n-dimensional system with constant coefficients (i.e., $J =$ constant matrix):

$$\frac{\partial \mathbf{u}}{\partial t} + J \frac{\partial \mathbf{u}}{\partial x} = 0, \quad \mathbf{u}(x, 0) = \bar{\mathbf{u}}(x). \tag{2.22}$$

At the end of the previous section we have mentioned the travelling wave solutions for a scalar conservation law. Note that these travelling waves could be either travelling pulses (as in Figs. 2.3, 2.1) or travelling fronts (as in Fig. 3.6), depending on the boundary conditions at the end points of the domain.

Definition 2.9 A *travelling wave* solution is a solution of the form $u(x, t) = U(x - \lambda t)$, where $\lambda \in \mathbb{R}$ is the travelling speed of the wave.

Next we show that the solution of the quasilinear system (2.22) with constant coefficients can be written as a sum of travelling waves. Assume that this system has n distinct real eigenvalues $\lambda_1 < \ldots < \lambda_n$. Using the basis of right and left eigenvectors defined by (2.15)–(2.16), and multiplying Eq. (2.22) on the left by l_i, $i = 1, \ldots, n$, one can decouple Eq. (2.22) into n scalar Cauchy problems [6]:

$$\frac{\partial u^i}{\partial t} + \lambda_i \frac{\partial u^i}{\partial x} = 0, \quad \text{with } u^i := l_i \mathbf{u}. \tag{2.23}$$

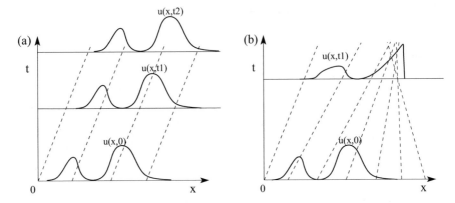

Fig. 2.3 (**a**) Propagation of initial data of linear system (2.22) along the characteristic lines; (**b**) Propagation of initial data for nonlinear or quasi-linear systems (2.9) or (2.10). The nonlinearity of the model can lead to the intersection between different characteristic lines, at which time-point a shock (discontinuity) forms. Moreover, the nonlinearity of the model can lead to changes in the shape of interacting waves [6]

This equation can be solved explicitly, and its solution has the form of a travelling wave (formed of sums of n waves, each travelling with one of the characteristic speeds λ_i):

$$\mathbf{u}(t, x) = \sum_{i=1}^{n} \bar{u}^i (x - \lambda_i t) r_i. \tag{2.24}$$

Simple calculations confirm that this function does indeed satisfy Eq. (2.22). Hence, this travelling wave solution is a superposition of the n travelling wave solutions $\bar{u}^i (x - \lambda_i t) r_i$. Since the hyperbolic system has constant coefficients, these travel waves propagate without distortion along the characteristic lines, with speeds λ_i (see Fig. 2.3a). We emphasise again that for the linear system (2.22) the characteristics are straight lines. However, for systems with flux terms (H) that depend explicitly on x (or systems with reaction terms), the characteristics are no longer straight lines.

Returning now to the quasi-linear system (2.10), we note that if the wave propagation speed depends on the density $\mathbf{u}$, the profile of the solution will change in time. In particular, the waves can interact with each other creating new waves. This nonlinear effect is explained in [6] with the help of the i-th component of the gradient of $\mathbf{u}$ with respect to the basis of eigenvectors $(r_1(\mathbf{u}), \ldots, r_n(\mathbf{u}))$: $u^i_x := \partial u^i/\partial x = l_i \cdot (\partial \mathbf{u}/\partial x)$. This allows to us to re-write the derivatives of $\mathbf{u}$ as follows:

$$\frac{\partial \mathbf{u}}{\partial x} = \sum_{i=1}^{n} u^i_x r_i(\mathbf{u}), \quad \frac{\partial \mathbf{u}}{\partial t} = - \sum_{i=1}^{n} \lambda_i(\mathbf{u}) u^i_x r_i(\mathbf{u}). \tag{2.25}$$

Differentiating first equation with respect to t and second equation with respect to x, and equating the terms leads to

$$\frac{\partial u_x^i}{\partial t} + \frac{\partial (\lambda_i u_x^i)}{\partial x} = \sum_{j>k} (\lambda_j - \lambda_k)(l_i \cdot ((\nabla r_k)r_j - (\nabla r_j)r_k)u_x^j u_x^k. \tag{2.26}$$

The difference in speed $\lambda_j - \lambda_k$ describes the rate at which the j-waves and k-waves interact with each other, $u_x^j u_x^k$ describes the product of the density of j-waves and k-waves, and $(\nabla r_k)r_j$ (or $(\nabla r_j)r_k$) is the directional derivative of r_k (r_j) in the direction of r_j (r_k). Overall, the term under the sum in (2.26) describes the amount of i-waves produced by the interactions of j-waves with k-waves [6].

Definition 2.10 A wave for which the characteristics with different slopes (that start from a line) fan out in the direction $t > 0$ is called a *rarefaction wave* (or expansion wave). A wave for which the characteristics (of different slopes) fan out from a point is called a *centred rarefaction wave*. A wave for which the characteristics (of different slopes) converge in the direction $t > 0$ is called a *compression wave*. A wave that forms a surface that separates zones of different density is called a *contact discontinuity*. A description of these types of waves is shown in Fig. 2.4.

Note that linearly degenerate fields lead to contact discontinuities, while genuinely nonlinear fields lead to rarefaction waves or shock waves (depending on whether the characteristics are divergent or convergent; see Fig. 2.4). Moreover, we emphasise that the nonlinear interactions in the model dynamics lead to a decay of the rarefaction waves, and to a steepness of the compression waves, which might eventually lead to the formation of *shocks* in finite time [6]. To discriminate between the shocks and the rarefaction waves one has to introduce some extra conditions, which will be discussed in the next section.

2.2.2 The Rankine-Hugoniot Jump Condition

To be able to understand the behaviour of the solutions of the nonlinear system (2.9) at discontinuities (particularly the relation between the solution states on both sides of a shock wave), we start with a simplified problem: the *Riemann problem*. We say that a function

$$\mathbf{U} = \mathbf{u}(x, t) = \begin{cases} \mathbf{u_l}, & \text{for } x < \lambda t, \\ \mathbf{u_r}, & \text{for } x > \lambda t \end{cases} \tag{2.27}$$

is a *weak solution* of (2.20a) if and only if the following condition holds [2]:

$$\lambda(\mathbf{u_r} - \mathbf{u_l}) = \mathbf{H}(\mathbf{u_r}) - \mathbf{H}(\mathbf{u_l}). \tag{2.28}$$

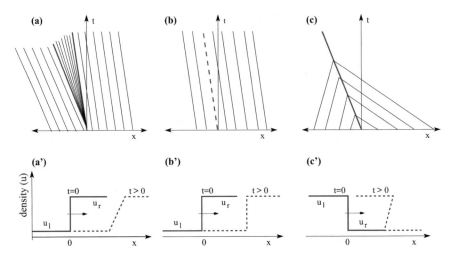

Fig. 2.4 Caricature description (in the (x, t) plane) of characteristic lines for the three types of wave solutions that could be exhibited by a Riemann problem (2.20), together with a visual description of these waves (in the (x, u) plane). (**a, a'**) centred rarefaction waves (where the characteristic lines of the i-th field diverge); (**b, b'**) contact discontinuity (where the characteristic lines travel parallel); (**c, c'**) shock waves (where the characteristic lines converge along a shock line)

Equation (2.28) is called the *Rankine-Hugoniot jump condition*. This condition connects the left ($\mathbf{u_l}$) and right ($\mathbf{u_r}$) states of the solution $\mathbf{u}$ with the speed ($\lambda = \lambda(\mathbf{u})$) of the discontinuity. We can re-write the jumps in $\mathbf{u}$ and $\mathbf{H(u)}$ (across the discontinuity) as follows: $[\mathbf{u}] = \mathbf{u_r} - \mathbf{u_l}$ and $[\mathbf{H}] = \mathbf{H(u_r)} - \mathbf{H(u_l)}$.

2.2.3 Admissibility Conditions: Entropy, Vanishing Viscosity and Speed Stability

As we have seen above, nonlinear hyperbolic equations and systems may loose their regularity: solutions that are initially smooth may become discontinuous at some finite time. After the characteristic lines cross (as in Fig. 2.3b), there are points where multiple characteristics lead back to $t = 0$, and thus the solution can take multiple values at that point. Therefore, the weak solutions are non-unique, and one has to use various additional conditions to select a unique physically realistic solution. The conditions used to select this unique solution are called *admissibility conditions*. In the following, we will discuss three such conditions [6]:

1. **Entropy Conditions.** To introduce these conditions, let us first define the entropy and the entropy flux:

Definition 2.11 A function $\eta : \mathbb{R}^n \to \mathbb{R}$ is called an *entropy* for system (2.9), with *entropy flux* $q : \mathbb{R}^n \to \mathbb{R}$, if for all $\mathbf{u} \in \mathbb{R}^n$ the following condition is true:

$$\nabla \eta(\mathbf{u}) \cdot \nabla \mathbf{H}(\mathbf{u}) = \nabla q(\mathbf{u}). \tag{2.29}$$

Entropy Condition A weak solution of (2.9) is entropy-admissible if the following inequality holds true (in the sense of distributions) for any pair (η, q) of convex entropy and corresponding entropy flux [6]:

$$\frac{\partial \eta(\mathbf{u})}{\partial t} + \frac{\partial q(\mathbf{u})}{\partial x} \leq 0. \tag{2.30}$$

This admissibility condition is useful only if one knows a nontrivial convex entropy solution of the conservation law (2.20a) [6]. In the following, we focus on two other types of admissibility conditions more amenable to direct applications.

Kružkov Entropy Condition A different type of entropy condition that is more convenient to work with is the Kružkov entropy condition, which combines the definition of an entropy condition with the one of a weak solution [23]. This condition was introduced in the context of scalar conservation laws:

$$\frac{\partial u}{\partial t} + \frac{\partial H(u)}{\partial \mathbf{x}} = 0, \qquad \text{with } (\mathbf{x}, t) \in D = \mathbb{R}^n \times (0, \infty),$$
$$u(x, 0) = u_0(x). \tag{2.31}$$

Definition 2.12 A function $u \in L^\infty(\mathbb{R}^n \times (0, \infty))$ is a *Kružkov entropy solution* for (2.31) if the following inequality holds for all constants k and all non-negative test functions $\phi \in C_c^\infty(\mathbb{R} \times (0, \infty))$:

$$\int \int \left(|u - k|\phi_t + sign(u - k)\big(H(u) - H(k)\big)\phi_x \right) dx \, dt \geq 0. \tag{2.32}$$

Equivalently, we can say that u is a Kružkov entropy solution if the following inequality holds in distributional sense for all constants k:

$$\frac{\partial}{\partial t}|u - k| + \frac{\partial}{\partial x} sign(u - k)\big(H(u) - H(k)\big) \leq 0. \tag{2.33}$$

Note that functions $\eta_K := |u - k|$ and $q_K := sign(u - k) \cdot \big(H(u) - H(k)\big)$ are similar to the (entropy,entropy flux)=(η, q) pairs defined above in the context of entropy solutions. However, the pair (η_K, q_K) is not of class C^1.

The first proof of uniqueness of the entropy solutions for scalar conservation laws was given by Kružkov [22] (and uses the Kružkov entropies together with a technical argument of doubling variables):

Theorem 2.1 *Consider u and v two entropy solutions for the system (2.31), such that $||u||_{L^\infty(D)}, ||v||_{L^\infty(D)} \leq M$. Then*

$$\int_{|x|<R} |u(x,t) - v(x,t)|dx \leq \int_{|x|<R+Lt} |u(x,0) - v(x,0)|dx, \qquad (2.34)$$

for every $R > 0$, $t > 0$, and with L the Lipschitz constant for $H(u)$: $|H(z) - H(w)| \leq L|z - w|$ for every $z, w \in [-M, M]$.

Therefore, if $u(x,0) = v(x,0)$ then $u = v$, and so the entropy solutions are unique if they exist.

2. **Vanishing Viscosity Admissibility Condition.** Based on the uniqueness results for solutions of parabolic equations, another approach to investigate the uniqueness of solutions for hyperbolic equations would be to transform the hyperbolic system (2.9) into a parabolic system by adding a small viscosity term $\epsilon(\partial^2\mathbf{u}/\partial x^2)$:

$$\frac{\partial\mathbf{u}}{\partial t} + \frac{\partial\mathbf{H}(\mathbf{u})}{\partial x} = \epsilon\frac{\partial^2\mathbf{u}}{\partial x^2}. \qquad (2.35)$$

Definition 2.13 A weak solution u of (2.9) is called *admissible in the vanishing viscosity sense* if there is a sequence of smooth solutions u^ϵ of (2.35) which converge to u in $\mathbb{L}_{loc}^1$ as $\epsilon \to 0$ [6].

It has been shown in [17] that the solutions of (2.35) are defined globally in time, satisfy uniform bounded variation estimates independent of ϵ, and depend continuously on the initial data (with a Lipschitz constant independent of time t and viscosity parameter ϵ). Moreover, in the limit $\epsilon \to 0$, the solutions of (2.35) converge to a unique limit that is a solution of (2.9) [17]. In the case where $J(\mathbf{u}) = \nabla\mathbf{H}(\mathbf{u})$ is the Jacobian of the flux $\mathbf{H}(\mathbf{u})$, the vanishing viscosity limits are the unique entropy weak solutions of the hyperbolic system (2.9) [17]. This way, the vanishing viscosity limit would help single out the unique "good" solution of the Cauchy problem that satisfies the entropy conditions.

Remark 2.1 The introduction of artificial viscosity into a hyperbolic model is useful not only from an analytical point of view, but also from a numerical point of view. Numerically, the goal of introducing artificial viscosity is to reduce the oscillations that form behind the shock, for most finite difference methods; see also Fig. 7.2. This approach has the advantage of being easy to incorporate into an existing numerical scheme, and preserves the high-order accuracy of the scheme [24].

3. **Speed Stability Conditions**. One can derive admissibility conditions by perturbing the initial piecewise state (2.20b) through the insertion of an intermediate state $u^* \in [u_l, u_r]$ (we focus for now on the one-equation model) [6]. Thus, the original shock is split into two other shocks, each satisfying the Rankine-Hugoniot conditions. The $\mathbb{L}^1$-distance between the original solution and the perturbed one does not increase in time if [6]:

speed of jump behind $u^* \geq$ speed of jump ahead u^*,

which translates into

$$\frac{H(u^*) - H(u_l)}{u^* - u_l} \geq \frac{H(u_r) - H(u^*)}{u_r - u^*}. \tag{2.36}$$

This condition was generalised by Liu [25] to hyperbolic systems by defining $s \rightarrow S_i(s)(\mathbf{u_l})$ to be the i-th shock curve through $\mathbf{u_l}$ (i.e., the curve of all states $\mathbf{u}$ that can be connected with $\mathbf{u_l}$ via a shock in the i-th field) [6]:

Liu Admissibility Condition [25] Consider a point $\mathbf{u_r}$ on the i-th shock curve through $\mathbf{u_l}$: $\mathbf{u_r} = S_i(s)(\mathbf{u_l})$, for some $s \in \mathbb{R}$. The shock with left state $\mathbf{u_l}$ and right state $\mathbf{u_r}$ satisfies the Liu admissibility condition if its speed is less or equal to the speed of any smaller shock that connects $\mathbf{u_l}$ to an intermediate state along the same i-th shock curve, $\mathbf{u^*} = S_i(\sigma)(\mathbf{u_l})$, for $\sigma \in [0, s]$:

$$\lambda_i(\mathbf{u_l}, \mathbf{u_r}) \leq \lambda_i(\mathbf{u_l}, \mathbf{u^*}). \tag{2.37}$$

A different admissibility condition for hyperbolic systems of conservation laws was introduced by Lax [6, 26]:

Lax Admissibility Condition [26] A shock of the i-th family that connects the states $\mathbf{u_l}$ and $\mathbf{u_r}$, and travels with speed $\lambda = \lambda_i(\mathbf{u_l}, \mathbf{u_r})$ satisfies the *Lax admissibility condition* if:

$$\lambda_i(\mathbf{u}_l) \geq \lambda_i(\mathbf{u_l}, \mathbf{u_r}) \geq \lambda_i(\mathbf{u_r}). \tag{2.38}$$

Note that for contact discontinuities, the Lax entropy condition holds with equalities:

$$\lambda_i(\mathbf{u}_l) = \lambda_i(\mathbf{u_l}, \mathbf{u_r}) = \lambda_i(\mathbf{u_r}). \tag{2.39}$$

We conclude this Chapter by giving the following uniqueness result for the solutions to the Riemann problem (2.20a)–(2.20b) [27].

Theorem 2.2 (Evans [27]) *Consider the Riemann problem (2.20a)–(2.20b), with* **H(u)** *a uniformly convex function. Then there exists a unique weak admissible solution for this Riemann problem. Moreover:*

- *if* $u_l > u_r$, *the admissible solution contains a shock curve of speed* $\lambda = [H]/[u]$. *In this case, the solution is given by*

$$\mathbf{u}(x, t) = \begin{cases} \mathbf{u_l}, & x < \lambda t, \\ \mathbf{u_r}, & x > \lambda t. \end{cases} \tag{2.40}$$

- *if* $u_l < u_r$, *the solution contains a rarefaction wave. In this case, the solution is given by*

$$\mathbf{u}(x, t) = \begin{cases} \mathbf{u_l}, & x < \mathbf{H}'(\mathbf{u_l})t, \\ (\mathbf{H}')^{-1}, & \mathbf{H}'(\mathbf{u_l})t < x < \mathbf{H}'(\mathbf{u_r}), \\ \mathbf{u_r}, & x > \mathbf{H}'(\mathbf{u_r})t. \end{cases} \tag{2.41}$$

References

1. P. Lax, *Hyperbolic Systems of Conservation Laws and the Mathematical Theory of Shock Waves* (SIAM, Philadelphia, 1973)
2. D. Serre, *Systems of Conservation Laws. 1. Hyperbolicity, Entropies, Shock Waves* (Cambridge University Press, Cambridge, 1999)
3. D. Serre, *Systems of Conservation Laws. 2. Geometric Structures, Oscillations, and Initial-Boundary Value Problems* (Cambridge University Press, Cambridge, 2000)
4. A. Bressan, *Hyperbolic Systems of Conservation Laws. The One-Dimensional Cauchy Problem* (Oxford University Press, Oxford, 2000)
5. A. Bressan, *Lecture Notes on Functional Analysis. With Applications to Linear Partial Differential Equations* (American Mathematical Society, Rhodes Island, 2013)
6. A. Bressan, *Hyperbolic Conservation Laws: An Illustrated Tutorial* (Springer, Berlin, 2013), pp. 157–245
7. C. Dafermos, *Hyperbolic Conservation Laws in Continuum Physics*. Grundlehren der Mathematischen Wissenschaften, vol. 325 (Springer, Berlin, 2000)
8. V. Sharma, *Quasilinear Hyperbolic Systems, Compressible Flows, and Waves* (CRC Press, Boca Raton, 2010)
9. M. Fey, R. Jeltsch (eds.), *Hyperbolic Problems: Theory, Numerics, Applications* (Birkhäuser Verlag, Basel, 1999)
10. B. Perthame, *Kinetic Formulation of Conservation Laws* (Oxford University Press, Oxford, 2002)
11. A. Bressan, D. Serre, M. Williams, K. Zumbrun, *Hyperbolic Systems of Balance Laws* (Springer, Berlin, 2007)
12. R. LeVeque, *Numerical Methods for Conservation Laws* (Birkhäuser, Basel, 1992)
13. M. Rosini, *Macroscopic Models for Vehicular Flows and Crowd Dynamics: Theory and Applications* (Springer, Berlin, 2013)
14. S. Goldstein, Quart. J. Mech. Appl. Math. **4**, 129 (1951)
15. A. Bressan, *Hyperbolic Systems of Balance Laws* (Springer, Berlin, 2007), pp. 1–78
16. J. Glimm, Commun. Pure Appl. Math. **18**, 697 (1965)
17. S. Bianchini, A. Bressan, Ann. Math. **161**, 223 (2005)

18. T. Liu, Commun. Math. Phys. **57**, 135 (1977)
19. C. Dafermos, J. Math. Anal. Appl. **38**, 33 (1972)
20. A. Bressan, J. Math. Anal. Appl. **170**, 414 (1992)
21. O. Oleinik, Amer. Math. Soc. Transl. **26**, 95 (1957)
22. S. Kružkov, Math. USSR Sb. **42**, 217 (1970)
23. H. Holden, N. Risebro, *Front Tracking for Hyperbolic Conservation Laws* (Springer, Berlin, 2011)
24. G. Sod, J. Comput. Phys. **27**, 1 (1978)
25. T. Liu, J. Math. Anal. Appl. **53**, 78 (1976)
26. P. Lax, Commun. Pure Appl. Math. **10**, 537 (1957)
27. L. Evans, *Partial Differential Equations* (American Mathematical Society, Rhodes Island, 1997)

Chapter 3
One-Equation Local Hyperbolic Models

3.1 Introduction

The simplest macroscopic hyperbolic models derived to investigate the movement of animal and human populations are advection and advection-reaction equations. These models describe the evolution of populations when random movement is negligible compared to directed movement. One-equation models have been employed to investigate the movement and growth of animal populations [1, 2], pedestrian and car traffic [3–7], or the formation of animal trails [8]. We mention here car traffic models, since they were the starting point for models for pedestrian dynamics. More precisely, while traffic models started being developed since the 1950s (see [3, 9]), the interest in developing models for crowd/pedestrian movement started about a decade later (see [10, 11]), being motivated by the acknowledgement of safety issues related to human traffic: evacuation of pedestrians in case of danger [12, 13], structural design of buildings, stadiums or bridges [14, 15]. We should note that more recent studies on pedestrian movement emphasised the need to consider individual social and psychological characteristics (e.g., physical and cognitive abilities, emotional condition, motivation) [16, 17] when investigating the collective movement of pedestrians. While these aspects could be easily incorporated into the kinetic theory of active particles (which will be reviewed in Chap. 6), they are not usually taken into consideration by the models investigated in this monograph.

The general advection-reaction equation that will be discussed throughout this chapter in various contexts related to animal, human, or car movement is given by:

$$\frac{\partial u}{\partial t} + H'(u)\frac{\partial u}{\partial x} = R(u), \quad u(x,0) = u_0(x). \tag{3.1}$$

Here, $u = u(x,t) : \mathbb{R} \times \mathbb{R}^+ \to \mathbb{R}$ is the population density (with $x, t \in \mathbb{R}$ and $t \geq 0$), $H(u)$ is a real nonlinear function describing the advective movement, and

© Springer Nature Switzerland AG 2018
R. Eftimie, *Hyperbolic and Kinetic Models for Self-organised Biological Aggregations*, Lecture Notes in Mathematics 2232,
https://doi.org/10.1007/978-3-030-02586-1_3

$R(u)$ is a real reaction term describing population growth as a result of birth and death processes. Note that the reaction term is used to model group behaviours when the movement occurs on a timescale similar or slower than the timescale of organisms growth [1, 2]. It can also model group behaviours when there is no population dynamics (i.e., no birth or death), but organisms can enter or leave a certain domain [8].

In the following we start reviewing simple models with no reaction ($R(u) = 0$). First, we present in more detail some examples of one-equation models for the evolution of the density of pedestrians/cars with different velocity functions: from constant or linear $H'(u)$ functions, to more complex multi-regime $H(u)$ functions describing different pedestrian behaviours. The traffic models in Sects. 3.2 and 3.3 are described by conservative advection equations. While the majority of examples discussed in Sects. 3.2 and 3.3 focus on single roads, we also mention briefly 1D models describing traffic on networks of roads. In Sects. 3.5 and 3.6 we review some advection-reaction models for car/pedestrian traffic and animal dynamics. In Sect. 3.7 we discuss in more detail an analytical approach used to investigate the speed of travelling wave patterns.

Remark 3.1 Before we continue with the discussion of different types of hyperbolic models for car/pedestrian traffic, we need to discuss the relevance of these continuous models at different spatial scales. The microscopic models consider the behaviour of individual cars/pedestrians under the influence of neighbouring cars/pedestrians. In these cases, the distance between cars/pedestrians is the same order of magnitude as the size of cars/pedestrians (and the spatial scale of the dynamics of the system). Note that these models could also incorporate different internal and external factors that might influence drivers/pedestrians behaviours (e.g., individual time pressure, presence of obstacles; see [18]). The mesoscopic models focus on the probability distribution of cars/pedestrians over a large spatial domain. These models investigate the impact of drivers/pedestrians attitudes and behaviours (e.g., their psychological states, or acceleration/braking and lane-changing behaviours, as given by probabilistic terms) on the overall dynamics of car/pedestrian aggregations, without distinguishing individual space-time dynamics [19]. Finally, the macroscopic models focus on the collective dynamics of cars/pedestrians, ignoring any individual behaviours. Again, the dynamics occurs on a spatial domain much larger than the magnitudes of cars/pedestrians and the distances between them. Returning to the discussion in Chap. 1 on the number of particles/cells/animals required for a mesoscopic/macroscopic model to be relevant, we note that even if the number of cars/pedestrians is usually small (e.g., in the order of tens or hundreds), continuum models have been used very often to describe the behaviours of these cars/pedestrians since existent analytical techniques can be applied to gain a better understanding of these models. It should be emphasised that there are a few studies which focus on the equivalence between microscopic models for car dynamics (in Lagrangian coordinates) and macroscopic models for the same car dynamics (in Eulerian coordinates) [20, 21]. While the fundamental diagram for the steady-state relation between speed and car density (discussed below in Fig. 3.3a, b) is usually the same for the two classes of models, other aspects are different (e.g., the string stability condition for a second-order discrete car-following model in [20] was shown to be different from the linear stability condition of an equivalent continuum model).

3.2 First-Order Traffic Models

The advection equations discussed in Chap. 2 were initially derived in the context of car traffic [3, 4, 22], but they were later applied to pedestrian traffic [4, 23–26], or to combined car and pedestrian traffic [6]. For this reason, we briefly discuss them in the following. The generic equations are

$$\frac{\partial u}{\partial t} + \frac{\partial (uv)}{\partial x} = 0, \tag{3.2}$$

where $u = u(x, t) : \mathbb{R} \times \mathbb{R}^+ \to \mathbb{R}$ describes the density of cars or pedestrians at position x and time t, and v is their velocity. If the velocity is a constant ($v \in \mathbb{R}$), or if it is a function of time and space only ($v = v(x, t) : \mathbb{R} \times \mathbb{R}^+ \to \mathbb{R}$) or a function of density only ($v = v(u(x, t))$), then Eq. (3.2) describes an instantaneous velocity adaptation. More complicated models with non-local velocities, which take into account the velocities of other cars further away have been investigated by Helbing [4] and Helbing et al. [27]. (We will briefly return to these nonlocal models in Chaps. 5 and 6.) If $v = v(x, t)$, applying the so-called total derivative (which describes the temporal changes in a coordinate system moving with velocity v) [4],

$$\frac{d_v}{dt} = \frac{\partial}{\partial t} + v\frac{\partial}{\partial x}, \tag{3.3}$$

leads to the following form of (3.2):

$$\frac{d_v u}{dt} + u\frac{\partial v(x, t)}{\partial x} = 0. \tag{3.4}$$

Equation (3.4) says that density increases in time when the velocity decreases along the domain (i.e., $\partial v/\partial x < 0$). These models are sometimes referred to as "first-order models", being described by a transport equation for the pedestrian/vehicle density and a closed equation for the instantaneous velocity $v(u)$.

In regard to the velocity $v(u)$, these hyperbolic models generally assume that there is a speed limit v_{max}. If there are no obstacles on the road, then cars/pedestrians travel at this maximum speed: $v = v_{max}$ [4, 33]. On crowded roads, the speed is reduced. These two assumptions are incorporated into the equation for the speed as

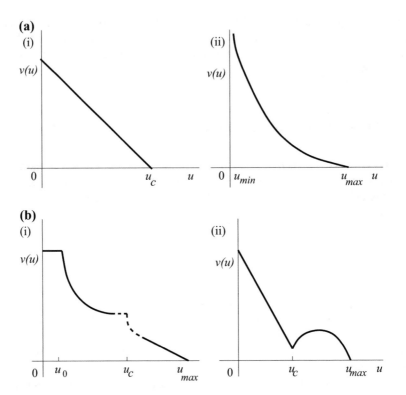

Fig. 3.1 Caricature description of various velocity functions $v(u)$. (**a**) Single stage velocities: (i) linear velocity, as introduced in [3, 28]—see also Eq. 3.5; (ii) logarithmic velocity, as given by Eq. (3.6); (**b**) Multi-stage velocity functions: (i) velocity function introduced in [29, 30]; (ii) velocity function introduced in [31]. The shape in (**b**)(ii) is consistent with experimental measurements in [32]

follows (see also Fig. 3.1a):

$$v(u) = v_{max}\left(1 - \frac{u}{u_{max}}\right), \quad \text{for } 0 \le u \le u_{max}, \tag{3.5}$$

with $u_{max} \in \mathbb{R}$ the maximum (car/pedestrian) density. A slightly different speed function, which also accounts for the slow-down at high densities is

$$v(u) = v_{max} \ln\left(\frac{u_{max}}{u}\right), \quad \text{for } u_{min} \le u \le u_{max}, \tag{3.6}$$

with $v_{max} = v_{max}^* / \ln(u_{max}/u_{min})$. Many models for traffic flow that incorporate such density-dependent speeds have been shown to exhibit shocks, which can propagate either upstream or downstream [4, 34].

While linear speeds are easier to investigate analytically, they cannot always fit observation data for pedestrian movement [14, 29, 32]. Therefore, the authors in

[14, 25] considered the following nonlinear velocity to account for the influence (on the walking velocity of pedestrians) of various physical, physiological and psychological factors represented by the geographical area and the purpose of travel (e.g., a rush hour vs. leisure time in Europe, USA or Asia):

$$v(u) = v_M\left(1 - e^{-\gamma\left(\frac{1}{u} - \frac{1}{u_M}\right)}\right). \tag{3.7}$$

Parameter γ is related to the purpose of travel, while the jam density u_M is given by $u_M = 1/(\beta_G S_m)$. Here, S_m is the mean surface occupied by a motionless pedestrian, and β_G is the coefficient for the geographical area occupied by a human body (e.g., $\beta_G = 1.075$ for European and American pedestrians, $\beta_G = 0.847$ for Asian pedestrians) [14]. Finally, v_M is the average speed, which depends on the geographic area and the purpose of travel. Note that S_m and β_G are related to the individual's repulsion area, an aspect which will be discussed again in Chap. 5 in the context of nonlocal models.

In general, for scalar conservation laws, if the initial density $u(x, 0) \in [u_0^{min}, u_0^{max}]$, for all $x \in \mathbb{R}$, then the solution stays within the same bounds: $u(x, t) \in [u_0^{min}, u_0^{max}]$ for all $t \geq 0$, $x \in \mathbb{R}$ [35]. In [31] this is referred to as the maximum principle for nonlinear hyperbolic equations. However, this principle cannot account for the people behaviour in panic situations, where the maximum density increases beyond the bounds for the initial density. To address this situation, researchers have introduced models with multi-stage velocity functions (see Fig. 3.1b(ii)). For example, Colombo and Rossini [31] constructed analytically a multi-stage velocity function similar to the one described in Fig. 3.1b(ii). This particular shape of $v(u)$ is the result of certain assumptions on the flow $H(u) = uv(u)$ [36]:

- $H(u) = 0$ for $u \in \{0, u_{max}\}$;
- $H(u) \in W^{1,\infty}([0, u_{max}] \times [0, \infty)$ (finite speed of propagation of waves);
- $H(u)$ is strictly concave for $u \in [0, u_c]$ and $u \in [u_c, u_{max}]$ (to avoid mixed waves);
- maximum flow $H(u)$ calculated during panic situations (i.e., $u \in (u_c, u_{max})$ is lower than the maximum flow calculated during normal situations ($u \in (0, u_c)$);
- $H(u)$ has a local minimum at $u = u_c$, describing an increase in the flow when entering a panic regime.

Colombo and Rossini [31] focused mainly on the analytical investigation of (3.2) and showed that in addition to classical shocks and rarefaction waves (see definitions in Chap. 2), this velocity function leads to the formation of nonclassical shocks (i.e., discontinuity solutions that satisfy the Rankine-Hugoniot conditions, but not necessarily Liu's entropy condition; for details see Chapter III in [35], and our discussion of these topics in Chap. 2). Figure 3.2 shows a caricature of a nonclassical shock solution, which can describe panic situations that arise when individuals face a higher density than the one they are in. Note that this solution does not satisfy the maximum principle as defined in [31].

Fig. 3.2 Example of a time
snapshot ($t = 1$) for a
non-classical shock solution,
as observed in [31], when the
change in density is described
by Eq. (3.2), the velocity $v(u)$
is given by a function similar
to the one depicted in
Fig. 3.1, and the initial
condition is $u(0, x) = u^l$ for
$x < 0$ and $u(0, x) = u^r$ for
$x > 0$. Note that for these
non-classical shocks to occur
it is necessary that $u^r > u^l$

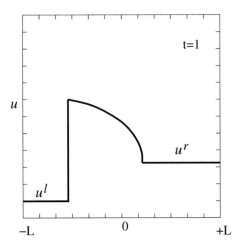

Hughes [29] also considered a multi-stage velocity function (see Fig. 3.1b) to
describe a particular situation when the speed is almost constant at low densities,
and concave-shaped for intermediate densities. Moreover, the velocity takes into
account pedestrian discomfort that arises at very high densities. Thus, Eq. (3.2) can
be modified as follows:

$$\frac{\partial u}{\partial t} + \frac{\partial}{\partial t}\left(ug(u)v^2(u)\bar{v}\right) = 0, \tag{3.8}$$

where $g(u)$ accounts for the discomfort, and the unit vector $\bar{v}$ gives the direction
of movement. The simplest case would be to choose $v(u)$ as described by (3.5),
and $g(u) = 1$. However, Hughes [29] chose a multi-stage velocity function (see
Fig. 3.1b)

$$v(u) = \begin{cases} A, & \text{for } u < u_0, \\ A(\frac{u_0}{u})^{1/2}, & \text{for } u_0 < u \le u_c, \\ A(\frac{u_0u_c}{u_{max}-u_c})^{1/2}(\frac{u_{max}-u}{u})^{1/2}, & \text{for } u_c < u \le u_{max}. \end{cases} \tag{3.9}$$

Under normal conditions, u varies between 0 and u_c. However, under exceptional
panic conditions, density u can cross the threshold u_c and increase up to u_{max} (the
maximum density in exceptional conditions of panic). Note that $v(u_{max}) = 0$, and
thus any movement stops at very high densities. Moreover, there is a discontinuity in
the gradient of the speed at $u = u_c$, which is not realistic. To model the discomfort
of pedestrians in dense crowds, Hughes [29] considered

$$g(u) = \begin{cases} 1, & \text{for } u < u_c, \\ \frac{u(u_{max}-u_c)}{u_c(u_{max}-u)}, & \text{for } u_c < u < u_{max}. \end{cases} \tag{3.10}$$

Di Francesco et al. [37] considered a model similar to (3.8), with the unit vector $\bar{v}$ assumed to be parallel to the gradient of a potential $\phi(x, t)$, which describes the direction of motion of pedestrians at each point in the domain. This potential is given by the Eikonal equation

$$|\nabla \phi| = \frac{1}{v(u)}. \tag{3.11}$$

Note that the choice $v(u) = 1 - u$ can lead to a possible blow-up of $|\nabla \phi|$ when $u \to 1$, rendering the analysis very difficult. Di Francesco et al. [37] tried to address this problem by adding a small viscosity term to the Eikonal equation. They proved the existence and uniqueness of entropy solutions for the regularised equation.

A different class of transport models focus on the concept of "apparent density" felt by pedestrians, which depends on the local gradient of pedestrians' density [38].

$$u^* = u\left(1 + \eta(1 - u)\nabla u \cdot \bar{v}\right), \tag{3.12}$$

where u^* denotes the apparent density, $\bar{v}$ is the direction vector and $\eta > 0$. If the local gradient of density is positive, then the apparent density is larger than the real density ($u^* > u$). This leads to traffic jams (associated with shocks). If, on the other hand, the local gradient is negative, then $u^* < u$. This leads to the formation of vacuum areas (associated with rarefaction waves) [14]. Although De Angelis [38] discussed possible numerical schemes to investigate the resulting models, numerical solutions were not shown, the results being mainly analytical.

The transport models (3.2) can be used to investigate the interactions between pedestrians and their environment, such as moving bridges. In this case, for example, velocity v can depend not only on the density u but also on the lateral acceleration of the bridge (z): $v = v(u, z)$ [14]. More complicated dynamics can include space dislocation (δ) and time delay (τ): $v = v(u(x + \delta, t))g(z(x, t - \tau))$. Note that these models (with space dislocation) are a particular case of nonlocal models, where the changes in velocity are given by nonlocal kernels; we will discuss nonlocal models in Chap. 5.

Many of the models described previously have been investigated in terms of existence, uniqueness and stability of solutions to the corresponding Cauchy problems (see Sect. 2.2 for a discussion of analytical approaches used to investigate these transport models). Moreover, some of those models have also been investigated in terms of optimal solutions that minimise the time of travel between two points, in the presence or absence of a time-dependent highway toll [39]. In fact, optimisation approaches are used very often in the context of car traffic on networks of roads, to choose the optimal travelling route between two points. The density-dependent speed models for traffic flow network have the general form [40, 41]:

$$\frac{du}{dt} + \frac{\partial v_{ij}(u)u}{\partial x} = 0, \tag{3.13}$$

where v_{ij} describe the nonnegative speed of cars along the arc γ_{ij} on the network (with the assumption that if two nodes, i and j, are not connected by a road then $v_{ij} = 0$). Many of these studies on flow networks focus on theoretical optimal control results, such as the minimisation of the cost incurred by different drivers with different departure and arrival times, and finding optimal and equilibrium solutions [41, 42], or the approximation of solutions of Riemann problems at the junctions [43, 44]. There are also a few studies that show numerical simulations of optimal solutions for traffic flow networks [42, 43, 45].

3.3   Second-Order Traffic Models

The models described in the previous section incorporate only the assumption that velocity changes in response to changes in pedestrian/car density. While such an assumption is not completely unrealistic [46], it does not fully describe the complexity of road and highway traffic, since velocity can also change in response to the traffic ahead (e.g., on-ramp highway situations). This leads to the so-called "second-order models": a conservation equation for the density $u \in \mathbb{R}$ of pedestrians/vehicles, coupled with a second equation for the changes in the average velocity $v \in \mathbb{R}^+$:

$$\frac{\partial u}{\partial t} + \frac{\partial (uv)}{\partial x} = 0 \qquad (3.14a)$$

$$\frac{\partial v}{\partial t} + v \frac{\partial v}{\partial x} = F(u). \qquad (3.14b)$$

Note that these two equations correspond to the conservation of mass and momentum, as used in continuum dynamics [47]. The term $F(u)$ describes (1) how the average velocity v adapts to an equilibrium velocity $v_e \in \mathbb{R}^+$, and (2) the drivers' (or pedestrians') awareness of the traffic conditions ahead. The simplest form of $F(u)$ includes only a velocity adaptation: e.g., $F(u) = \frac{1}{\tau}(v_e(u) - v)$, where τ is a relaxation time scale, and v_e could be a constant [48] or could depend on the car density [49]. More complex (and more realistic) terms $F(u)$ can incorporate traffic pressure $(1/u)\partial p/\partial x$, with $p = p(u)$ being the pressure, which is an increasing function of the density u: $F(u) = -\frac{1}{u}p_x + \frac{1}{\tau}(v_e - v)$ [50]. This pressure term describes preventive driving in response to road conditions [50]. In addition, $F(u)$ could also include a small viscosity component $v_0 \partial^2 v/\partial x^2$, which is introduced to smear out sharp shocks and to allow for a continuous description of freeway traffic flow that can exhibit stop-and-go waves [51]: e.g., $F(u) = -\frac{1}{u}p_x + \frac{1}{\tau}(\bar{v}-v) + v_0 v_{xx}$. In terms of pattern formation, model (3.14) can exhibit shock waves, rarefaction waves and clustering [46, 50, 52, 53]. The existence of shock waves has been confirmed both analytically (via the Rankine-Hugoniot conditions) and numerically [50, 52, 53].

In the following we will discuss briefly a few examples of second-order models, by focusing on the various terms that they incorporate. For a more general review of equations describing the changes in the velocity of these models, see [20]. A first example that we focus on here is the model introduced by Payne [54], where the author proposed the following equation for the changes in vehicles speed:

$$\frac{\partial v}{\partial t} + v\frac{\partial v}{\partial x} = \frac{v_e(u) - v}{\tau} - \frac{c_0^2}{u}\frac{\partial u}{\partial x}. \tag{3.15}$$

Parameter τ is the relaxation constant for the vehicles to approach the equilibrium speed v_e, and c_0 is an "anticipation constant" (describing driver's anticipation of the traffic ahead). Investigation of the Riemann problem associated with this model showed the existence of two types of shock waves and two types of rarefaction waves; see Sect. 2.2 for a more detailed discussion of different types of analytical solutions exhibited by hyperbolic equations/systems. Numerical investigation of model (3.14a)–(3.15) confirmed these shock waves and rarefaction waves, as well as the formation of free-flow regions and formation of clusters [52, 53].

One of the drawbacks of the model in [54] is the lack of preservation of the anisotropic nature of traffic: vehicles can move against the flow, with negative speeds [55, 56]. To address this problem, Aw and Rascale [56] introduced the following equation for the changes in speed:

$$\frac{\partial(v + p(u))}{\partial t} + v\frac{\partial(v + p(u))}{\partial x} = 0, \tag{3.16}$$

with $p(u)$ a smooth increasing function of the form $p(u) = u^\gamma$, $\gamma > 0$. Analytical results showed that model (3.14a) with speed (3.16) and stepwise initial conditions (i.e., $(u(x,0), v(x,0)) = (u_l, v_l)$ for $x < 0$ and $(u(x,0), v(x,0)) = (u_r, v_r)$ for $x > 0$) can exhibit shocks, rarefaction waves and contact discontinuities. However, in contrast to the model (3.14a)+(3.15) which can exhibit negative velocities, model (3.14a)+(3.16) exhibits only non-negative velocities.

A generalisation of the model (3.14a)–(3.16) was proposed in [33] to investigate the phase transitions (i.e., jump discontinuities—see Sect. 2.2) between un-congested and congested pedestrian regions. The model is given as follows:

$$\frac{\partial u}{\partial t} + \frac{\partial(uv)}{\partial x} = 0, \tag{3.17a}$$

$$\frac{\partial(uw)}{\partial t} + \frac{\partial(uvw)}{\partial x} = 0, \tag{3.17b}$$

$$w = v + p(u), \tag{3.17c}$$

with $u(x,t) \in \mathbb{R}$ being pedestrian density, $v \in \mathbb{R}^+$ their velocity, $w(x,t) \in \mathbb{R}^+$ the desired velocity in the absence of any obstacles, and p being the offset velocity between the desired and actual velocities (where p is an increasing function of pedestrian density). If one assumes that the desired velocity is a constant $w = V \in \mathbb{R}^+$, then the actual velocity is given by $u = V - p(u)$. This model was then generalised to incorporate congestion constraints (as controlled by parameter ϵ that appears in the correction term Q^ϵ added to pedestrian pressure):

$$\frac{\partial u^\epsilon}{\partial t} + \frac{\partial (u^\epsilon v^\epsilon)}{\partial x} = 0 \tag{3.18a}$$

$$\frac{\partial u^\epsilon w^\epsilon}{\partial t} + \frac{\partial u^\epsilon w^\epsilon v^\epsilon}{\partial x} = 0, \tag{3.18b}$$

$$w^\epsilon = v^\epsilon + p(u^\epsilon) + Q^\epsilon(u^\epsilon), \quad Q^\epsilon(u) = \frac{\epsilon}{\left(\frac{1}{u} - \frac{1}{u^*}\right)^g}, \quad g > 1, \tag{3.18c}$$

with $p(u)$ a convex function describing pedestrian pressure, $p(0) = 0$, $p'(0) \geq 0$ and $p(u) \to \infty$ as $u \to u^*$. The ϵ-correction term (Q^ϵ) for the background pressure p of the pedestrians is turned on when the density is close to the congestion density (i.e., $u \to u^*$). The authors investigate this perturbation problem as $\epsilon \to 0$, and show that the transition from un-congested movement ($u < u^*$) to congested movement ($u = u^*$) corresponds to a phase transition from a compressible to an incompressible flow regime [33]. The interface between congested and un-congested regions could be investigated with the help of Rankine-Hugoniot conditions. However, such an analysis was not performed in [33]. (For a brief discussion of the Rankine-Hugoniot conditions, see Sect. 2.2.)

To understand the relation between the traffic flux uv (i.e., vehicles/pedestrians per hour) and traffic density u (i.e., vehicles/pedestrians per surface area), one uses *fundamental diagrams*. The majority of traffic models in the literature that show phase transitions [3], display a fundamental diagrams similar to the one shown in Fig. 3.3a. However, empirical data similar to Fig. 3.3b suggest that the free-flow can be represented by a curve in the flow-density plane, while the congested traffic is described by a broad spreading of measurement points [57]. This led researchers to try to develop more realistic models that can exhibit two qualitatively-different behaviours [58]: (1) free vehicular traffic at low densities; (2) congested vehicular traffic at higher densities, with one more degree of freedom (thus covering a 2-dimensional domain). However, before we discuss these models in more detail, we need to mention that for pedestrian movement, the fundamental diagrams depend on the psychological status of individuals (i.e., normal versus panic conditions; see Fig. 3.3c), on various infrastructural elements (e.g., upwards/downwards stairs; see Fig. 3.3d), or on whether the flow is unidirectional or bidirectional [32, 59, 60]. Note in Fig. 3.3c that some fundamental diagrams for pedestrian movement show increased flows not only at small/intermediate densities, but also at higher densities (e.g., in panic situations), which is different from the car traffic flows.

A new class of models that investigate phase transitions in vehicular traffic were proposed by Colombo [58, 61]. These models assume that the free-flow and the congested phases are modelled by different equations defined on two different domains (corresponding to the two domains shown in Fig. 3.3b): a first-order model

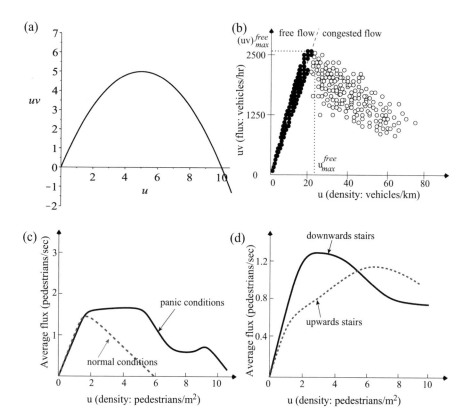

Fig. 3.3 (a) Fundamental traffic diagram showing traffic flux (uv) versus vehicle density (u), for the general flux function $uv_{max}(1 - u/u_{max})$; here $u_{max} = 10$, $v_{max} = 2$. (b) Caricature description of averaged traffic data based on real data from [57, 62–64]. As noted in [64], data in the fundamental diagram depends on the freeway location where the fundamental diagram is measured, and on the traffic demand. The diagrams show that the higher density in traffic flow corresponds to lower average vehicle speed. (c) Caricature description of averaged pedestrian traffic data based on real data shown in [32, 65]. The dotted blue curves approximate pedestrian movement under *normal conditions* [32, 65], while the continuous black curves approximate pedestrian movement under *panic conditions* [32]. (d) Caricature description of averaged pedestrian traffic data based on real data shown in [60, 66]. The dotted blue curves describe the movement on *upward stairs*, while the continuous black curves describe the movement on *downward stairs*

for the free-flow (on domain $\Omega_f \subseteq \mathbb{R}^+ \times \mathbb{R}^+$), and a second-order model for the congestion (on domain $\Omega_c \subseteq \mathbb{R}^+ \times \mathbb{R}^+$). For example, the model in [58] is described by the following equations:

$$\frac{\partial u}{\partial t} + \frac{\partial(uv_f(u))}{\partial x} = 0, \quad \text{for } (u, p) \in \Omega_f \text{ (free flow)}, \tag{3.19a}$$

$$\begin{cases} \frac{\partial u}{\partial t} + \frac{\partial(uv_c(u,p))}{\partial x} = 0, \\ \frac{\partial p}{\partial t} + \frac{\partial((p-Q)v_c(u,p))}{\partial x} = 0, \end{cases} \quad \text{for } (u, p) \in \Omega_c \text{ (congested flow)}, \tag{3.19b}$$

with $v_f(u)$ the velocity for the free phase and $v_c(u, p)$ the velocity for the congested phase:

$$v_f(u) = \left(1 - \frac{u}{u_{max}}\right)v^*, \quad v_c(u, p) = \left(1 - \frac{u}{u_{max}}\right)\frac{p}{u}, \tag{3.20}$$

where v^* is the fee-flow speed and u_{max} is the maximum vehicle density. The variable p describes a perturbation or deviation from the equilibrium state [67]. Finally, parameter Q depends on the road conditions and characterises the phenomenon of wide traffic jams. The domains for the two phases are defined as

$$\Omega_f = \{(u, p) \in [0, u_{max}] \times [0, +\infty] | v_f(u) \geq v_f^*, q = pv^*\}, \tag{3.21a}$$

$$\Omega_c = \left\{(u, p) \in [0, u_{max}] \times [0, +\infty] | v_c(u, p) \leq v_c^*, \frac{p - Q}{u}\right.$$

$$\left. \in \left[\frac{Q_- - Q}{u_{max}}, \frac{Q_+ - Q}{u_{max}}\right]\right\}, \tag{3.21b}$$

with $v_f^* < v_c^*$ two threshold speeds (above v_f^* the flow is free, below v_c^* the flow is congested). Moreover, parameters $Q_- \in (0, Q)$ and $Q_+ \in (Q, +\infty)$ depend on the environment conditions and define the width of the congested region [68]. These two phase domains are invariant, i.e., if the initial data is in the free (respectively congested) phase, then the solution stays in the free (respectively congested) phase [58]. Moreover, the solution of this model (with initial conditions $(u(x, 0), v(x, 0)) = (u_l, p_l)$ for $x < 0$, and $(u(x, 0), v(x, 0)) = (u_r, p_r)$ for $x > 0$) has been shown to be formed by phase-transition waves or rarefaction waves in Ω_f, and by shocks, contact discontinuities or rarefaction waves in Ω_c [69].

A slightly different two-phase model was proposed in [70], where the authors started with the model in [61] and assumed that $Q = 0$, while the free-flow and congestion vehicle speeds are described by:

$$v = \begin{cases} v_f(u) := v^*, & \text{for } u \in \Omega_f \\ v_c(u, p) := \left(1 - \frac{u_{max}}{u}\right)\left(a(u - u_c) + \frac{u_c v^*}{u_c - u_{max}}\right)(1 + p), & \text{for } (u, p) \in \Omega_c. \end{cases} \tag{3.22}$$

Here, u_c is a critical vehicle density (marking the transitions between the two phases), and parameter $a \in [-A, 0)$ (with $A = u_c v^*/(u_c - u_{max})^2$). See, in Fig. 3.4 two examples of congested velocities for two different values of parameter a. The Riemann problem associated with model (3.19) and speeds (3.22) (i.e., the Cauchy problem with piecewise constant initial condition $(u(x, 0), p(x, 0)) = (u_l, p_l)$ for $x < 0$ and $(u(x, 0), p(x, 0)) = (u_r, p_r)$ for $x > 0$; see also the discussion in

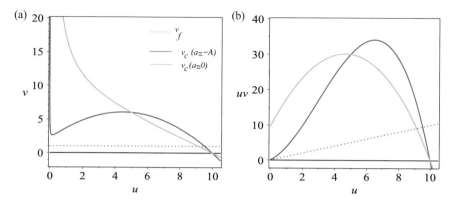

Fig. 3.4 Examples of (**a**) free-flow speed (v_f) and congested speed (v_c); (**b**) the flux function $H(u) = uv$, for the second-order traffic model (3.19)–(3.22) in [70]. The parameters are: $u_c = 5$, $u_{max} = 10$, $v^* = 0$, $A = 2$, and two values of parameter a: $a = -0.19$ (red curve), $a = -0.01$ (green curve)

Sect. 2.2) can exhibit different types of solutions when the initial conditions belong to the different phase domains [70, 71]:

- For $u_l, u_r \in \Omega_f$, the solution is represented by a contact discontinuity from u_l to u_r [71].
- For $u_l, u_r \in \Omega_c$, the solution is represented by a shock/rarefaction wave from u_l to u_m, and a 2-contact discontinuity from u_m to u_r, where u_m is the solution to the system: $p_l/u_l = p_m/u_m$, and $v_c(u_r, p_r) = v_c(u_m, p_m)$ [71].
- For $u_l \in \Omega_c, u_r \in \Omega_f$, the solution is represented by a 1-rarefaction wave from u_l to u_m, and a contact discontinuity from u_m to u_r [71].
- For $u_l \in \Omega_f, u_r \in \Omega_c$, the solution is represented by a shock from u_l to u_{m-}, and a 2-contact discontinuity from u_{m-} to u_r, where u_{m-} is the solution of the system $Q_-/u_{max} = p_{m-}/u_{m-}$ and $v_c(u_{m-}, p_{m-}) = v_c(u_r, p_r)$ [71] (Fig. 3.5).

This class of two-phase models has been further developed in other studies that used different functions for the congested speed. For example, Goatin [72] assumed that the congested speed is similar to the one in the Aw-Rascale (AR) model [56] (see Eq. (3.16)). By generalising the Aw-Rascale model to a two-phase model, the author aimed to correct some drawbacks of the original AR model, namely: lack of well-posedness near vacuum (i.e., when density is close to zero), and dependence on initial data for the maximum speed reached by vehicles on empty roads. The new two-phase model in [72] was described by the following equations:

$$\begin{cases} \frac{\partial u}{\partial t} + \frac{\partial (uv(u))}{\partial x} = 0, & \text{for } (u, v) \in \Omega_f \text{ (free flow),} \\ v(u) := v_f(u) = \left(1 - \frac{u}{u_{max}}\right)v^* \end{cases} \quad (3.23a)$$

$$\begin{cases} \frac{\partial u}{\partial t} + \frac{\partial (uv_c(u,p))}{\partial x} = 0, & \text{for } (u, v) \in \Omega_c \text{ (congested flow),} \\ \frac{\partial u(v+p(u))}{\partial t} + \frac{\partial (uv(v+p(u)))}{\partial x} = 0, \\ p(u) = v_{ref} \ln\left(\frac{u}{u_{max}}\right). \end{cases} \quad (3.23b)$$

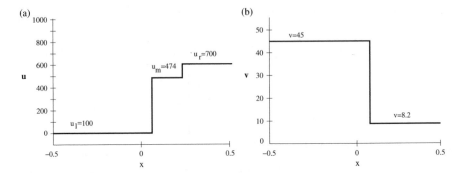

Fig. 3.5 Example of (**a**) density and (**b**) velocity profiles for a phase-transition model described in [71]. The graphs, which are re-drawn from [71], correspond to the case $u_l \in \Omega_f$, $u_r \in \Omega_c$, discussed above, where a new equilibrium state u_m appears between u_l and u_r. The parameters are: $v_* = 45$, $u_{max} = 1000$

Here, p is a pressure function related to drivers anticipation of the traffic ahead of them. The invariant domains for the free-flow and congested flow are described by:

$$\Omega_f = \{(u, v) \in [0, u_{max}^f] \times [v_f^*, v^*] | v = v_f(u)\},$$

$$\Omega_c = \{(u, v) \in [0, u_{max}] \times [0, v_c^*] | p(Q) \le v + p(u) \le p(u_{max})\},$$

with v_f^* and v_c^* are two threshold speeds (above v_f^* the flow is free, and below v_c^* the flow is congested; $v^* > v_f^* > v_c^*$). Parameter $Q \in (0, u_{max})$ depends on environmental conditions and determines the width of the congested region, and the maximal free-flow density u_{max}^f and reference velocity v_{ref} satisfy condition $v_f^* + v_{ref} \ln(u_{max}^f / u_{max}) = 0$. Analytical results for this model showed the existence of shocks, rarefaction waves and contact discontinuities depending on the initial data [72].

3.4   Third-Order Traffic Models

A third-order model was introduced in [73, 74], where in addition to an equation for the conservation of the density ($u \in \mathbb{R}$) and an equation for the evolution of the velocity ($v \in \mathbb{R}^+$), the model contained also an equation for the velocity variance $\theta \in \mathbb{R}^+$. To this end, Helbing [73, 74] started with the following gas-kinetic traffic model for the phase-space density of vehicles $\rho(w, t, x)$ (moving with velocity w, but having a desired velocity w_e),

$$\frac{\partial \rho}{\partial t} + \partial \rho w \partial x + \frac{\partial}{\partial w}\left(\rho \frac{dw}{dt}\right) + \frac{\partial}{\partial w_e}\left(\rho \frac{dw_e}{dt}\right) = \left(\frac{\partial \rho}{\partial t}\right)_{fluctuation}$$

$$+ \left(\frac{\partial \rho}{\partial t}\right)_{interaction}. \quad (3.24)$$

This equation includes (on the right-hand side) a velocity fluctuation term due to imperfect driving, and a deceleration term due to interactions between different cars. By considering the moment equations associated with this kinetic model, Helbing [73, 74] derived the following third-order traffic model for the spatial density $u(x, t) = \int \rho(w, t, x)dv$, the average velocity $v(t, x) = \int w(\rho(w, t, x)/u(t, x))dw$, and the velocity variance $\theta(t, x) = \int (w - v(t, x))^2(\rho(w, t, x)/u(t, x))dw$:

$$\frac{\partial u}{\partial t} + \frac{\partial(uv)}{\partial x} = 0, \tag{3.25a}$$

$$\frac{\partial v}{\partial t} + v\frac{\partial v}{\partial x} + \frac{1}{u}\frac{\partial(u\theta)}{\partial x} = \frac{1}{\tau}(v_e(u, v, \theta) - v), \tag{3.25b}$$

$$\frac{\partial \theta}{\partial t} + v\frac{\partial \theta}{\partial x} + 2\theta\frac{\partial v}{\partial x} = \frac{2}{\tau}(\theta_e(u, v, \theta) - \theta). \tag{3.25c}$$

Here, $v_e(u, v, \theta)$ is an equilibrium velocity (of the stationary and spatially homogeneous traffic flow), $\theta_e(u, v, \theta)$ is an equilibrium variance, and τ is the relaxation time (for the adaptation in the average velocity and average variance to the equilibrium velocity and variance). In the equation for velocity, the term $v\partial(v)/\partial x$ describes velocity changes at x caused by average vehicle motion, while the term $(1/u)(\partial u\theta)/(\partial x)$ accounts for driver's awareness of the traffic ahead [73]. (Note that the number "2" in the equation for the velocity variance arises during the calculation of the second moment $m_2 = \int dw \int w^2(w_e)\rho(w, w_e, t, x)dw_e$.)

As we have previously discussed, one of the most studied characteristics of vehicular traffic is equilibrium solutions. Since the number of vehicles is conserved, the equilibrium traffic of all these models (including model (3.25)) is uniquely determined by the averaged spatial density $\bar{u} = \int u(x, t)dx$. The equilibrium velocity and equilibrium variance can be obtained after integrating the previous equations [73]. This particular model added more realism to the literature of complex highway traffic: for small variance in speed, the cars travelled more or less at the same speed, while for large variance in the speed some vehicles travelled faster than others, causing lane changes.

3.5 Traffic Models that Include Reaction Terms

For all these different types of traffic models, reaction terms can be added to describe the rates at which pedestrians (or cars) enter or leave a particular domain (see [4, 8] and the references therein). For example, the following equation was derived by Helbing et al. [8] to describe the movement of pedestrians belonging to a sub-population u_a (part of the whole population u):

$$\frac{\partial u_a}{\partial t} + \frac{\partial(u_a v(x, t))}{\partial x} = R_a^+(x, t) - R_a^-(x, t). \tag{3.26}$$

Here, $R_a^{\pm}(x, t)$ are the rates at which pedestrians join and leave sub-population u_a. (In the context of car traffic, $R^{\pm}(x, t)$ usually denote the inflow/outflow of cars on ramps.) In [8], this reaction-advection equation was then coupled with an equation for the environmental changes (i.e., trail formation) caused by these pedestrians:

$$\frac{dG}{dt} = G_0(x) - G(x, t) + \left(1 - \frac{G(x, t)}{G_{max}(x)}\right) \sum_a K(x) u_a(x, t). \qquad (3.27)$$

Here G is a ground potential that measures the comfort of walking (with G_{max} the maximum clearing of the trail from vegetation), $G_0(x)$ describes the existent trails (at positions x), and $K(x)$ measures the attractiveness of the trail (at different spatial positions x). Numerical simulations showed the formation of a simple trail system (i.e., network patterns; see also Fig. 1.11), the shape of which depended on the magnitude of K.

Models for pedestrian traffic (with pedestrians entering/leaving the domain) exhibit various types of human behaviours, depending on the density of individuals [4]. For example, at low densities the pedestrian flow is similar to streamlines of fluids, while at higher densities the pedestrians organise themselves into different lines of uniform walking direction [24]. Similar to car traffic, pedestrian traffic can lead as well to jams [24]. When the rates $R_{\pm}(u)$ are nonzero, models (3.26) can also exhibit patterns with increasing amplitude, such as localised clusters and stop-and-go waves (which are sequences of traffic jams that alternate with free traffic) [75]. Other pedestrian models (of Langevin type) can incorporate social interactions as well as boundary forces [23]. In particular, these models assume that pedestrians are attracted/repelled by other persons or by boundary objects (e.g., buildings). In Chap. 5, we will discuss in more detail the importance of these social interactions to the formation of various group patterns.

As we have previously seen in Sect. 3.3, reaction terms can be also added to second-order traffic models, to describe highway entries and exits, or local changes in traffic flow caused by inhomogeneities in the road. The simplest and probably one of most widely-used reaction terms is the relaxation velocity shown in Eq. (3.15). In the following we discuss the more general model introduced in [76], which incorporates different reaction terms with different physical meanings:

$$\frac{\partial u}{\partial t} + \frac{\partial u v(u, p)}{\partial x} = s_u(t, x, u, p), \qquad (3.28a)$$

$$\frac{\partial p}{\partial t} + \frac{\partial (p - Q) v(u, p)}{\partial x} = s_p(t, x, u, p), \qquad (3.28b)$$

with the velocity $v(u, p) = (1 - u/u_{max})(p/u)$. The explicit form of the source functions depends on the assumptions being modelled:

- vehicles entries and exits along an interval $[a, b]$ [76]:

$$s_u(t, x, u, p) = a_{in}(t, x)\left(1 - \frac{u}{u_{max}}\right) - a_{out}(t, x)\frac{u}{u_{max}},$$

$$s_p(t, x, u, p) = -\left(\frac{a_{in}(t, x)}{u_{max}} + \frac{a_{out}(t, x)}{u_{max}}\right)(p - Q).$$

Here, $a_{in}(t, x) = g_{in}(t)\chi_{[a,b]}(x)$ and $a_{out}(t, x) = g_{out}(t)\chi_{[a,b]}(x)$, with $g_{in}(t)$ and $g_{out}(t)$ the fraction of the traffic density (per unit time) that enters and exits the road, respectively. The source term s_p describes a "stabilising" effect caused by the entries and exits on vehicle flow [76].

- relaxation term added to the second equation, to model acceleration towards an equilibrium velocity $v_e(u)$ [27, 76]:

$$s_u(t, x, u, p) = a(t, x)\left(1 - \frac{u}{u_{max}}\right),$$

$$s_p(t, x, u, p) = -\frac{a(t, x)}{u_{max}}(p - Q) + \frac{u_{max}}{u_{max} - u}\frac{u(v_e - v)}{\tau}.$$

with τ the relaxation time corresponding to the average acceleration time. This can be a constant, or can be density-dependent ($\tau = \tau(u)$).

- local changes in traffic speed (while the total vehicle density is conserved) [76]:

$$s_u(t, x, u, p) = 0,$$

$$s_p(t, x, u, p) = \chi(x)ua(t, u, p),$$

where $\chi(x)$ gives the location of a descent along the highway (between some points x_1 and x_2), and $a(t, u, p) \geq 0$ is the mean acceleration.

For more examples of source terms see [76] and the references therein. The general model (3.28) has been shown to be well posed (for suitable conditions on the source terms), thus admitting a unique solution defined for all $t > 0$ [76]. In contrast to models without a source, these models have been shown to capture the formation of queues (e.g., as for $s_u = 0$ and $s_p(t, x, u, p) = \chi(x)ua(t, u, p)$ in [76]). In this case, the increase in vehicle speed (triggered by a decrease in density associated with the exit) leads to an accumulation of vehicles into a wide moving traffic jam.

Remark 3.2 The positive and negative aspects of different traffic models have been discussed in various studies (see, for example, [55, 77, 78]), and thus we will not review them here. However, we would like to emphasise that the continuous development of all these different traffic models was triggered by the lack of realistic patterns exhibited by some earlier models. For example, since many of first-order models assumed that the average speed (v) was in equilibrium with the density

(u), they could not describe non-equilibrium situations taking place at on-ramps, or stop-and-go traffic (where the speed changes). Therefore, second-order models were introduced to add a new dynamic equation for the average velocity v. Even some of these models do not always describe the self-organised phenomena of stop-and-go waves when the density increases above a critical value [54], which lead to further refining of models [79, 80].

We will return to traffic models in Chap. 4, when we will discuss not only the patterns generated by multiple populations of car drivers/pedestrians, but also the patterns generated by ants moving along pheromone trails, and the application of systems of hyperbolic equations to the traffic of molecules along filaments.

3.6   Advection-Reaction Equations for Animal Population Dynamics

In the context of animal population growth and movement, the most common types of reaction terms $R(u)$ that appear in the general advection equation:

$$\frac{\partial u}{\partial t} + h(u)\frac{\partial u}{\partial x} = R(u), \tag{3.29}$$

are [81]:

1. quadratic (logistic) growth: $R(u) = ru(1 - u)$;
2. cubic growth ("Allee" effect): $R(u) = ru(1 - u)(u - \alpha)$, with $\alpha \in (0, 1)$.

Mickens [1] used a simpler version of (3.1), with constant velocity ($h(u) = 1$) and logistic growth ($R(u) = r_1u(1 - r_2u)$), to describe the dynamics of a generic population u. Because of the simplicity of this model, it was possible to find exact time-dependent and time-independent solutions. The time-independent solutions (i.e., the stationary solutions) were shown to be either bounded or unbounded, depending on the initial condition. Another type of solution discussed by Mickens [1] was the travelling front solution (see Fig. 3.6b). Again, because of the simplicity of the model it was possible to find an exact expression for these travelling fronts.

A slightly more complex model was investigated by Lika and Hallam [2]. When the advective velocity was described by a linear term $h(u) = -ku$ (and the growth was logistic), model (2.4) could exhibit shock solutions (see Fig. 3.6a). In particular, advective waves with low speed were shown to steepen and lead to shock formation. Note that these patterns are very common in large human aggregations which result in panic stampede. The model in [2] can also exhibit travelling front patterns (see Fig. 3.6b). These solutions, which exist only for speeds $c > 1$, are stable to certain semi-finite domain perturbations. These authors also showed that it is not possible to have travelling front patterns when the initial conditions have compact support. This is in contrast with the classical Fisher's equation, where initial conditions with

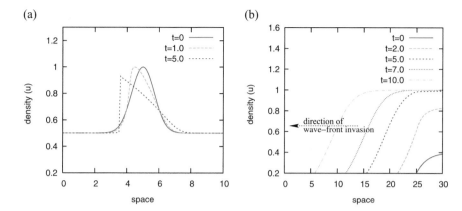

Fig. 3.6 (a) Shock patterns exhibited by the advection model (3.2) with density-dependent speeds $v = u(1 - u)$. The initial condition is $u(0) = 0.5 + 0.5e^{-(x-5)^2}$. (b) Travelling front patterns exhibited by the advection reaction model (2.4), with $h(u) = u$, and $f(u) = u(1 - u)$. The initial data is similar to the data used by Lika and Hallam [2]: if $x \le x_0$ then $u(0) = 0.2e^{0.5(x-x_0)}$, and if $x > x_0$ then $u(0) = 0.2(2 - e^{-0.5(x-x_0)})$. We take $x_0 = 5L/6$, where $L = 30$ is the domain length

compact support lead to travelling front solutions moving with the minimum speed c (where $c = 2$) [82]. In Sect. 3.7, we will discuss in more detail the approach to find the speed and shape of travelling-wave solutions.

To reproduce the intricate animal population dynamics during self-organised behaviours, the models derived in the recent years contain more complex assumptions—such as nonlocal attractive-repulsive velocities [83]. We will discuss these models in more detail in Chap. 5, where we will also show the numerical patterns displayed by them.

We conclude this section by observing that some hyperbolic models consider the change of the population with respect to age $a \in \mathbb{R}^+$ (and not necessarily with respect to space x). The classical example is represented by the McKendrick–Von Foerster equation [84, 85]:

$$\frac{\partial u}{\partial t} + \frac{\partial u}{\partial a} = -\mu(a)u(a, t), \quad a > 0. \tag{3.30}$$

Here $\mu(a) \in \mathbb{R}^+$ represents the age-dependent death rate of population $u \in \mathbb{R}$. Therefore, this equation models the changes in a population as a result of becoming older and eventually dying. If $\mu(a) = 0$ then population change is only the result of getting older. The birth of the population is usually incorporated into the initial conditions:

$$u(0, t) = \int_0^\infty b(a)u(a, t)da, \quad t > 0, \tag{3.31}$$

where $b(a) \in \mathbb{R}^+$ is the birth rate for age a. The problem is completed by the specification of the initial age distribution: $u(a, 0) = u_0(a)$. This equation can be solved exactly along the characteristic lines to obtain a closed-form solution [86, 87]:

$$u(a, t) = \begin{cases} u(0, t - a)e^{-\int_0^a \mu(s)ds}, & \text{when } a < t, \\ u_0(a - t)e^{-\int_{a-t}^a \mu(s)ds}, & \text{when } a \geq t. \end{cases} \tag{3.32}$$

Note that these models could also include changes in population size with respect to the space variable x:

$$\frac{\partial u}{\partial t} + \frac{\partial u}{\partial a} + \frac{\partial u}{\partial x} = -\mu(a)u(a, t). \tag{3.33}$$

A slightly different type of age-structured model with local birth terms will be presented at the end of Sect. 4.6.

3.7 Analytical Approaches for the Investigation of Patterns: Speed of Travelling Waves

Since the travelling wave solutions are one of the simplest (but biologically very important) solutions exhibited by the transport models for collective movement of animals (as well as models for car/pedestrian movement), in the following we present briefly the analytical approaches used to calculate the propagating speed and shape of these solutions for a simple 1-equation hyperbolic model introduced in Lika and Hallam [2]. For a more detailed discussion of travelling waves (and their propagation speeds) in various biological populations we refer the reader to [88–90] (although all these studies focus on parabolic equations).

Lika and Hallam [2] investigated the existence and stability of travelling wave solutions for the following (non-dimensionalised) advection-reaction model describing the movement and growth of a population with density $u(x, t)$:

$$\frac{\partial u}{\partial t} - u\frac{\partial u}{\partial x} = u(1 - u). \tag{3.34}$$

Travelling wave solutions for Eq. (3.34) are nonnegative functions $0 \leq u(x, t) \leq 1$ (which connect the two spatially homogeneous steady states $u = 0$ and $u = 1$). Writing $u(x, t) = w(z)$, with $z = x + ct$ the travel-wave coordinate and $c > 0$ the wave speed to be determined, leads to the following differential equation

$$(c - w)\frac{dw}{dz} = w(1 - w), \tag{3.35}$$

together with the boundary conditions

$$\lim_{z \to -\infty} w(z) = 0, \quad \lim_{z \to +\infty} w(z) = 1. \tag{3.36}$$

The travelling wave solution should also satisfy $\frac{dw}{dz} > 0$. By linearising Eq. (3.35) about the steady states $w = 0$ and $w = 1$, one obtains that the trivial state $w = 0$ is unstable for all $c > 0$ while the state $w = 1$ is stable for $c > 1$ and unstable for $0 < c < 1$.

To investigate whether it is possible to have travelling waves with speeds $c < 1$ or $c > 1$, the authors integrate Eq. (3.35) to obtain a closed-form solution:

$$\ln \left(\frac{w^c}{(1 - w)^{c-1}} \right) = z + C^*, \quad C^* = const. \tag{3.37}$$

Due to the invariance of the travelling wave solution to any shift in the origin of the coordinate system, one can assume that for $z = 0$ one has $w = 1/2$, and thus the solution is $\ln(2w^c/(1 - w)^{c-1}) = z$. Moreover, the steepness of the wave at $z = 0$ is $\frac{dw(0)}{dz} = 1/(4c - 2)$, suggesting that the wave steepness decreases as the speed c increases; see also Fig. 3.7a.

The boundary conditions (3.36) are then used to determine the existence/absence of travelling wave solutions. For example, the boundary condition at $z = -\infty$ is satisfied by solution of (3.37) for any $c > 0$ and any constant C^*. In contrast, the boundary condition at $z = +\infty$ is (1) satisfied by the solution of (3.37) for any constant C^* when $c > 1$; (2) not satisfied by the solution of (3.37) for any constant C^* when $c < 1$. The authors in [2] conclude that for any $c > 1$ there is a travelling wave $u(x, t) = w_c(x + ct)$ that satisfies $0 \le w \le 1$ and $w(0) = 1/2$, while for

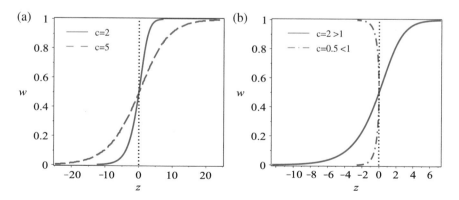

Fig. 3.7 (a) Description of two possible travelling wave profiles, as given by Eq. (3.37) with $C^* = 0$, for two speed values: $c = 2$ and $c = 5$. Note that lower speed is associated with a very steep profile. (b) Comparison between the solution of (3.37) with $C^* = 0$, when the speed $c = 0.5 < 1$ (and no travel front exists) and $c = 2 > 1$ (and a travel front exists)

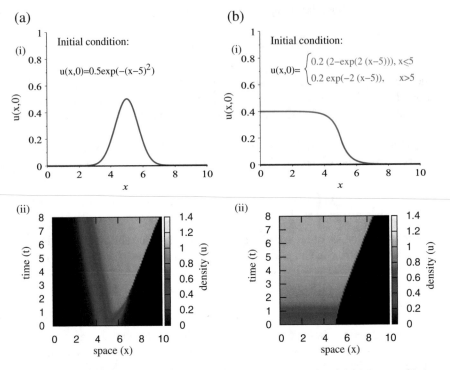

Fig. 3.8 (i) Initial density profile and (ii) spatio-temporal patterns for Eq. (3.34) when considering two types of initial conditions: (a) $u(x, 0) = 0.5 \exp(-(x - 5)^2)$, and (b) $u(x, 0) = 0.2 * (2 - \exp(2 * (x - 5)))$ for $x \le 5$ and $u(x, 0) = 0.2 \exp(-2(x - 5))$. The simulations are performed on a finite domain $[0, L] = [0, 10]$ with: (a) periodic boundary conditions, and (b) no-flux boundary conditions.

$c \le 1$ there is no such solution; see also Fig. 3.7b. (From (3.37) note that for $c = 1$, $w(z)$ is increasing exponentially and reaches the threshold $w = 1$ at some finite z.)

Figure 3.8 shows the numerically-simulated travelling wave/front solutions for two different types of initial conditions:

(a) a Gaussian-like function $u(x, 0) = 0.5e^{-(x-5)^2}$;
(b) a composite function, as in [2]:

$$u(x, 0) = \begin{cases} 0.2(2 - e^{2(x-5)}), & x \le 5, \\ 0.2e^{-2(x-5)}, & x > 5. \end{cases} \tag{3.38}$$

Note that the logistic proliferation of population u leads to the spread of the front (in Fig. 3.8a(ii) the spread can occur both ahead and behind the initial aggregation), which adds to the transport of the population to the right, via the advection term.

Lika and Hallam [2] also investigated the stability of the travelling front to small semi-finite domain perturbations. The authors have re-written $u(x, t) = y(z, t)$ with

$z = x + ct$, and then added a small perturbation v to the travel wave solutions $w_c(z)$ (i.e., $y(z, t) = w_c(z) + v(z, t)$) and investigated when the perturbations decayed towards 0. The perturbations were defined on a semi-domain as $v(z, t) = 0$ for $z \in (-\infty, a)$ with some $a \in \mathbb{R}$. For more details, we refer the reader to [2].

Remark 3.3 We need to emphasise that for parabolic PDEs, the existence and stability of travelling front solutions is proven rigorously with the help of a classical comparison argument based on upper and lower solutions; see [91, 92]. Since for hyperbolic PDEs a similar comparison principle does not hold (see [93]), proving the existence and uniqueness of such travelling waves/fronts for kinetic transport equations (even 1-population models) is more complicated, and is not the aim of this study. Nevertheless, we need to emphasise here that all these theoretical aspects are still open problems for the great majority of the local and nonlocal kinetic and hyperbolic models presented throughout this monograph (as the majority of hyperbolic/kinetic models that investigate the travelling wave/front patterns focus mainly on the speed at which the front of the population wave invades new territories).

3.8 Numerical Approaches

For numerical simulations of these local models, one can use various numerical schemes: from first order dissipative finite difference upwind and Lax-Friedrichs schemes, to second-order dispersive finite-difference Lax-Wendroff and MacCormack schemes, and even finite volume schemes (e.g., Godunov, Roe or Engquist-Osher schemes). To minimise the dispersion that appears in second-order schemes, various flux limiters (minmod, superbee, monotonized central—MC) are used. A more detailed discussion of these numerical schemes can be found in LeVeque [47, 94]. For hyperbolic models describing traffic on networks, in the past the focus was mainly on first-order or second-order numerical schemes [95], although more recent studies have started to focus on the construction of higher order finite volume schemes [96, 97]. For an overall discussion of these various numerical schemes (and other numerical schemes derived to discretise kinetic models), see Chap. 7. Note that these numerical schemes can be applied also to systems of local and nonlocal equations (described in Chaps. 4 and 5), and for this reason in the next two chapters we will not discuss again these numerical approaches.

Regarding the specific model discussed in Sect. 3.7, the numerical simulations shown in Fig. 3.8 were obtained by discretising the transport equation using a second order MacCormack finite difference scheme (which includes also the reaction terms). The same MacCormack scheme was used to create the travelling wave and shock profiles shown in Fig. 3.6.

References

1. R. Mickens, SIAM Rev. **30**(4), 629 (1988)
2. K. Lika, T. Hallam, J. Math. Biol. **38**, 346 (1999)
3. M. Lighthill, G. Whitham, Proc. R. Soc. Lond. Ser. A. **229**(1178), 317 (1955)
4. D. Helbing, Rev. Mod. Phys. **73**, 1067 (2001)
5. D. Helbing, P. Monar, I. Farkas, K. Bolay, Environ. Plann. B. Plann. Des. **28**, 361 (2001)
6. R. Borsche, A. Meurer, Discret. Contin. Dyn. Syst. Ser. S **7**(3), 363 (2014)
7. S. Göttlich, C. Harter, Netw. Heterog. Media **11**(3), 447 (2016)
8. D. Helbing, F. Schweitzer, J. Keltsch, P. Molnar, Phys. Rev. E **56**, 2527 (1997)
9. I. Prigogine, R. Herman, *Kinetic Theory of Vehicular Traffic* (Elsevier, New York, 1971)
10. L. Henderson, Nature **229**, 381 (1971)
11. L. Henderson, Transp. Res. **8**, 509 (1975)
12. R. Colombo, M. Garavello, M. Lécureux-Mercier, Math. Models Methods Appl. Sci. **22**(4), 1150023 (2012)
13. D. Yanagisawa, A. Kimura, R. Nishi, A. Tomoeda, K. Nishinari, Distrib. Auton. Robot. Syst. **8**, 227 (2009)
14. F. Venuti, L. Bruno, Phys. Life Rev. **6**(3), 176 (2009)
15. F. Venuti, L. Bruno, Eng. Struct. **56**, 95 (2013)
16. C. Schäfer, R. Zinke, L. Künzer, G. Hofinger, R. Koch, Transp. Res. Proc. **2**, 636 (2014)
17. A. Sieben, J. Schumann, A. Seyfried, PLoS ONE **12**(6), e0177328 (2017)
18. S. Hoogendoorn, P. Bovy, Transp. Res. Board **1710**, 28 (2000)
19. S. Hoogendoorn, P. Bovy, Proc. Inst. Mech. Eng. Pt. I J. Syst. Control Eng. **215**(4), 283 (2001)
20. W. Jin, Transp. Res. B Methodol. **93**(A), 543 (2016)
21. R. Abeyaratne, Int. J. Mech. Eng. Educ. **42**(3), 185 (2014)
22. B. Piccoli, M. Garavello, *Traffic Flow on Networks* (American Institute of Mathematical Sciences, San Jose, 2006)
23. D. Helbing, P. Molnar, Phys. Rev. E **51**(5), 4282 (1995)
24. D. Helbing, Complex Syst. **6**, 391 (1992)
25. F. Venuti, L. Bruno, N. Bellomo, Math. Comput. Model. **45**(3–4), 252 (2007)
26. N. Bellomo, C. Dogbé, Math. Models Methods Appl. Sci. **18**, 1317 (2008)
27. D. Helbing, A. Hennecke, V. Shvetsov, M. Treiber, Math. Comput. Model. **35**(5–6), 517 (2002)
28. P. Richards, Oper. Res. **4**, 42 (1956)
29. R. Hughes, Transp. Res. B **36**, 507 (2002)
30. R. Hughes, Annu. Rev. Fluid Mech. **35**, 169 (2003)
31. R. Colombo, M. Rosini, Math. Method Appl. Sci. **28**(13), 1553 (2005)
32. D. Helbing, A. Johansson, H.Z. Al-Abideen, Phys. Rev. E **75**, 046109 (2007)
33. C. Appert-Rolland, P. Degond, S. Motch, Netw. Heterog. Media **6**(3), 351 (2011)
34. G. Schütz, Exactly solvable models for many-body systems far from equilibrium, in *Phase Transitions and Critical Phenomena*, vol. 19 (Academic Press, London, 2001), pp. 1–251
35. P. Lefloch, *Hyperbolic Systems of Conservation Laws. The Theory of Classical and Nonclassical Shock Waves*. Lectures in Mathematics. ETH Zürich (Birkhäuser, Basel, 2002)
36. R. Colombo, P. Goatin, M. Rosini, GAKUTO Int. Ser. Math. Sci. Appl. **32**, 255 (2010)
37. M.D. Francesco, P. Markowich, J.F. Pietschmann, M.T. Wolfram, Math. Models Methods Appl. Sci. **250**(3), 1334 (2011)
38. E.D. Angelis, Math. Comput. Model. **29**, 83 (1999)
39. A. Bressan, K. Han, SIAM J. Math. Anal. **43**, 2384–2417 (2011)
40. G. Coclite, B. Piccoli, SIAM J. Math. Anal. **36**(6), 1862 (2005)
41. A. Bressan, K. Han, Netw. Heterog. Media **8**, 627 (2013)
42. M. Herty, S. Moutari, M. Rascale, Netw. Heterog. Media **1**(2), 275 (2006)
43. G. Bretti, R. Natalini, B. Piccoli, Netw. Heterog. Media **1**(1), 57 (2006)
44. A. Bressan, S. Canić, M. Garavello, M. Herty, B. Piccoli, EMS Surv. Math. Sci. **1**, 47 (2014)
45. M. Gugat, M. Herty, A. Klar, G. Leugering, J. Optim. Theory Appl. **126**(3), 589 (2005)

46. D. Helbing, Phys. Rev. E **53**, 2366 (1996)
47. R. LeVeque, *Numerical Methods for Conservation Laws* (Birkhäuser, Basel, 1992)
48. P. Ross, Transp. Res. **22**(6), 421 (1988)
49. W. Phillips, Transp. Plan. Technol. **5**(3), 131 (1979)
50. M. Flynn, A. Kasimov, J.C. Nave, R. Rosales, B. Seibold, Phys. Rev. E **79**(5), 056113 (2009)
51. R. Kühne, M. Rödiger, *Proceedings of the 1991 Winter Simulation Conference* (1991), pp. 762–770
52. W. Jin, H. Zhang, Solving the Payne-Whitham traffic flow model as a hyperbolic system of conservation laws with relaxation. Technical report, University of California, Davis, 2001
53. A. Delis, I. Nikolos, M. Papageorgiou, Transp. Res. C.: Emerg. Technol. **44**, 318 (2014)
54. H. Payne, *Mathematical Models of Public Systems*, vol. 28 (Simulation Council, La Jolla, 1971), pp. 51–61
55. C. Daganzo, Transp. Res. B **28**, 35 (1995)
56. A. Aw, M. Rascale, SIAM J. Appl. Math. **60**, 916 (2000)
57. B. Kerner, Math. Comput. Model. **35**, 481–508 (2002)
58. R. Colombo, P. Goatin, Flow Turbul. Combust. **76**(4), 383 (2006)
59. F. Navin, R. Wheeler, Traffic Eng. **39**, 31 (1969)
60. L. Vanumu, K. Rao, G. Tiwari, Eur. Transp. Res. Rev. **9**, 49 (2017)
61. R. Colombo, SIAM J. Appl. Math. **63**, 708 (2003)
62. B.S. Kerner, P. Konhäuser, Phys. Rev. E **50**, 54 (1994)
63. B. Kerner, Phys. A: Stat. Mech. Appl. **333**, 379 (2004)
64. B. Kerner, arXiv prerint cond-mat/0309018 (2003)
65. A. Seyfried, B. Steffen, W. Klingsch, M. Boltes, J. Stat. Mech. **2005**(10), P10002 (2005)
66. V. Predtechenskii, A. Milinskii, *Planning for Foot Traffic Flow in Buildings* (Amerind Publishing, New Delhi, 1978). Translation of: Proekttirovanie Zhdanii s Uchetom Organizatsii Dvizheniya Lyuddskikh Potokov. Stroiizdat Publishers, Moscow
67. B. Piccoli, K. Han, T. Friesz, T. Yao, J. Tang, Transp. Res. C **52**, 32 (2015)
68. R. Colombo, P. Goatin, B. Piccoli, J. Hyperbolic Differ. Equ. **7**(1), 85 (2010)
69. M. Garavello, B. Piccoli, Netw. Heterog. Media **4**(1), 107 (2009)
70. S. Blandin, P. Goatin, B. Piccoli, A. Bayen, D. Work, Proc. Soc. Behav. Sci. **54**, 302 (2012)
71. S. Blandin, D. Work, P. Goatin, B. Piccoli, A. Bayen, SIAM J. Appl. Math. **71**(1), 107 (2011)
72. P. Goatin, Math. Comput. Model. **44**(3–4), 287 (2006)
73. D. Helbing, Phys. A **219**, 375 (1995)
74. D. Helbing, in *A Perspective Look at Nonlinear Media*, ed. by J. Parisi, S. Müller, W. Zimmermann. Lecture Notes in Physics, vol. 503 (Springer, Berlin, 1998), pp. 122–139
75. M. Schönhof, D. Helbing, Transp. Sci. **41**(2), 135 (2007)
76. P. Bagnerini, R. Colombo, A. Corli, Math. Comput. Model. **44**(9–10), 917 (2006)
77. J.P. Lebacque, J.B. Lesort, *Proceedings of the 14th International Symposium on Transportation and Traffic Theory, Jerusalem* (1999)
78. J.B. Lesort, E. Bourrel, V. Henn, *Proceedings of Traffic and Granular Flow?03, Delft* (2003), pp. 125–139
79. T. Li, Phys. D: Nonlinear Phenom. **207**(1–2), 41 (2005)
80. H. Yeo, A. Skabardonis, *Proceedings of the 18th International Symposium on Transportation and Traffic Theory, Hong-Kong* (2009), pp. 99–115
81. K. Hadeler, Reaction transport systems in biological modelling, in *Mathematics Inspired by Biology*. Lecture Notes in Mathematics (Springer, Berlin, 1999), pp. 95–150
82. A. Kolmogorov, I. Petrovsky, N. Piscounov, Mosc. Univ. Bull. Math. **1**, 1 (1937)
83. K. Fellner, G. Raoul, Math. Models Methods Appl. Sci. **20**, 2267 (2010)
84. A. McKendrick, Proc. Edinb. Math. Soc. **44**, 98 (1926)
85. H.V. Foerster, *The Kinetics of Cell Proliferation* (Grune and Stratton, New York, 1959), pp. 382–407
86. G. Webb, *Theory of Nonlinear Age-Dependent Population Dynamics* (Marcel Dekker, New York, 1985)
87. B. Keyfitz, N. Keyfitz, Math. Comput. Model. **26**(6), 1–9 (1997)

88. J.D. Murray, *Mathematical Biology* (Springer, New York, 1989)
89. A. Volpert, V. Volpert, V. Volpert, *Travelling Wave Solutions of Parabolic Systems* (American Mathematical Society, Providence, 2000)
90. V. Volpert, S. Petrovskii, Phys. Life Rev. **6**, 267 (2009)
91. P. Fife, J. Diff. Equ. **40**, 168 (1981)
92. P. Fife, J. McLeod, Arch. Ration. Mech. Anal. **65**(4), 335 (1977)
93. A. Bressan, *Hyperbolic Systems of Conservation Laws. The One-Dimensional Cauchy Problem* (Oxford University Press, Oxford, 2000)
94. R. LeVeque, *Finite Volume Methods for Hyperbolic Problems* (Cambridge University Press, Cambridge, 2002)
95. G. Bretti, R. Natalini, B. Piccoli, J. Comput. Appl. Math. **210**, 71 (2007)
96. R. Borsche, J. Kall, J. Comput. Phys. **327**, 678 (2016)
97. Y. Shi, Y. Guo, Appl. Numer. Math. **108**, 21 (2016)

Chapter 4
Local Hyperbolic/Kinetic Systems in 1D

4.1 Introduction

The models discussed in Chap. 3 assumed that individuals can move in only one direction, either left or right, depending on the sign of the velocity. Of course, the individual's velocity can be density-dependent and change its sign as time progresses, leading to changes in the movement direction of the whole population. However, recent approaches in experimental ecology have started to focus on collecting data on the speed and turning rates of different animals exhibiting collective behaviours [1–3]. Similarly, studies in cell biology have started to investigate the individual movement of cells and bacteria, some of which exhibit for example "run-and-turn" behaviours [4]. These behaviours are similar to the "run-and-tumble" and "run-reverse-turn" behaviours in bacterial communities [5, 6]; see also Fig. 4.1a–c. While the turning events of bacteria are usually random, the turning events of amoebae cells can exhibit short-term memory with cells having a tendency of turning away from previous turns [4]; see also Fig. 4.1d. The running speeds, turning rates and turning angles during changes in movement direction of cells and bacteria can all be measured experimentally [2, 4, 5], thus allowing for a better quantitative description of cell/bacterial/animal movement. However, when cells/bacteria/animals interact with conspecifics within a community [7], they might change their velocity and turning in response to the behaviour of their conspecifics. This adds another layer of complexity to the dynamics of the whole system, as individual cell/bacteria/animal behaviours are difficult to predict emergent group-level behaviours.

A different type of transport phenomena occurs in cellular biology, during the bidirectional movement of molecules/particles along polarised microtubules or axons [10–12] (see also Fig. 4.2). These particles (e.g., mitochondria, pigment granules, lipid droplets, viruses [13]) are transported with the help of molecular motors such as cytoplasmic dynein and conventional kinesin. Usually the kinesin

© Springer Nature Switzerland AG 2018
R. Eftimie, *Hyperbolic and Kinetic Models for Self-organised Biological Aggregations*, Lecture Notes in Mathematics 2232,
https://doi.org/10.1007/978-3-030-02586-1_4

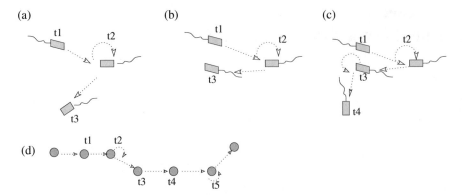

Fig. 4.1 Caricature description of three different bacterial movement strategies, and one cell movement strategy. (**a**) The "run-and-tumble" strategy, where the turning angle during the "tumble" is random. This strategy has been observed, for example, in *Escherichia coli, Salmonella typhimurium, Bacillus subtilis* [8]; (**b**) The "run-and-reverse" strategy, where the bacteria undergo a 180° reorientation. This strategy has been observed, for example, in *Shewanella putrefaciens, Pseudoalteromonas haloplanktis* and *Deleya marina* [8]; (**c**) The "run-reverse-flick" strategy, which combines the run-and-reverse with a random and fast tumble described by the flick. This strategy has been observed in *Vibrio Alginolyticus* [8, 9] or in *Pseudomonas aeruginosa* [5]. (**d**) The "run-and-turn" strategy exhibited by crawling cells [4], which is similar to the run-and-tumble strategy in bacteria. Note the tendency of cells to reverse the turning direction from one step to another (as emphasised in time steps $t2$ and $t5$)

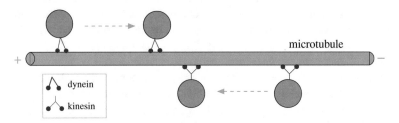

Fig. 4.2 Caricature description of bidirectional movement of particles along a microtubule filament (or axon), which displays polarity (see the "+" and "−" signs at the end of the microtubule). The transported particles are bound to (and transported along) the microtubule by molecular motors such as dynein and kinesin

moves to the microtubules' "+" end, while the dynein moves to the "−" end [12, 13]. Since many molecules have both types of motors, it has been suggested that the actual transport could be the results of a "tug-of-war" between molecular motors for the "+" and "−" directions, with a "pause" state being reached when a particle has activated both types of motors [12]. The transported particles can reverse their direction every few seconds, depending on the motors that dominate the dynamics [12, 13]. This movement-turning behaviour (towards the +/− ends of the microtubules) gives rise to a biased random walk, which characterises the transport of fish and frog pigment granules inside cells, or the transport of adenoviruses and herpes viruses through the cytoplasm [13].

It should be mentioned that bi-directional movement has also been observed in various ecological contexts: e.g., during ants movement along pheromone trails [14], or during pedestrian movement at crosswalks [15] or along corridors [16].

To be able to incorporate the directionality of movement into the 1D mathematical models for self-organised biological behaviours (to further investigate emergent group-level phenomena, also by using available experimental data), one can split the initial population (u) into two subpopulations of left-moving (u^-) and right-moving (u^+) individuals (or cells/bacteria/etc.). This leads to systems of equations describing the movement and interactions of left-moving and right-moving individuals as they change their direction of travel and/or their speeds. Because these models incorporate detailed individual-level information regarding movement directions (left or right), they are in fact kinetic (i.e. mesoscale) models, being often referred to as "two-speed kinetic models".

In the following, we review briefly some hyperbolic systems derived to investigate the movement of organisms in response to local conspecifics. We start with the simplest hyperbolic models, which assume that left-moving and right-moving individuals travel at a constant speed and have constant turning rates. Since these models could not explain the complexities of some observed biological aggregations (e.g., the splitting and merging of groups, or the increase in population size), more complicated models have been derived. These new models incorporate density-dependent turning rates, density-dependent speeds, or reaction terms describing population growth and decay. Throughout this section, the underlying assumption is that the behaviour of individuals is influenced only by the local density of their conspecifics.

As for the one-population models in Chap. 3, one can use a variety of simple first order and higher order finite difference schemes or finite volume schemes to discretise and simulate the solutions of these hyperbolic systems. Since the numerical approaches are similar to those mentioned in Sect. 3.8, we will not discuss them here again. However, for a more detailed discussion of these various numerical schemes, see Chap. 7.

4.2 Derivation of Local Models in 1D

The derivation of local hyperbolic models for self-organised movement follows the classical Goldstein-Kac theory for correlated random walk [17, 18]. This approach has been reviewed, for example, in [19–21]. In brief, denote by $u^+(x, t) \in \mathbb{R}$ and $u^-(x, t) \in \mathbb{R}$ the densities of right-moving and left-moving individuals at position $x \in \mathbb{R}$ and time $t \in \mathbb{R}^+$. To derive the equations describing the evolution of these densities, let us first consider the behaviour of an individual organisms (in a population of size N). Denote by $p^+(x, t)$ ($p^-(x, t)$) the probability that a randomly chosen right-moving (left-moving) individual is found inside the interval

$[x - \Delta x/2, x + \Delta x/2]$ at time t. These probabilities are defined as follows [22, 23]:

$$p^{\pm}(x, t) = \frac{1}{N} \int_{x-\Delta x/2}^{x+\Delta x/2} u^{\pm}(s, t)ds \longrightarrow \frac{\Delta x u^{\pm}(x, t)}{N} \quad \text{when } \Delta x \to 0. \quad (4.1)$$

We denote by Δx and Δt the space and time steps, respectively. At each time step, the individual changes direction with probability $\lambda^{+} \Delta t$ (if it was initially moving right) or $\lambda^{-} \Delta t$ (if it was initially moving left); see also Fig. 4.3. This leads to the following *master equations*:

$$p^{+}(x, t + \Delta t) = (1 - \lambda^{+} \Delta t)p^{+}(x - \Delta x, t) + (\lambda^{-} \Delta t)p^{-}(x + \Delta x, t), \quad (4.2a)$$

$$p^{-}(x, t + \Delta t) = (\lambda^{+} \Delta t)p^{+}(x - \Delta x, t) + (1 - \lambda^{-} \Delta t)p^{-}(x + \Delta x, t). \quad (4.2b)$$

Expanding Eqs. (4.2) in Taylor series, taking the limits $\Delta t, \Delta x \to 0$, such that $\gamma = \Delta x/\Delta t$, and using (4.1), leads to the following hyperbolic equations for the evolution of densities of right- and left-moving individuals:

$$\frac{\partial u^{+}}{\partial t} + \gamma \frac{\partial u^{+}}{\partial x} = -\lambda^{+} u^{+} + \lambda^{-} u^{-}, \quad (4.3a)$$

$$\frac{\partial u^{-}}{\partial t} - \gamma \frac{\partial u^{-}}{\partial x} = \lambda^{+} u^{+} - \lambda^{-} u^{-}. \quad (4.3b)$$

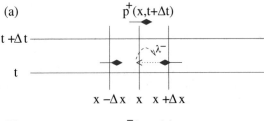

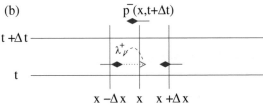

Fig. 4.3 The movement and turning behaviour of (**a**) a right-moving individual, and (**b**) a left-moving individual, in response to *local neighbours*. A right-moving individual can be positioned at x at time $t + \Delta t$, if at the previous time step t it was at $x - \Delta x$ and kept moving in the same direction, or it was at $x + \Delta x$ and moving left, and changed direction at the end of the time step Δt. A similar explanation holds for the movement of a left-moving individual (see (**b**)). The change in direction occurs at rates $\lambda^{\pm}$

Here, $\lambda^\pm$ are the constant turning rates, and γ is the constant speed. Note that Eq. (4.1) helps us connect the random movement of an individual (given by the position probabilities $p^\pm(x, t)$) to the population distribution (described by the density functions $u^\pm(x, t)$).

Remark 4.1 As mentioned above, models (4.3) are often referred to as "two-speed kinetic models", since they incorporate individual-level characteristics (left and right velocities $\pm\gamma$, and turning probabilities $\lambda^\pm$) into the description of population variables. For example, we could have defined $u^\pm(x, t) := u^\pm(x, t; \pm\gamma)$, to make it more clear that the population densities depend on the discrete speeds $\pm\gamma$. We will return to this aspect in Chap. 6, in the context of 2D kinetic Boltzmann-type equations that can be reduced in 1D to equations similar to (4.3), by assuming that the direction angle ϕ can take only two values, $\pm\pi$ (which reduces the continuous speed/orientation model to a discrete speed/orientation model).

The hyperbolic system (4.3) can be reduced to a 1D telegraph equation by considering the total population density $u = u^+ + u^-$ and the population flow $v = u^+ - u^-$. These two variables satisfy the following equations

$$\frac{\partial u}{\partial t} + \gamma \frac{\partial v}{\partial x} = 0, \tag{4.4a}$$

$$\frac{\partial v}{\partial t} + \gamma \frac{\partial u}{\partial x} = \Lambda_1 u - \Lambda_2 v. \tag{4.4b}$$

Here, $\Lambda_1 = \lambda^- - \lambda^+$ and $\Lambda_2 = \lambda^- + \lambda^+$. Differentiating Eqs. (4.4) with respect to x and t, and eliminating the derivatives $\frac{\partial^2 v}{\partial t \partial x}$ and $\frac{\partial v}{\partial x}$ (a process known as the Kac trick [18]) leads to the following telegraph equation:

$$\frac{\partial^2 u}{\partial t^2} + \Lambda_2 \frac{\partial u}{\partial t} = \gamma^2 \frac{\partial^2 u}{\partial x^2} - \gamma \Lambda_1 \frac{\partial u}{\partial x}. \tag{4.5}$$

A similar equation (with no drift $\Lambda_1 = 0$ and with a reaction term) was investigated by Holmes [19] in the context of population dispersal. Holmes [19] showed that this telegraph equation can predict dispersal patterns that are very similar to the ones obtained with parabolic equations.

Remark 4.2 Parabolic equations can be obtained as a limit of Eqs. (4.3), following a rescale of the speed and turning rates: $\gamma = \gamma_0/\epsilon$, $\lambda^\pm = \lambda_0/\epsilon^2$ (i.e., individuals travel extremely fast and turn very rapidly), or equivalently a rescale of the space and time variables: $x = x^*/\epsilon$, $t = t^*/\epsilon^2$. The limit $\epsilon \to 0$ applied in the context of the previous rescaling is known as the *parabolic limit*.

Remark 4.3 Hyperbolic equations of the form (2.1) (with $R(u) = 0$) can be obtained as a limit of these two-speed kinetic equations (4.3), following a slightly different rescaling: $x = x^*/\epsilon$, $t = t^*/\epsilon$. The limit $\epsilon \to 0$ applied in the context of the previous rescaling is known as the *hydrodynamic limit*.

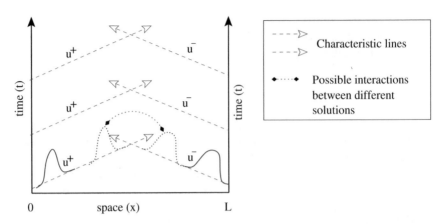

Fig. 4.4 Solutions of hyperbolic systems transported along characteristic lines, on a finite domain $[0, L]$. If the left-moving and right-moving solutions do not interact with each other, they maintain their shapes until they reach the other end of the domain. Otherwise, the left-moving and right-moving solutions might change their shapes following interactions with each other

Since the majority of biological populations live in restricted areas, the mathematical models that investigate these populations are usually defined on finite domains $[0, L]$. For hyperbolic systems, the solutions are transported along the characteristic lines, and thus the boundary conditions have to be prescribed only at the points where the characteristics are directed inward [20] (see also Fig. 4.4). Briefly, the three biologically realistic boundary conditions are: Dirichlet conditions (for domains on which the population density is known at the boundary; including the case where populations vanish on the boundaries), Neumann conditions (for closed domains where no individuals/cells/particles can leave) and periodic boundary conditions (for ring domains, or approximations of infinite domains). A detailed discussion of these boundary conditions and their role on the existence of weak and classical solutions for system (4.3) can be found in [20, 24]. Similar boundary conditions can be derived for the corresponding telegraph equations that model population movement [20]. We will return to these boundary conditions in Sect. 7.3, in the context of numerical simulations for kinetic and hyperbolic systems.

In the following, we discuss briefly a few examples of hyperbolic models that have been derived to investigate interactions between prey and predators moving in opposite directions [25], or the formation and movement of bacterial aggregations, such as colonies of *Myxobacteria* [26], colonies of *Dictyostelium discoideum* [27], or even the intracellular active transport of particles along microtubules [28]. While many of these models consider constant speeds and turning rates, a few others consider turning rates and speeds that are influenced by conspecifics (either directly through their density, or indirectly through chemical signals produced by these conspecifics). We will also discuss a few examples of models that have different reaction terms (i.e., the right-hand-sides of Eqs. (4.3) do not contain only transition terms between the left-moving/right-moving states, but also transitions between moving/stationary states, as well as birth/death processes). Although we discuss separately models with density-dependent speeds, density-dependent turning rates, reaction terms, multiple-population models, as well as first and second order traffic models, we need to emphasise that many of these models can fall into multiple categories (e.g., could have density-dependent speeds and reaction terms, or could be traffic-like models with multiple populations). Also, while a few models are described exclusively by transport and/or reaction-transport equations, other models combine reaction-transport equations with parabolic equations or ordinary differential equations.

4.3 Density-Dependent Turning Rates

Model (4.3) describes the movement of right-moving and left-moving individuals when their turning rates ($\lambda^{\pm}$) are constant. While this assumption is supported by experimental data on individual cells, bacteria or animals [1–3], it does not tell the whole story since individuals in a group can also turn in response to interactions with their neighbours. These interactions can be direct [25, 26, 29], or indirect through chemicals produced by these neighbours [27, 30].

For the hyperbolic models (4.3) with constant speed and very simple density-dependent turning rates (e.g., $\lambda^{+}(u^{+}, u^{-}) = 0$ and $\lambda^{-}(u^{+}, u^{-}) = u^{+}$ as in [25]), it is possible to find exact analytical solutions. For more general models, however, only numerical and analytical methods (such as existence and asymptotic results) could be used to investigate the various types of solutions. For example, Lutscher and Stevens [26] studied the rippling behaviour observed in *Myxobacteria* colonies. The hyperbolic model that they used incorporated turning rates that have a constant random component (μ) and a directed density-dependent component ($\mu^{\pm}$): $\lambda^{\pm} = \mu + \mu^{\pm}(u^{+}, u^{-})$, with $\mu^{\pm}(u^{+}, u^{-}) = 2 \tanh u^{-} \sqrt{u^{-}}$ or $\mu^{\pm}(u^{+}, u^{-}) = \frac{3}{1+(u^{+}+u^{-})^{3}}(u^{-})^{2}$. First, the authors showed the existence of unique solutions (in $\mathbb{C}^{1}$) for the local hyperbolic system introduced in [26]. (Note that the proof used a contraction argument; see also Chap. 8.) Then, using analytical and numerical results, they investigated the role of these nonlinear turning functions $\mu^{\pm}$ on the overall group patterns. For the numerical results, the authors used a first-order upwind scheme with periodic boundary conditions, as well as a second-order Lax-Wendroff scheme. The results showed that when the turning functions are equal and depend only on the individuals moving in the opposite direction (see the $\mu^{\pm}(u^{+}, u^{-})$ functions above), the model exhibits ripples and moving aggregations (travelling pulses). These two patterns are shown in Fig. 4.5; see also Table 1.2 for a description of these patterns.

Very recently, Kang et al. [31] identified parameter regions of global existence for ripples and waves, and parameter regions of existence of finite-time blow-up patterns, in a local hyperbolic model with density-dependent turning rates. The authors used the same model (4.3) with the turning rates given by $\lambda^{+}(u^{+}, u^{-}) =$

(a)

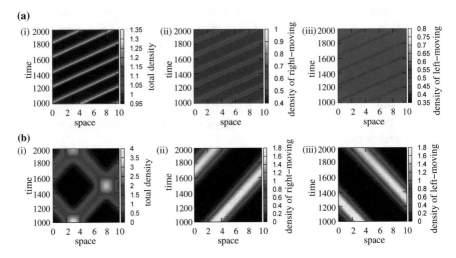

Fig. 4.5 Reproduction of spatio-temporal patterns observed in [26] for a hyperbolic system of the form (4.3) with constant speed γ and turning rates $\lambda^{\pm} = \mu + 2\tanh u^- \sqrt{u^-}$. (**a**) travelling pulses (or waves); (**b**) ripples (or standing waves). Panels (i) show the total population density $u = u^+ + u^-$; panels (ii) show the density of right-moving subpopulation u^+; panels (iii) show the density of left-moving sub-population u^-

$g(u^-)$ and $\lambda^-(u^+, u^-) = g(u^+)$, where

$$g(u) = \mu + \frac{u^p}{1 + \nu u^q}, \quad \text{for } u \in \mathbb{R}^+. \tag{4.6}$$

As before, μ is a constant spontaneous reversal rate. Parameter p is a rate of reversal increase for small population densities, while parameter $p - q$ is a rate of reversal increase/decrease for large population densities [31]. Finally, ν describes a saturation level where the reversal rates stop increasing at the same rates as for small populations [31]. For specific values of p, q, and μ the authors have shown the existence of nontrivial wave patterns when $\nu \in (0, 1/8) \cup \{0\}$, existence of finite-time blow-up solutions for $\nu = 0$, and stable spatially homogeneous (or equidistributed [31]) steady states for $\nu \in [1/8, \infty)$.

We note that the rippling patterns obtained by Lutscher and Stevens [26], Kang et al. [31] occurred when individuals changed their movement direction only in response to conspecific moving in the opposite direction. Actually, this biological mechanism seems to be crucial for the formation of ripples, as observed in discrete models [32, 33], parabolic models [34–36], as well as nonlocal hyperbolic models [37]. (We will revisit this pattern in Chap. 5.) Mathematically, it was shown that the formation of ripples is associated with Hopf bifurcations [26]. For a review of these bifurcations see the discussion in Chap. 8.

In contrast to these hyperbolic models where the turning rates depend on the local density of conspecifics, Hillen and Stevens [27] derived a slightly different model to investigate the effect of indirect interactions among organisms. These interactions were mediated by an external signal S, which was produced by individuals at a rate α, decayed at a rate β, and diffused into the environment at a rate D:

$$\frac{\partial u^+}{\partial t} + \gamma \frac{\partial u^+}{\partial x} = -\lambda^+ \left(S, \frac{\partial S}{\partial x}\right) u^+ + \lambda^- \left(S, \frac{\partial S}{\partial x}\right) u^-, \tag{4.7a}$$

$$\frac{\partial u^-}{\partial t} - \gamma \frac{\partial u^-}{\partial x} = \lambda^+ \left(S, \frac{\partial S}{\partial x}\right) u^+ - \lambda^- \left(S, \frac{\partial S}{\partial x}\right) u^-, \tag{4.7b}$$

$$\tau \frac{\partial S}{\partial t} = D \frac{\partial^2 S}{\partial x^2} + f(S, u^+ + u^-). \tag{4.7c}$$

Here, function $f(S, u^+ + u^-) = -\beta S + \alpha(u^+ + u^-)$ describes the production/decay of signal S. Moreover, the turning rates depend not only on the signal alone but also on the gradient of this signal ($\partial S/\partial x$), suggesting some long-distance interactions. Focusing on existence (and uniqueness) results, the authors showed that solutions exist globally (i.e., there are no blow-up solutions) when signal production is linear (i.e., $f(S, u^+ + u^-) = \alpha(u^+ + u^-)$) and the turning rates ($\lambda^\pm(S, \partial S/\partial x)$) are bounded. However, blow-up solutions are possible when the production of signal S is nonlinear (i.e., $f(S, u^+ + u^-) = S(u^+ + u^-)$). Hillen and Levine [38] investigated a similar hyperbolic model and showed that for their case, the formation of blow-up patterns is preceded by the turning rates becoming negative. This causes the densities $u^\pm$ to become negative, meaning that the hyperbolic model becomes unrealistic just before the blow-up [38]. Even if these density blow-up patterns are not biologically realistic, they can indicate that the investigated organisms tend to form very dense aggregations. Examples of blow-up solutions are shown in Figs. 1.10 and 5.22, and discussed in Table 1.2.

Note that these hyperbolic systems could be seen as displaying another type of pattern: spatially homogeneous solutions with the majority of individuals aligned in one direction. Since individuals are spread over the entire domain, they do not actually form heterogeneous patterns. However, because the population can split into two subpopulations of left-moving and right-moving individuals, this splitting this could be seen as a type of group pattern. When the majority of individuals are moving in one direction, this pattern is the precursor of lane patterns observed in some 2D models [39]; see also Fig. 1.11.

4.4 Analytical Approaches for the Investigation of Patterns: Stability of Homogeneous States and Travelling Waves for Models with Density-Dependent Turning Rates

In the following we focus on a simple local hyperbolic model introduced by Lutscher [29] to describe the alignment of animals in one spatial dimension, and discuss the stability of spatially homogeneous steady states and the stability of

waves of perfect alignment (i.e., all individuals are aligned in one direction, causing
the second population to become zero; which is in contrast to the case depicted in
Fig. 4.5a, where the second population persists at some very low densities). The
model takes the form [29]

$$\frac{\partial u^+}{\partial t} + \gamma \frac{\partial u^+}{\partial x} = \mu(u^+, u^-)(u^+ - u^-), \tag{4.8a}$$

$$\frac{\partial u^-}{\partial t} - \gamma \frac{\partial u^-}{\partial x} = \mu(u^+, u^-)(u^- - u^+), \tag{4.8b}$$

with

$$\mu(u^+, u^-) = \frac{a}{(u^+ + u^-)^3} u^+ u^- - \frac{\mu_*}{2}. \tag{4.9}$$

The first term in the above equation is a density-dependent turning rate (with
$a > 0$) that leads to alignment, while the second term (with $\mu_* \geq 0$) is a constant
random tuning rate. It is clear from Eqs. (4.8) that this model can have spatially
homogeneous steady states with $u^+ = u^- = \bar{u}$, or states with $u^+ \neq u^-$ (that
satisfy $\mu(u^+, u^-) = 0$); see Fig. 4.6. It was shown in [29] that small spatial
perturbations of spatially homogeneous steady states $u^+(x, t) = u^-(x, t) = \bar{u}$
(i.e., $u^\pm = \bar{u} + c_\pm e^{\sigma t + ikx}$, with $0 < c_\pm \ll 1$) lead to a dispersion relation with
eigenvalues given by:

$$\sigma_{1,2} = \left(\frac{a}{4}\bar{u}^2 - \frac{\mu_*}{2}\right) \pm \frac{1}{2}\sqrt{\left(\frac{a}{4}\bar{u}^2 - \frac{\mu_*}{2}\right)^2 - 4\gamma^2 k^2}. \tag{4.10}$$

When μ_* is large, $Re(\sigma_{1,2}) < 0$ and all modes are stable. However, as we decrease
μ_* below a critical threshold $\mu_* = a\bar{u}^2/2$ all modes become unstable; see Fig. 4.7.
(Note that similar dispersion relations, where eigenvalues can be unstable for an

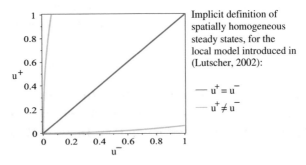

Fig. 4.6 Examples of spatially homogeneous steady states (i.e., states which satisfy $\partial u^\pm/\partial t = \partial u^\pm/\partial x = 0$) in the (u^+, u^-) plane, for the local hyperbolic model (4.8). One can have states with
$u^+ = u^-$ (magenta line), or states with $u^+ \neq u^-$ (green curve). Note the symmetry of the steady
states with respect to the $u^+ \leftrightarrow u^-$ interchange

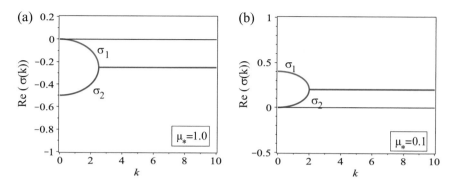

Fig. 4.7 Real part of the dispersion relation as given by (4.10); the red curves describe σ_1, while the blue curves describe σ_2. Note that **(a)** for large μ_* (e.g., $\mu_* = 1.0$) all wavenumbers k are stable, while **(b)** for small μ_* (e.g., $\mu_* = 0.1$) all wavenumbers k are unstable

infinite range of wavenumbers, have been shown to exist in a relatively similar local hyperbolic system introduced in [40].) This spectral result impacts the existence of a centre manifold reduction—as it will be discussed in Chap. 8.

Lutscher [29] also showed that simple waves of total alignment (i.e., waves of the form $u^+(x, t) = u_0^+(x - \gamma t)$ and $u^-(x, t) = 0$, or $u^+(x, t) = 0$ and $u^-(x, t) = u_0^-(x + \gamma t)$) are linearly stable with respect to the positive cone. To this end, the author assumed that $u^+ \geq 0$ and $u^- = 0$, and applied small perturbations $w^\pm$. These perturbations satisfy the following linearised system:

$$\frac{\partial w^+}{\partial t} + \gamma \frac{\partial w^+}{\partial x} = a(u)(u^+)^2 w^-,$$

$$\frac{\partial w^-}{\partial t} - \gamma \frac{\partial w^-}{\partial x} = -a(u)(u^+)^2 w^-. \tag{4.11}$$

Since the second equation (for w^-) decouples from the first equation, one can integrate it to obtain

$$\frac{d}{dt} \int_D w^-(x, t)dx = -\int_D a(u)u^+(x, t)^2 w^-(x, t)dx \leq 0, \tag{4.12}$$

which implies that w^- converges to 0, and thus the simple wave ($u^+ > 0$, $u^- = 0$) is stable.

To conclude the discussion of the model in [29] we note that the author also showed that for all initial data $u_0^\pm \in \mathbb{L}^\infty(\mathbb{R})$ there exists a unique mild solution (see definition below) $u^\pm \in \mathbb{L}^\infty(\mathbb{R} \times [0, T))$ of the initial value problem, for some $T > 0$. Moreover, if the initial data is in $C^k(\mathbb{R})$ then all solutions are in $C^k(\mathbb{R})$ provided that $\mu \in C^{k,1}$. The proof of this existence result uses a standard contraction argument [29] (see also Chap. 8).

Definition 4.1 A *mild solution* is a pair of functions $(u^+, u^-) \in \mathbb{L}^\infty(\mathbb{R} \times [0, T))^2$ that satisfies the following integral equation (obtained by integrating system (4.8) along the characteristic lines $x \pm \gamma t = const$ and using a new variable $z = z_0 \pm \gamma t$):

$$u^\pm(z) = u_0^\pm \pm \int_{z_0}^z \mu(u^+, u^-)(u^+ - u^-)(y)dy. \tag{4.13}$$

4.5 Density-Dependent Speeds

The models mentioned in the previous section assume that the speed is constant, while the turning behaviour is the result of local alignment interactions. However, it is well known that some organisms change also their speed in response to interactions with neighbours [41]. This can lead to an additional equation for the evolution of the speed.

One of the first generalisations of models (3.1) to two different populations u and v (or "two-phase" models) was proposed by Bick and Newell [42] in the context of two-lane traffic:

$$\frac{\partial u}{\partial t} + \frac{\partial}{\partial x}(u(1 - u - \beta v)) = 0, \tag{4.14a}$$

$$\frac{\partial v}{\partial t} + \frac{\partial}{\partial x}(-v(1 - \beta u - v)) = 0. \tag{4.14b}$$

Here, $\beta > 0$ models the changes in the velocities of the two populations as a result of their interactions. The authors then investigated analytically the possibility of having shock curves that connect different states.

Chertock et al. [43] derived a local hyperbolic model for pedestrian movement from a microscopic model on a lattice. The authors considered the probabilities of left-moving ($p_x^-(t)$) and right-moving ($p_x^+(t)$) pedestrians to be in a lattice cell x at time t, and defined four different velocities which depend on the presence or absence of pedestrians in the adjacent lattice cells (at $x \pm \Delta x$). For example, if right-moving pedestrians are found in a cell x (i.e., $p_x^+ = 1$, but $p_{x+\Delta x}^+ = 0$), then their velocities depend on the presence/absence of left-moving pedestrians:

1. velocity=c_0 if $p_x^- = p_{x+\Delta x}^- = 0$ (no left-moving pedestrians in cells at x or $x + \Delta x$),
2. velocity=c_1 if $p_x^- = 1$, $p_{x+\Delta x}^- = 0$ (left-moving pedestrians in the cell at x),
3. velocity=c_2 if $p_x^- = 0$, $p_{x+\Delta x}^- = 1$ (left-moving pedestrians in the cell at $x + \Delta x$),
4. velocity=c_3 if $p_x^- = p_{x+\Delta x}^- = 1$ (left-moving pedestrians in cells at x and $x + \Delta x$),

with $c_3 < c_2 \approx c_1 < c_0$. Similar velocities can be defined for left-moving pedestrians when $p_x^- = 1$ (and their velocities depend on the presence/absence of

right-moving pedestrians). With these definitions, the probabilities of moving from
cell x to cell $x + \Delta x$ (for p_x^+) or from cell x to cell $x - \Delta x$ (for p_x^-) in one time
step Δt are

$$
\begin{aligned}
P_{x \to x+\Delta x}^+ = \Delta t \Big(&c_0 p_x^+ (1 - p_{x+\Delta x}^+)(1 - p_x^-)(1 - p_{x+\Delta x}^-) \\
&+ c_1 p_x^+ (1 - p_{x+\Delta x}^+) p_x^- (1 - p_{x+\Delta x}^-) \\
&+ c_2 p_x^+ (1 - p_{x+\Delta x}^+)(1 - p_x^-) p_{x+\Delta x}^- \\
&+ c_3 p_x^+ (1 - p_{x+\Delta x}^+) p_x^- p_{x+\Delta x}^- \Big),
\end{aligned}
\tag{4.15a}
$$

$$
\begin{aligned}
P_{x \to x-\Delta x}^- = \Delta t \Big(&c_0 p_x^- (1 - p_{x-\Delta x}^-)(1 - p_x^+)(1 - p_{x-\Delta x}^+) \\
&+ c_1 p_x^- (1 - p_{x-\Delta x}^-) p_x^+ (1 - p_{x-\Delta x}^+) \\
&+ c_2 p_x^- (1 - p_{x-\Delta x}^-)(1 - p_x^+) p_{x-\Delta x}^+ \\
&+ c_3 p_x^- (1 - p_{x-\Delta x}^-) p_x^+ p_{x-\Delta x}^+ \Big).
\end{aligned}
\tag{4.15b}
$$

In the limit $\Delta x, \Delta t \to 0$, this microscale model becomes the following macroscale
model:

$$
\frac{\partial u^+}{\partial t} + \frac{\partial}{\partial x}\Big(f(u^+) g(u^-) \Big) = 0,
\tag{4.16a}
$$

$$
\frac{\partial u^-}{\partial t} - \frac{\partial}{\partial x}\Big(f(u^-) g(u^+) \Big) = 0,
\tag{4.16b}
$$

with

$$
f(w) = w(1-w), \quad g(w) = (c_3 - c_2 - c_1 + c_0) w^2 + (c_2 + c_1 - 2c_0) u + c_0.
\tag{4.17}
$$

The authors used numerical simulations to compare the dynamics of the microscale
and macroscale models. For the simulation of the macroscopic model, they
employed a semi-discrete, second-order, central-upwind scheme [44], with a
nonlinear "minmod" limiter to avoid oscillations in the reconstruction of the
solution (see a description of this limiter in Chap. 7). The simulations showed
that both models could exhibit rippling behaviour, with pulses of left- and right-
moving pedestrians passing through each other. The strength of the slow-down
interactions (which is determined by the values of the velocities $c_j, \, j = 0, 1, 2, 3$)
determines how well the numerics for the microscopic model match the numerics
for the macroscopic model: lower slow-down interactions are associated with an
almost perfect match. Model (4.16)–(4.17) can be non-hyperbolic in some regions
of the parameter space, since the Jacobian matrix

$$
J = \begin{pmatrix} f'(u^+) g(u^-) & f(u^+) g'(u^-) \\ -f(u^-) g'(u^+) & -f'(u^-) g(u^+) \end{pmatrix}
\tag{4.18}
$$

has real eigenvalues (which ensures hyperbolicity) only if the following condition holds true:

$$\left(f'(u^-)g(u^+) + f'(u^+)g(u^-)\right)^2 - 4f(u^-)f(u^+)g'(u^-)g'(u^+) > 0.$$

Hence it is possible to have a non-hyperbolic regime for some specific values of c_0, c_1, c_2 and c_3, for which the above inequality does not hold (and this happens when the left-moving and right-moving pedestrians were both present at a particular location in space). In this regime, the solution of the model (4.16) was shown to exhibit spurious oscillations. To address this numerical problem, the authors considered also a diffusive correction which described the presence of pedestrians moving in the opposite directions:

$$\frac{\partial u^+}{\partial t} + \frac{\partial}{\partial x}\left(f(u^+)g(u^-)\right) = \frac{\epsilon}{2}\frac{\partial}{\partial x}\left(g(u^-)\frac{\partial u^+}{\partial x}\right), \tag{4.19a}$$

$$\frac{\partial u^-}{\partial t} - \frac{\partial}{\partial x}\left(f(u^-)g(u^+)\right) = \frac{\epsilon}{2}\frac{\partial}{\partial x}\left(g(u^+)\frac{\partial u^-}{\partial x}\right), \tag{4.19b}$$

Numerical simulations for this new model showed solutions corresponding to rarefaction waves and stop-and-go behaviours in pedestrian dynamics [43]; see also Fig. 1.10b.

Model (4.16) can be generalised to include a desired pedestrian velocity V. Appert-Rolland et al. [45] proposed a model for two-way one-lane pedestrian traffic which assumes that pedestrians adapt their speed in response to pressure from their neighbours:

$$\frac{\partial u^+}{\partial t} + \frac{\partial}{\partial x}\left(u^+(V - p(u^+, u^-))\right) = 0, \tag{4.20a}$$

$$\frac{\partial u^-}{\partial t} - \frac{\partial}{\partial x}\left(u^-(V - p(u^+, u^-))\right) = 0. \tag{4.20b}$$

Here, function p is an increasing function of both u^+ and u^-. Generally, p can be obtained from the data. However, in [45], the authors propose a particular form for function p that allows for the investigation of congestion effects (see also Eqs. (3.18)):

$$p(u^+, u^-) = p^\epsilon(u^+, u^-) = P(u^+ + u^-) + Q^\epsilon(u^+, u^-), \tag{4.21a}$$

$$P(u^+ + u^-) = M(u^+ + u^-)^m, \quad m \geq 1, \tag{4.21b}$$

$$Q^\epsilon(u^+, u^-) = \frac{\epsilon}{q(u^+)\left(\frac{1}{u^+ + u^-} - \frac{1}{u^*}\right)^d}, \quad d > 1, \tag{4.21c}$$

with q an increasing function. Here, congestion occurs when $u^+ + u^- \to u^*$. Note that also for this model there are regions in the parameter space where (4.20) is non-hyperbolic. As for the previous example, the authors considered a diffusive perturbation of the model to stabilise the large wavenumbers that arise in the parameter region where the model is non-hyperbolic. The numerical simulations (performed with a central difference scheme [46]) have shown the existence of travelling pulses, where both subpopulations $u^\pm$ travel in the same direction. For large diffusive perturbations, the dynamics of the system approaches a spatially homogenous steady state.

A slightly different hyperbolic model with density-dependent speeds ($\gamma^+ = \gamma^+(u^+, \partial u^+/\partial x)$ and $\gamma^- = \gamma^-(u^-, \partial u^-/\partial x)$) and turning rates ($\lambda^+ = \lambda^- = \lambda(u^+, u^-)$) was introduced by Lutscher [47]:

$$\frac{\partial u^+}{\partial t} + \frac{\partial\big(u^+\gamma^+(u^+, \partial u^+/\partial x)\big)}{\partial x} = -\lambda(u^+, u^-)u^+ + \lambda(u^+, u^-)u^-, \qquad (4.22a)$$

$$\frac{\partial u^-}{\partial t} - \frac{\partial\big(u^-\gamma^-(u^-, \partial u^-/\partial x)\big)}{\partial x} = \lambda(u^+, u^-)u^+ - \lambda(u^+, u^-)u^-. \qquad (4.22b)$$

The speeds $\gamma^\pm$ depend not only on the local density of individuals, but also on the gradient of this density. In particular, these speeds satisfy the following additional elliptic equations

$$\beta\frac{\partial^2\gamma^+}{\partial x^2} = \gamma^+ - E\Big(u^+, \frac{\partial u^+}{\partial x}\Big), \quad \beta\frac{\partial^2\gamma^-}{\partial x^2} = \gamma^- - E\Big(u^-, -\frac{\partial u^-}{\partial x}\Big), \qquad (4.23)$$

where E describes an expected speed. If this expected speed is constant (i.e., $E = \gamma$), then the speed of the individuals is also constant ($\gamma^\pm = \gamma$). Analytical results were used to discuss the stability of spatially homogeneous steady states. Numerical simulations were performed only for right-moving individuals, with the speed satisfying a parabolic equation of the form

$$\tau\frac{\partial\gamma^+}{\partial t} = \beta\frac{\partial^2\gamma^+}{\partial x^2} - \gamma + E(u^+, \frac{\partial u^+}{\partial x}). \qquad (4.24)$$

The results showed that model (4.22)–(4.24) can exhibit travelling pulse solutions.

The density-dependent speeds could also be the result of indirect interactions among organisms. These indirect interactions can be mediated by signals S produced by the organisms themselves: $\gamma^\pm = \gamma(S, \frac{\partial S}{\partial t}, \frac{\partial S}{\partial x})$, where S satisfies Eq. (4.7c) [27]. These signals could also be produced by other sources not investigated explicitly. Such a situation is common in tumour immunology, where the tumour-immune interactions are mediated by cytokines produced by various types of immune cells, not all of them being investigated in the models (see, for example, the tumour-immune interactions discussed in [48]). Mathematically, this situation could be modelled by incorporating into Eq. (4.7c) an external source of communication signals. In Chap. 5 we will discuss a nonlocal model that considers such an approach.

A different situation that could be modelled by density-dependent speeds is the interaction of multiple species, each species influencing directly the velocity of the other species. To our knowledge, this "cross-advection" mechanism has not received much attention in the context of hyperbolic systems (although it has been recently investigated in parabolic models for population dynamics [49]). To model such a situation, the individuals' speed would be described by $\gamma^\pm(u^\pm, v^\pm)$, where $u^\pm$ and $v^\pm$ are the two species. More generally, each species could produce signals S that would influence indirectly the speed of the other species: $\gamma^\pm(S(u^\pm, v^\pm))$.

Finally we note that the organisms' velocities could also depend on their age (a) as well as their location in space (x): $\gamma = \gamma(a, x)$ [20]. We will revise this aspect in the next section, in the context of reaction hyperbolic systems.

We conclude this discussion on density-dependent speeds, by mentioning that in Sect. 5.8 we will consider a more general situation where the speed of individuals depends on the density of neighbours located further away (i.e., nonlocal speeds).

4.6 Models that Include Reaction Terms

When the hyperbolic models describe the evolution of organisms over long periods of time, they incorporate also population dynamics (i.e., birth and death processes). In this case, one has to be careful when modelling the death terms, since the left- and right-moving particles (organisms) die as left- and right-moving particles (organisms). On the other hand, the newly formed particles (organisms) can be either left-moving or right-moving. For detailed discussions of such models see [19, 20, 50–53]. The general equations for reaction-transport models with population dynamics are

$$\frac{\partial u^+}{\partial t} + \gamma \frac{\partial u^+}{\partial x} = -\lambda^+ u^+ + \lambda^- u^- + \frac{1}{2}m(u)u - g(u)u^+, \tag{4.25a}$$

$$\frac{\partial u^-}{\partial x} - \gamma \frac{\partial u^-}{\partial x} = \lambda^+ u^+ - \lambda^- u^- + \frac{1}{2}m(u)u - g(u)u^-. \tag{4.25b}$$

Here $u = u^+ + u^- \in \mathbb{R}$ is the total density, $m(u)$ models the production (multiplication) of particles/individuals/cells, and $g(u)$ models the death of these particles/individuals/cells. Note that while the production term depends on the total density u (and the new-born particles/individuals/cells become left-moving or right-moving with similar probabilities $1/2$), the death terms depend on u^+ and u^-. These equations have been used, for example, to study animal dispersal [19], or to model epidemiological problems when the movement of infected individuals can be described by a correlated random walk process [20, 54].

Solutions for the reaction-transport system (4.25) have been investigated extensively: from existence and uniqueness results [24, 55, 56], to asymptotic behaviour of solutions [56, 57] and invariance results [51]. System (4.25) was also shown to exhibit travelling front solutions [19, 20, 53, 56, 58, 59]. The minimal speed of the travelling fronts was proven to depend on the nonlinear functions $m(u)$ and $g(u)$ [58, 59].

The reaction-transport models (4.25) can be further generalised by adding density-dependent speeds ($\gamma(u)$) and turning rates ($\lambda^{\pm}(u)$) [60, 61]. Such models had been shown to exhibit travelling front solutions [61].

Finally, these reaction-transport models could be used to study populations structured by location in space (x) and age ($a \geq 0$) [20], or by some other internal state (especially relevant in case of cells, where signalling pathways inside cells influence cells' movement) [62]. For example, in [20] the authors assumed that age can influence the speed (γ) as well as the turning rates ($\lambda^{\pm}$) of organisms:

$$\frac{\partial u^+}{\partial t} + \frac{\partial u^+}{\partial a} + \frac{\partial \gamma(a, x) u^+}{\partial x} = -\lambda^+(a) u^+ + \lambda^-(a) u^- + \frac{1}{2} m(u) u - g(u) u^+,$$
(4.26a)

$$\frac{\partial u^-}{\partial x} + \frac{\partial u^-}{\partial a} - \frac{\partial \gamma(a, x) u^-}{\partial x} = \lambda^+(a) u^+ - \lambda^-(a) u^- + \frac{1}{2} m(u) u - g(u) u^-.$$
(4.26b)

Here, $u^{\pm}(a, x, t)$ describes the density of left- and right-moving individuals of age a, positioned at $x \in \mathbb{R}$ at time $t \in \mathbb{R}^+$. In contrast to the nonlocal birth term in model (3.30), here the birth rate does not depend explicitly on age.

In the context of bacterial colonies, Xue et al. [62] considered a hyperbolic model for the movement of an *E. coli* bacterial population, coupled with parabolic equations for the dynamics of chemicals consumed (i.e., aspartate $S(x, t)$) and secreted (i.e., succinate $F(x, t)$) by cells, and ordinary differential equations for the dynamics of internal cell variables. The density of left-moving and right-moving bacterial population is described by functions $u^{\pm}(x, \mathbf{y}, \mathbf{z}, t)$. Here $[\mathbf{y}, \mathbf{z}]$ is an internal state with variable $\mathbf{y}$ describing signal transduction (where the coordinates $y_i, i = 1, \ldots q$, of this vector represent the concentrations of various proteins in the cell and receptor states involved in signal transduction) and variable $\mathbf{z}$ describing cellular metabolism (where the coordinates $z_i, i = 1, 2$ include the concentration of the components in the TCA cycle and ATP). The equations for the evolution of the internal variables $\mathbf{y} = (y_1, \ldots, y_q)$ and $\mathbf{z} = (z_1, z_2)$ are [62]:

$$\frac{dy_i}{dt} = \mathbf{f}(y_i, S), \quad i = 1, \ldots, q, \quad \text{and} \quad \mathbf{f} = (f_1, \ldots, f_q),$$
(4.27)

$$\frac{dz_1}{dt} = g_1(\mathbf{z}, F), \quad \text{with} \quad g_1(\mathbf{z}F) = \frac{F(x, t) - z_1}{t_f},$$
(4.28)

$$\frac{dz_2}{dt} = g_2(\mathbf{z}, F), \quad \text{with} \quad g_2(\mathbf{z}, F) = \frac{z_1 - z_2}{t_m}.$$
(4.29)

Here t_f is the characteristic time scale for the generation of the metabolic variable z_1 (e.g., fumarate in the TCA cycle; it is measured in seconds), while t_m is the characteristic time scale for the generation of variable z_2 (e.g., associated with the starving state of bacteria; it is measured in tens of minutes).

The equations for cells' movement are [62]:

$$
\frac{\partial u^+}{\partial t} + s\frac{\partial u^+}{\partial x} + \sum_{i=1}^{q}\frac{\partial}{\partial y_i}\Big(f_i(\mathbf{y}, S)u^+\Big) + \sum_{i=1}^{2}\frac{\partial}{\partial z_i}\big(g_i(\mathbf{z}, F)u^+\big)
$$
$$
= \lambda(\mathbf{y})[-u^+ + u^-] + k(\mathbf{z}u^+), \tag{4.30}
$$

$$
\frac{\partial u^-}{\partial t} - s\frac{\partial u^-}{\partial x} + \sum_{i=1}^{q}\frac{\partial}{\partial y_i}\Big(f_i(\mathbf{y}, S)u^-\Big) + \sum_{i=1}^{2}\frac{\partial}{\partial z_i}\big(g_i(\mathbf{z}, F)u^-\big)
$$
$$
= \lambda(\mathbf{y})[u^+ - u^-] + k(\mathbf{z}u^-). \tag{4.31}
$$

It is assumed that bacteria travel with a constant speed s, and the turning frequency of bacteria depends on the signal transduction variable ($\lambda(\mathbf{y})$), while the proliferation rate depends on the metabolic variables ($k(\mathbf{z})$). Finally, the equations for the evolution of the two chemicals, aspartate (S) and succinate (F) are [62]:

$$
\frac{\partial S}{\partial t} = D_S\frac{\partial^2 S}{\partial x^2} + \alpha F \int\int h(z_2)\big(u^+(\mathbf{y}, \mathbf{z}) + u^-(\mathbf{y}, \mathbf{z})\big)d\mathbf{y}d\mathbf{z}
$$
$$
-\beta S \int\int \big(1 - h(z_2)\big)\big(u^+(\mathbf{y}, \mathbf{z}) + u^-(\mathbf{y}, \mathbf{z})\big)d\mathbf{y}d\mathbf{z} - \gamma S, \tag{4.32}
$$

$$
\frac{\partial F}{\partial t} = D_F\frac{\partial^2 F}{\partial x^2} - \beta F \int\int h(z_2)\big(u^+(\mathbf{y}, \mathbf{z}) + u^-(\mathbf{y}, \mathbf{z})\big)d\mathbf{y}d\mathbf{z}, \tag{4.33}
$$

where it was assumed that $\alpha, \beta > 0$, $\gamma \geq 0$ and $h(z_2)$ is an increasing function of z_2 (e.g., $h(z_2) = az_2$). It was shown in [62] that this complex 1D mathematical model exhibits global weak solutions. This analytical result was complemented by numerical simulations that have shown stationary aggregations (which were also very slowly diffusing), and travelling band solutions (i.e., travelling pulses or rotating waves). It should be noted that for the numerical simulations, the authors used zero-flux boundary conditions. The hyperbolic equations for bacterial movement were discretised using an upwind scheme combined with a Van Leer flux limiting approach to preserve the positivity of solutions. The diffusion terms were approximated with the help of a central difference scheme. The integral terms in Eqs. (4.32)–(4.33) were discretised using a trapezoidal scheme. For more details regarding these numerical approaches, see Chap. 7.

The models discussed until now assumed that the individuals/particles/cells are moving either left ($-$) or right ($+$). However, in many situations it is possible to have also third state, where the individuals/particles/cells are not moving. Such a class of models was discussed in the review article by Bressloff and Newby [28], in the context of active intracellular transport. Particles moving along a single microtubular filament (of length L) can be bound to the filament and moving left (u^-) or right (u^+), or can be unbound from the filament and (maybe) diffusing (u_0); see also Fig. 4.2. The equations describing the dynamics of these different types of particles could be written as follows:

$$\frac{\partial u^+}{\partial t} + v\frac{\partial u^+}{\partial x} = -\beta^+ u^+ + \alpha u_0, \tag{4.34a}$$

$$\frac{\partial u^-}{\partial t} - v\frac{\partial u^-}{\partial x} = -\beta^- u^- + \alpha u_0, \tag{4.34b}$$

$$\frac{\partial u_0}{\partial t} = D_0\frac{\partial^2 u_0}{\partial x^2} + \beta^+ u^+ + \beta^- u^- - 2\alpha u_0. \tag{4.34c}$$

In the above equations $v \in \mathbb{R}^+$ is the speed of bound left-moving/right-moving particles, D_0 is the diffusion rate of unbound particles, while $\beta^\pm \in \mathbb{R}^+$ and $\alpha \in \mathbb{R}^+$ are the transition rates between the moving and stationary states. Note that in this model a filament-bound left-moving particle can become right-moving only after becoming stationary and unbound (i.e., there is no direct transition between $u^- \leftrightarrow u^+$, and so the reaction terms are slightly different from those in the models above). Since this model was defined on a finite domain $[0, L]$, boundary conditions were included in the model description: for example, a reflecting boundary at $x = 0$ (i.e., $u^-(0, t) = u^+(0, t)$), and an absorbing boundary at $x = L$ (i.e., $u^-(L, t) = 0$) [28].

This class of models (with/without diffusion for u_0 and even for $u^\pm$) for motor-assisted transport of particles along microtubules/axons have been investigated analytically and numerically in terms of: finding exact analytical solutions (using Laplace transforms) [63–65], showing the existence of travelling waves [66], performing numerical simulations for the time-evolution of solutions [63, 65, 67, 68]. Some studies even compared the numerical simulated solutions with experimental data [67, 68].

4.7 Traffic Models

4.7.1 First-Order Models

Local hyperbolic systems are also used to model traffic or pedestrian flows characterised by two distinct velocities $\gamma_{1,2} \in \mathbb{R}$. However, in this case, the "reaction" terms are actually speed-adaptation terms. A two-velocity model was first introduced by Ruijgrok and Wu [69] in the context of gas dynamics:

$$\frac{\partial u^+}{\partial t} + \gamma_1\frac{\partial u^+}{\partial x} = -\alpha u^+ + \beta u^- + \mu u^+ u^-, \tag{4.35a}$$

$$\frac{\partial u^-}{\partial t} + \gamma_2\frac{\partial u^-}{\partial x} = \alpha u^+ - \beta u^- - \mu u^+ u^-, \tag{4.35b}$$

As opposed to the models in the previous section, where the two velocities were $\gamma_1 = -\gamma_2$, here both velocities $\gamma_{1,2}$ could be positive. The last terms on the right-hand-side of Eqs. (4.35) describe the transition from velocity γ_2 to velocity γ_1 following a binary collision with intensity μ. If $\gamma_2 > \gamma_1$, these terms model the slowing-down process that takes place when a car of velocity γ_2 travels behind a car with velocity γ_1.

A different class of multi-equation models describes the heterogeneous traffic exhibited by multi-groups of drivers (i.e., cars, trucks, or vehicles with different trips [70]). These types of models (see, for example, [71, 72]) denote by $u_i(t, x)$, $i = 1, \ldots, n$, the density of vehicles in the i-th class of drivers, and by $v_i(u1, .., u_n)$ the average speed of the i-th class:

$$\frac{\partial u_i}{\partial t} + \frac{\partial(u_i v_i)}{\partial x} = 0, \quad i = 0, \ldots, n. \tag{4.36}$$

As for single-population first-order models, the speed has the general form $v_i(u) = \psi(\bar{u})v_i^*$, with v_i^* the maximum speed for drivers in the i-th class, and $\psi(\bar{u})$ a decreasing function of the total population $\bar{u} = u_1 + \ldots + u_n$, with $\psi(0) = 1$ (or some positive constant) and $\psi(u_{max}) = 0$ (with u_{max} the maximum possible density for the road) [70]. These types of models (with these assumptions on the speed variable) admit entropy admissible shock wave solutions [70, 72].

4.7.2 Second-Order Models

Similar to the one-population models, the two-population models (for the dynamics of left-moving/right-moving individuals) can be cast as second-order models by introducing explicit formulas for the velocity of left-moving and right-moving individuals. Appert-Roland et al.[45] introduced the following model for the movement of two pedestrian subpopulations $u^\pm \in \mathbb{R}$ on one lane:

$$\frac{\partial u^+}{\partial t} + \frac{\partial(u^+ v^+)}{\partial x} = 0, \tag{4.37a}$$

$$\frac{\partial u^-}{\partial t} + \frac{\partial(u^- v^-)}{\partial x} = 0, \tag{4.37b}$$

$$\frac{\partial(u^+ w^+)}{\partial t} + \frac{\partial(u^+ w^+ v^+)}{\partial x} = 0, \tag{4.37c}$$

$$\frac{\partial(u^- w^-)}{\partial t} + \frac{\partial(u^- w^- v^-)}{\partial x} = 0, \tag{4.37d}$$

$$w^+ = v^+ + p(u^+, u^-), \tag{4.37e}$$

$$w^- = -v^- + p(u^-, u^+). \tag{4.37f}$$

As mentioned in Chap. 3, where we discussed a similar one-population model, $(v^+, v^-) \in \mathbb{R}^+ \times \mathbb{R}^-$ describe the actual velocities for the two populations (with $v^+ > 0$ for right-moving individuals and $v^- < 0$ for left-moving individuals), while $(w^+, w^-) \in \mathbb{R}^+ \times \mathbb{R}^+$ describe the desired pedestrian velocities in the absence of any obstacles. Because the offset velocities $p(u^\pm, u^\mp) > 0$ can be very large (since p is not bounded; see Chap. 3), the actual velocities can reverse their sign. This way, under the pressure of large right-moving (left-moving) crowds, the left-moving (right-moving) individuals will reverse direction and will follow the majority of the population, thus leading to the formation of travelling pulses. The model can be easily generalised to consider congestion dynamics: $u^\pm \to u_\epsilon^\pm$, $w^\pm \to w_\epsilon^\pm$, $v^\pm \to v_\epsilon^\pm$ and $p(u^\pm, u^\pm) \to p^\epsilon(u_\epsilon^\pm, u_\epsilon^\mp)$, with p^ϵ described by (4.21).

Model (4.37) can be further generalised to describe the movement of pedestrians via multiple lanes. Thus, if one denotes by k the lane index, $k = 1, 2, \ldots K < \infty$, then the model reads [45]:

$$\frac{\partial u_k^+}{\partial t} + \frac{\partial (u_k^+ v_k^+)}{\partial x} = S_k^+, \tag{4.38a}$$

$$\frac{\partial u_k^-}{\partial t} + \frac{\partial (u_k^- v_k^-)}{\partial x} = S_k^-, \tag{4.38b}$$

$$\frac{\partial (u_k^+ w_k^+)}{\partial t} + \frac{\partial (u_k^+ w_k^+ v_k^+)}{\partial x} = R_k^+, \tag{4.38c}$$

$$\frac{\partial (u_k^- w_k^-)}{\partial t} + \frac{\partial (u_k^- w_k^- v_k^-)}{\partial x} = R_k^-, \tag{4.38d}$$

$$w_k^+ = v_k^+ + p_k(u_k^+, u_k^-), \tag{4.38e}$$

$$w_k^- = -v_k^- + p_k(u_k^-, u_k^+). \tag{4.38f}$$

Here, $S_k^\pm$ and $R_k^\pm$ are source terms that model the transition rates between different lanes. Generally, pedestrians change lanes from k to $k \pm 1$. In [45] the authors denoted by $\lambda_{k \to k\pm1}^+$ and $\lambda_{k \to k\pm1}^-$ the transition rates for right-moving and left-moving pedestrians, and assumed that $\lambda_{k \to k\pm1}^\pm$ are decreasing with $u_{k\pm1} = u_{k\pm1}^+ + u_{k\pm1}^-$, while the rates are zero at congestion: $u_{k\pm1} = u^*$. With these assumptions, the transition rates can be written as:

$$S_{k,\alpha} = \lambda_{k+1 \to k}^\alpha u_{k+1}^\alpha + \lambda_{k-1 \to k}^\alpha - \left(\lambda_{k \to k+1}^\alpha + \lambda_{k \to k-1}^\alpha\right) u_k^\alpha, \tag{4.39a}$$

$$R_k^\pm = \lambda_{k+1 \to k}^\alpha u_{k+1}^\alpha w_{k+1}^\alpha + \lambda_{k-1 \to k}^\alpha u_{k-1}^\alpha w_{k-1}^\alpha - \left(\lambda_{k \to k+1}^\alpha + \lambda_{k \to k-1}^\alpha\right) u_k^\alpha w_k^\alpha, \tag{4.39b}$$

$$\alpha = \pm. \tag{4.39c}$$

Note that model (4.38) does not actually describe the spontaneous segregation of pedestrians into different lanes. It only describes pedestrian dynamics after the lanes are formed. Unfortunately, the authors in [45] do not present any numerical simulations for these multi-lane models.

4.8 Multiple Population Models

In contrast to the models presented in the previous section (which described the dynamics of one population formed of left-moving/right-moving individuals travelling on one lane (4.37) or on different lanes (4.38)), we now consider the case of multiple populations moving through a one-lane domain. To this end, we generalise models (4.3)–(4.25) to multiple species by considering a vector $u^\pm = [u_i^\pm] \in \mathbb{R}^n$, $i = 1..n$ [20, 26]. Here, $u_i^\pm$ denotes the left-moving $(-)$ and right-moving $(+)$ particles of species i. Such models have been used to investigate typical species interactions, such as predator-prey, competition, or mutualism [20], Turing aggregation patterns [52], as well as epidemics spread [54].

As an example, Lutscher and Stevens [26] introduced the following two-population model to describe the interactions between two cell types that differ in their turning behaviour (due to different inter-cellular signalling):

$$\frac{\partial u_1^+}{\partial t} + \gamma \frac{\partial u_1^+}{\partial x} = -(\mu + \lambda^+)u_1^+ + (\mu + \lambda^-)u_1^-, \tag{4.40a}$$

$$\frac{\partial u_1^-}{\partial t} - \gamma \frac{\partial u_1^-}{\partial x} = (\mu + \lambda^+)u_1^+ - (\mu + \lambda^-)u_1^-, \tag{4.40b}$$

$$\frac{\partial u_2^+}{\partial t} + \gamma \frac{\partial u_2^+}{\partial x} = -(\mu + \nu^+)u_1^+ + (\mu + \nu^-)u_1^-, \tag{4.40c}$$

$$\frac{\partial u_2^-}{\partial t} - \gamma \frac{\partial u_2^-}{\partial x} = (\mu + \nu^+)u_1^+ - (\mu + \nu^-)u_1^-, \tag{4.40d}$$

where $\mu \geq 0$ is a constant turning rate and $\lambda^\pm(u_1^\pm, u_2^\pm)$ and $\nu^\pm(u_1^\pm, u_2^\pm)$ are the density-dependent turning rates that result from cell-cell interactions. The authors also assumed various symmetry conditions in regard to the turning rates (conditions that affect also the symmetry of the entire system):

$$\lambda^\pm(u_1^+, u_1^-, u_2^+, u_2^-) = \lambda^\pm(u_1^+, u_1^-, u_2^-, u_2^+), \tag{4.41a}$$

$$\nu^\pm(u_1^+, u_1^-, u_2^+, u_2^-) = \nu^\pm(u_1^+, u_1^-, u_2^-, u_2^+), \tag{4.41b}$$

$$\nu^+(u_1^+, u_1^-) = \nu^-(u_1^-, u_1^+), \tag{4.41c}$$

The authors in [26] investigated the stability of the spatially homogeneous steady states $u_1^\pm = c_1$ and $u_2^\pm = c_2$, and showed analytically that at the bifurcation point all eigenvalues cross the imaginary axis. Thus, the instability arises via a Hopf bifurcation. Although the model likely displays rotating waves (i.e., travelling pulses—as a result of Hopf bifurcations), no numerical simulations were presented to confirm the type of patterns obtained with model (4.40).

A somehow similar two-population model was introduced by Watmough and Edelstein-Keshet [73] to describe the dynamics of ant leaders/pioneers and followers, in response to the density of a pheromone trail. However, in contrast to the model in [26], here the hyperbolic equations for the dynamics of left-moving and right-moving leaders $(L^{\pm})$ and right-moving and left-moving followers $(F^{\pm})$, are coupled with an ODE for the degradation and production of pheromones (T):

$$\frac{\partial T(x, t)}{\partial t} = -\gamma T(x, t) + \tau_f F(x, t) + \tau_l L(x, t), \quad \text{with}$$

$$F = F^+ + F^-, \quad L = L^+ + L^-, \tag{4.42}$$

and

$$\frac{\partial F^+}{\partial t} + v \frac{\partial F^+}{\partial x} = -\epsilon F^+ + \alpha L^+ T - \rho^+ F^+ + \rho^- F^-, \tag{4.43a}$$

$$\frac{\partial F^-}{\partial t} - v \frac{\partial F^-}{\partial x} = -\epsilon F^- + \alpha L^- T + \rho^+ F^+ - \rho^- F^-, \tag{4.43b}$$

$$\frac{\partial L^+}{\partial t} + s \frac{\partial L^+}{\partial x} = \epsilon F^+ - \alpha L^+ T - \lambda L^+ + \lambda L^-, \tag{4.43c}$$

$$\frac{\partial L^-}{\partial t} - s \frac{\partial L^-}{\partial x} = \epsilon F^- - \alpha L^- T + \lambda L^+ - \lambda L^-. \tag{4.43d}$$

In the pheromone equation (4.42) γ is the degradation rate, while τ_f and τ_l are the production rates. In Eqs. (4.43) for the ants movement, λ and $\rho^{\pm}$ describe the turning rates, s and v are the speeds of leaders and followers, while ϵ and α describe the exchanges between leaders and followers. Note that the pheromone density acts as a communication mechanisms between leaders and followers (with leaders becoming followers, at a rate αT, when they encounter and follow the pheromone trails). The authors reduced this 5-equation system to a 3-equation system for the total follower (F) and leader (L) populations, and the density T of pheromones. Then they studied the stability of the spatially-homogeneous steady states displayed by the reduced model, and used the results to show qualitatively the existence of travelling waves as heteroclinic orbits connecting the steady states. Numerical simulations confirmed the existence of these travelling waves.

We decided to present this model for ants movement along pheromone trails, as a simple example of local communication between different populations via a chemical produced by the members of both populations. In contrast to the model by Hillen and Stevens [27] discussed above, where the turning rates depended also on the gradient of the external chemical (thus suggesting nonlocal interactions via the chemical signal, among the members of the same population), in model (4.43) the interactions between the two populations depend on the local density of the chemical. In the next chapter we will discuss more complex communication mechanisms, which can lead to short-range and/or long-range interactions among individuals (i.e., repulsion, attraction and alignment). In that context, we will return to the investigation of multiple populations which use different animal communication mechanisms to interact with each other. Furthermore, in Chap. 6 we will discuss another model for ant-trail formation through pheromone deposition, where the pheromone sensing will be assumed nonlocal.

Remark 4.4 The local hyperbolic models presented in these last two chapters form the basis of modelling directional movement in animal/human communities. Moreover, the majority of models presented here were simple enough to be amenable to analytical and numerical investigation of the solutions. However, the approaches taken in these studies did not identify a large variety of patterns displayed by these local models. It is not clear whether this is the result of an insufficient investigation of the parameter spaces for these local models, or the models themselves cannot exhibit more exotic patterns (due to their bifurcation and symmetry structure). In the next two chapters, the focus will be on hyperbolic/kinetic models that include nonlocal interactions between conspecifics. Particular emphasis will be on the complex types of patterns that can be displayed by these nonlocal models. This way, we aim to emphasise the contribution of nonlocal interactions (as a result of nonlocal animal-animal and cell-cell communication) to the complex behaviours observed in animal and cell aggregations.

Remark 4.5 Since many of the models discussed above have been derived using random walk approaches, they incorporate intrinsically various stochastic aspects (e.g., random initial positions of particles, transitions between moving/stationary states governed by Markov processes, etc.; Note that a Markov process is a stochastic process in which the future is independent of the past, given that we know the present [74]). For a more in-depth discussion of stochastic processes in relation to transport equations in cell biology, we refer the reader to [28, 75]. We will return to the aspect of stochasticity at the end of Chap. 5, where we will discuss the explicit incorporation of environmental noise into hyperbolic/kinetic models for collective dynamics.

References

1. A. Zienkiewicz, D. Barton, M.D. Bernardo, Eur. Phys. J. Spec. Top. **224**, 3343 (2015)
2. V. Mwaffo, S. Butail, M. di Bernardo, M. Porfiri, Zebrafish **12**(3), 250 (2015)
3. J. Killeen, H. Thurfjell, S. Ciuti, D. Paton, M. Musiani, M. Boyce, Mov. Ecol. **2**(1), 15 (2014)
4. T. Yang, J.S. Park, Y. Choi, W. Choi, T.W. Ko, K. Lee, PLoS ONE **6**(6), e20255 (2011)
5. C. Qian, C. Wong, S. Swarup, K.H. Chiam, Appl. Environ. Microbiol. **79**(15), 4734 (2013)
6. A. Patterson, A. Gopinath, M. Goulian, P. Arratia, Sci. Rep. **5**, 15761 (2015)
7. S. Yazdi, A. Ardekani, Biomicrofluidics **6**, 044114 (2012)
8. R. Stocker, Proc. Natl. Acad. Sci. USA **108**(7), 2635 (2011)
9. L. Xie, T. Altindal, S. Chattopadhyay, X.L. Wu, Proc. Natl. Acad. Sci. USA **105**, 4209 (2011)

10. M. Segal, I. Soifer, H. Petzold, J. Howard, M. Elbaum, O. Reiner, Biol. Open **1**, 1–12 (2012)
11. E. Reese, L. Haimo, J. Cell Biol. **151**, 155 (2000)
12. M. Müller, S. Klumpp, R. Lipowsky, Proc. Natl. Acad. Sci. **105**(12), 4609 (2008)
13. M. Welte, Curr. Biol. **14**, R525 (2004)
14. A. John, A. Schadschneider, D. Chowdhury, K. Nishinari, J. Theor. Biol. **231**(2), 279 (2004)
15. W. Alhajyaseen, H. Nakamura, M. Asano, Proc. Soc. Behav. Sci. **16**, 526 (2011)
16. C. Feliciani, K. Nishinari, Phys. Rev. E **94**, 032304 (2016)
17. S. Goldstein, Quart. J. Mech. Appl. Math. **4**, 129 (1951)
18. M. Kac, Rocky Mt. J. Math. **4**, 497 (1974)
19. E.E. Holmes, Am. Nat. **142**, 779 (1993)
20. K. Hadeler, Reaction transport systems in biological modelling, in *Mathematics Inspired by Biology*. Lecture Notes in Mathematics (Springer, Berlin, 1999), pp. 95–150
21. E. Codling, M. Plank, S. Benhamou, J. R. Soc. Interface **5**(25), 813 (2008)
22. D. Grünbaum, A. Okubo, in *Frontiers in Mathematical Biology*, ed. by S.A. Levin. Lecture Notes in Biomathematics, vol. 100 (Springer, Berlin, 1994), pp. 296–325
23. M.A. Lewis, Theor. Popul. Biol. **45**, 277 (1994)
24. T. Hillen, Can. Appl. Math. Q. **18**(1), 1 (2010)
25. H. Hasimoto, Proc. Jpn. Acad. Ser. A Math. Sci. **50**, 623 (1974)
26. F. Lutscher, A. Stevens, J. Nonlinear Sci. **12**, 619 (2002)
27. T. Hillen, A. Stevens, Nonlinear Anal.: Real World Appl. **1**, 409 (2000)
28. P. Bressloff, J. Newby, Rev. Mod. Phys. **85**(1), 135 (2013)
29. F. Lutscher, J. Math. Biol. **45**, 234 (2002)
30. L.A. Segel, SIAM J. Appl. Math. **32**, 653 (1977)
31. K. Kang, A. Scheel, A. Stevens, ArXiv (2018)
32. U. Börner, A. Deutsch, H. Reichenbach, M. Bär, Phys. Rev. Lett. **89**, 078101 (2002)
33. U. Börner, A. Deutsch, M. Bär, Phys. Biol. **3**, 138 (2006)
34. O. Igoshin, A. Mogilner, R. Welch, D. Kaiser, G. Oster, Proc. Natl. Acad. Sci. USA **98**, 14913 (2001)
35. O.A. Igoshin, R. Welch, D. Kaiser, G. Oster, Proc. Natl. Acad. Sci. USA **101**, 4256 (2004)
36. O.A. Igoshin, G. Oster, Math. Biosci. **188**, 221 (2004)
37. R. Eftimie, G. de Vries, M.A. Lewis, Proc. Natl. Acad. Sci. USA **104**(17), 6974 (2007)
38. T. Hillen, H. Levine, Z. Angew. Math. Phys. **54**, 1 (2003)
39. Y.L. Chuang, M. D'Orsogna, D. Marthaler, A. Bertozzi, L. Chayes, Phys. D **232**, 33 (2007)
40. A. Scheel, A. Stevens, J. Math. Biol. **75**, 1047 (2017)
41. J.R. Hunter, Anim. Behav. **17**, 507 (1969)
42. J. Bick, G. Newell, Q. Appl. Math. **18**, 191 (1960)
43. A. Chertock, A. Kurganov, A. Polizzi, I. Timofeyev, Math. Models Methods Appl. Sci. **81**, 1947 (2003)
44. A. Kurganov, C.T. Lin, Commun. Comput. Phys. **2**, 141 (2007)
45. C. Appert-Rolland, P. Degond, S. Motch, Netw. Heterog. Media **6**(3), 351 (2011)
46. A. Kurganov, E. Tadmor, J. Comput. Phys. **160**, 240 (2000)
47. F. Lutscher, Eur. J. Appl. Math. **14**, 291 (2003)
48. R. Eftimie, J. Bramson, D. Earn, J. Theor. Biol. **265**, 467 (2010)
49. E. Zemskov, K. Kassner, M. Tsyganov, M. Hauser, Eur. Phys. J. B. **72**, 457 (2009)
50. K. Hadeler, Math. Comput. Model. **31**(4–5), 75 (2000)
51. T. Hillen, J. Math. Anal. Appl. **210**, 360 (1997)
52. T. Hillen, J. Math. Biol. **35**, 49 (1996)
53. K. Hadeler, Can. Appl. Math. Q. **2**, 27 (1994)
54. K. Hadeler, in *Proceedings of the Thirteenth Dundee Conference*, ed. by R. Jarvis (1996), pp. 18–32
55. T. Hillen, Nichtlineare hyperbolische systeme zur modellierung von ausbreitungsvorgängen und anwendung auf das turing modell. Ph.D. thesis, Universität Tübingen, 1995
56. D. Needham, J. Leach, IMA J. Appl. Math. **73**, 158 (2008)

57. T. Hillen, Qualitative analysis of hyperbolic random walk systems. Technical report, SFB 382, Report No. 43, 1996
58. K. Hadeler, Nonlinear propagation in reaction transport systems, in *Differential Equations with Applications to Biology*. Fields Institute Communications (American Mathematical Society, Providence, 1998), pp. 251–257
59. K. Hadeler, J. Math. Sci. **149**(6), 1658 (2008)
60. K. Hadeler, Reaction-telegraph equations with density-dependent coefficients, in *Partial Differential Equations. Models in Physics and Biology*. Mathematical Research, vol. 82 (Akademie-Verlag, Berlin, 1994), pp. 152–158
61. K. Hadeler, in *Differential Equations and Applications to Biology and Industry. Proceedings of the Claremont International Conference*, ed. by M. Martelli, K. Cooke, E. Cumberbatch, B. Tang, H. Thieme (1996), pp. 145–156
62. C. Xue, H. Hwang, K. Painter, R. Erban, Bull. Math. Biol. **73**, 1695 (2011)
63. A. Kuznetsov, A.A. Avramenko, Proc. R. Soc. A **464**, 2867 (2008)
64. A. Kuznetsov, Proc. R. Soc. A **468**, 3384 (2012)
65. I. Kuznetsov, A. Kuznetsov, J. Biol. Phys. **40**, 41 (2014)
66. A. Friedman, G. Craciun, SIAM J. Math. Anal. **38**(3), 741 (2006)
67. P. Jung, A. Brown, Phys. Biol. **6**(4), 046002 (2009)
68. P. Monsma, Y. Li, J. Fenn, P. Jung, A. Brown, J. Neurosci. **34**(8), 2979 (2014)
69. T. Ruijgrok, T. Wu, Phys. A **113**, 401 (1982)
70. M. Garavello, B. Piccoli, Netw. Heterog. Media **4**(1), 107 (2009)
71. G. Wong, S. Wong, Transp. Res. A **36**, 827 (2002)
72. S. Benzoni-Gavage, R. Colombo, Eur. J. Appl. Math. **14**, 587 (2003)
73. J. Watmough, L. Edelstein-Keshet, J. Math. Biol. **33**, 459 (1995)
74. E. Dynkin, *Markov Processes*. Die Grundlehren der Mathematischen Wissenschaften (In Einzeldarstellungen mit Besonderer Berücksichtigung der Anwendungsgebiete), vol. 121/122 (Springer, Berlin, 1965)
75. P. Bressloff, *Stochastic Processes in Cell Biology* (Springer, Cham, 2014)

Chapter 5
Nonlocal Hyperbolic Models in 1D

5.1 Introduction

The hyperbolic models discussed in Chaps. 3 and 4 assumed that the interactions among organisms are only local. However, many organisms can receive information about conspecifics that are further away using, for example, acoustic or visual signals. An eloquent example is offered by migrating birds that fly during the night in very loose flocks, with individuals 200–300 m apart from each other [1]. Despite these large inter-individual distances, radar tracking data indicates that all birds move with the same speed and in the same direction [1]. The persistence of these migrating flocks of birds can be explained by the use of long-distance acoustic communication, to share "in-flight" information about movement speed and directions [1, 2]. These nonlocal interactions among individuals belonging to the same group are not restricted only to flocks of birds. Minnows, for example, have the tendency to approach the larger of the two shoals of conspecifics when exposed to predatory treat [3]. Their ability to decide which shoal is larger suggests the possibility of long-distance communication through visual signals. Monkeys have also been shown to use long-distance calls to communicate with conspecifics or in response to predators [4, 5]. More recently, experimental studies on the collective movement of cells have shown that cells can coordinate their behaviours across many cell rows—through mechano-transduction pathways [6]. Thus, even cells display nonlocal (i.e., long-distance) interactions. We need to emphasise that "long distance" is a relative term, which depends on the physiological characteristics of each species: it can range from micrometers for cells, to a few body lengths for fish, and meters and even kilometres for birds or monkeys.

In general, animal communication is the result of various signals (visual, chemical, acoustic, electrical or tactile signals) that organisms use to receive or to send information about their position and direction of movement [7]. Some signals (i.e., tactile and electrical signals) act on short spatial distances, while other signals

© Springer Nature Switzerland AG 2018   107
R. Eftimie, *Hyperbolic and Kinetic Models for Self-organised Biological
Aggregations*, Lecture Notes in Mathematics 2232,
https://doi.org/10.1007/978-3-030-02586-1_5

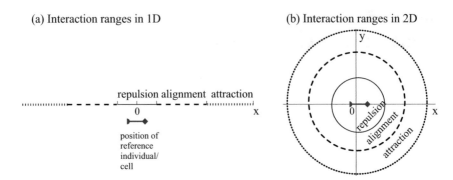

Fig. 5.1 A simplified depiction of the spatial ranges for three social interactions: repulsion, alignment and attraction. (**a**) one-dimensional (1D) case; (**b**) two-dimensional (2D) case. A reference individual is positioned at $x = 0$

(i.e., visual and acoustic signals) act on large spatial distances; see also Fig. 1.6. These communication signals are used to influence the social interactions among group members by causing organisms to change their travel direction and speed. For example, individuals can turn to move away from nearby neighbours (repulsion), to move towards conspecifics that are further away (attraction), or to align with other organisms that are at intermediate distances (alignment). They can also slow-down to avoid collision with other individuals nearby (repulsion), or speed-up to approach those that are further away (attraction). The spatial ranges for these three social interactions are depicted in Fig. 5.1.

Incorporating nonlocal interactions into the mathematical models which describe biological aggregations is a natural step in the modelling process. The majority of these nonlocal models were derived to explain complex biological processes that could not be reproduced by local models. For example, in the context of insect aggregations, a well-known nonlocal parabolic model for swarm formation was developed by Mogilner and Edelstein-Keshet [8] to address the lack of cohesive swarms (i.e., lack of travelling pulse solutions) displayed by local aggregation models [9]. In the context of cell aggregations, another nonlocal continuous model was developed by Armstrong et al. [10] to address the lack of cell segregation/mixing patterns observed experimentally with different cross-adhesion and self-adhesion strengths of cells during cell-cell interactions, but not reproduced by local continuous models.

Another natural step in the modelling process is to investigate the structure of the group patterns when individuals in a group communicate via different mechanisms. Generally, the self-organised biological aggregations are very heterogeneous, with individuals having different physiological characteristics (e.g., different age [11] or different health states [12]). This heterogeneity also translates into a heterogeneity of inter-individual communication, as discussed in [11] in the context of age-related alarm calls in vervet monkeys (see also Fig. 1.8). Moreover, the differences in communication could also affect the patterns exhibited by the animal groups, which in some extreme cases could mean speciation (thus spatial segregation of animal communities), as discussed in [13].

In this chapter, we extend the previous hyperbolic models to account for the long-distance communication that takes place among organisms belonging to the same group. We first discuss some general 1-equation nonlocal hyperbolic models. Then we focus on systems of hyperbolic equations, and discuss the group patterns that arise when individuals use only one communication mechanism. Next, we investigate the patterns that form in the presence of two different communication mechanisms. In this context of animal communication, we also discuss the role of particular social interactions (attraction, repulsion and alignment) on the resulting group patterns. We conclude with a discussion on the effect of environmental noise on inter-individual communication and aggregation patterns.

5.2 One-Equation Nonlocal Models with Density-Dependent Speeds

The simplest way to incorporate nonlocal social interactions into a mathematical model is to assume that individuals ($u \in \mathbb{R}$) speed-up or slow-down in response to neighbours detected ahead of them (3.2) [14, 15]:

$$\frac{\partial u}{\partial t} + \frac{\partial (\Gamma[u]u)}{\partial x} = 0, \quad \text{with } x \in \mathbb{R},\ t \in \mathbb{R}^+. \tag{5.1}$$

The speed $\Gamma[u] \in \mathbb{R}$ is usually a functional (as described by [·]) of the nonlocal repulsive and attractive interactions (with "$\star$" denoting the convolution symbol):

$$\Gamma[u] = G(K \star u) \equiv G\left(\int_{-\infty}^{\infty} K(x-y)u(y)dy \right). \tag{5.2}$$

The simplest choice for the speed function is a linear function: $G(y) = y$ (see, for example [14, 15]). The kernel $K(x)$ is usually defined to be the difference between a repulsive (K_r) and an attractive (K_a) force, which have opposite effects—hence the opposite sign [14, 16–18]:

$$K(x) = q_r K_r(x) - q_a K_a(x). \tag{5.3}$$

If we consider also alignment interactions, then we have a linear combination of repulsive (K_r), attractive (K_a) and alignment (K_{al}) forces [19–21]:

$$K(x) = q_r K_r(x) - q_a K_a(x) + q_{al} K_{al}(x). \tag{5.4}$$

Here, q_r, q_a and q_{al} describe the strength of the repulsive, attractive and alignment interactions, respectively (with $q_{r,al,a} \in \mathbb{R}^+$). The following are the most com-

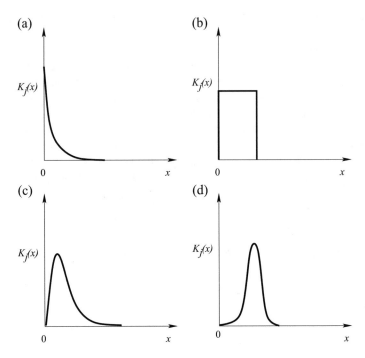

Fig. 5.2 Examples of long-distance interaction functions $K_j(x)$, $j = r, a$ (where K_r describes the spatial range of repulsive interactions, and K_a describes the spatial range of attractive interactions). (**a**) Morse interaction (described by a decaying exponential); (**b**) Piecewise constant; (**c**) Gaussian; (**d**) Translated Gaussian. The equations for these functions are given in the text below

mon choices for the repulsive (K_r) and attractive (K_a) kernels (depicted also in Fig. 5.2):

(a) Morse interaction kernel: $K_j(s) = e^{|s|/s_j}$, $j = r, a$, with s_j describing the length scale of the interactions [14, 16, 17]. See Fig. 5.2a;

(b) Piecewise constant kernel: $K_j(s) = 1/2d$ if $|s| \leq s_r$ and $K_j(s) = 0$ otherwise [15]. See Fig. 5.2b;

(c) Gaussian kernel: $K_j(s) = s e^{-s^2/(2s_j^2)}/2s_j^2$, $j = r, a$, with s_j describing the length scale of the interactions [8, 15, 19]. See Fig. 5.2c;

(d) Translated Gaussian kernel: $K_j(s) = \frac{1}{2\pi m_j^2} e^{-(s-s_j)^2/(2m_j^2)}$, $j, r, a, m_j = s_j/8$ [19–21]. See Fig. 5.2d.

In Fig. 5.3 we graph the combined repulsive-attractive functions $K(x) = K_r(x) - K_a(x)$ for the four types of interaction kernels discussed above (and shown in Fig. 5.2). We note that the last two functions in Fig. 5.3c and d are continuous everywhere. This implies that on very short inter-individual distances (i.e., $x \approx 0$), the neighbours are not repelled very strongly. While this assumption is not particularly realistic from a biological point of view (although one could argue that

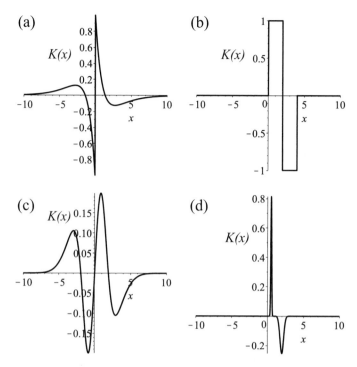

Fig. 5.3 Examples of long-distance repulsive-attractive interaction kernels $K(x) = K_r(x) - K_a(x)$. (**a**) Morse kernel; (**b**) Piecewise constant kernel; (**c**) Gaussian kernel; (**d**) Translated Gaussian kernel

the repulsion should be stronger at points x on the surface of the animal/cell where $0 < x \ll 1$, to avoid pilling-up before animals/cells start pressing on each other), it is convenient mathematically since it means infinite differentiability and bounded norms. A more biologically realistic assumption is described by the Morse kernel (see Fig. 5.3a),

$$K(s) = \text{sgn}(s)\left(-q_a e^{-|s|/s_a} + q_r e^{-|s|/s_r}\right). \tag{5.5}$$

Note that this kernel has the same qualitative behaviour as the atomic force microscopy measurements of cell-cell adhesion interactions [22]. We also emphasise that the Morse interaction kernel has a jump discontinuity at zero, which means that individuals are repelled more strongly at very short distances (i.e., distances close to zero). The discontinuity of kernel (5.5) seems to contribute to the formation of blow-up patterns, as shown by Leverentz et al. [14]. In particular, when the social interactions are attractive on long distances (i.e., $s_a > 1$, $s_r = 1$), model (5.1)+(5.5) can display a single-point blow-up. The blow-up time is given by $T = \left(b(0) - a(0)\right)/2Mq(0)$, where $b(0)$ and $a(0)$ are the endpoints of an initial aggregation shaped like a sufficiently narrow pulse, and $Mq(0)$ is the minimum speed at which the two endpoints travel towards each other; for details, see [14]. In contrast, when the social interactions are repulsive on long distances ($s_a < 1$, $s_r = 1$), the model can display two-point blow-ups (i.e., the singularities form at two different points in space). Note that these blow-up patterns are caused by the structure of the kernel $K(s)$ at the origin (i.e., the discontinuity). In addition, when the social interactions are attractive on long distances (e.g., $s_a = 4$, $s_r = 1$), but the repulsion force is stronger than the attraction force (e.g., $q_a = 0.4$, $q_r = 1$), these models can exhibit stationary pulses [14].

Topaz and Bertozzi [15] considered a more general velocity function

$$\Gamma[u(x,t)] = F(u) + G_1(u)(K_1 \star H_1(u)) + G_2(u)(K_2 \star H_2(u)), \tag{5.6}$$

where $F(u)$ describes local density-dependence of the velocity, $K_{1,2}$ are attractive-repulsive interaction kernels, $G_{1,2}$ describe the strength of the interactions, and $H_{1,2}$ describe the dependence of convolutions on population density u. (As above, we used $[\cdot]$ to emphasise that Γ is a functional, which depends on the nonlocal interactions described with the help of the convolution symbol "$\star$".) By considering the particular case $F = 0$ and $G_2 = 0$ in (5.6), the authors showed that in 1D, purely nonlocal velocities cannot lead to the formation of travelling pulses when the kernels are biologically realistic (i.e., $dK_1/d|z| \leq 0$) [15]. To this end, they considered the constant-density travelling band/pulse ansatz:

$$u(x,t) = u_0 W_L(x - ct), \quad W_L(x - ct)\Gamma[u(x,t)] = cW_L(x - ct), \tag{5.7}$$

with

$$W_L = \begin{cases} 1, & x \in [0, L], \\ 0, & \text{otherwise.} \end{cases} \tag{5.8}$$

Substituting these equations into (5.1), Topaz and Bertozzi [15] showed that the nonlocal equation can admit constant-density travelling band solutions only if the kernel K_1 is a L-periodic kernel on $[-L, L]$. However, such a kernel cannot be biologically realistic (since it does not decay with the distance). Moreover, when $G_2 = 0$ and $F, G_1 \neq 0$, it can be shown that the nonlocal model admits constant-density travelling band solutions for any choice of kernels K_1.

A slightly different class of nonlocal mathematical models assumes that the population velocity depends on the gradient of the interaction potential $K(x)$, and eventually on an external potential $V(x)$ [23–25]:

$$\frac{\partial u}{\partial t} = \frac{\partial}{\partial x}\left(u\frac{\partial}{\partial x}(K \star u) + V\right), \tag{5.9}$$

This model describes the movement of a population u in the direction of increasing density of neighbours. In regard to the qualitative behaviour of these models, it has been shown in [24] that for a symmetric singular interaction potential K (i.e., the kernel K is such that $K'(0^+) \neq 0$, and $K(x) = K(-x)$ for all $x \in \mathbb{R}$), the stable stationary states (pulses) are finite sums of Dirac masses:

$$\bar{u} = \sum_{i=1}^{n} \bar{u}_i \delta_{\bar{u}_i}(x), \quad \text{with } \bar{u}_i > 0, \ \sum_{i=1}^{n} \bar{u}_i = 1. \tag{5.10}$$

Moreover, in [24] the authors showed analytically that these steady states (i.e., stationary pulses) are linearly stable. Numerical simulations (using an implicit Euler scheme) confirmed the shape of the steady states (as a sum of approximations of Dirac masses), and their stability.

A similar one-equation nonlocal model introduced in [26] was shown to display different types of steady states, depending on the type of the potential $K(x)$: periodic steady states with period r (if $K(x) = (r - |x|)_+$), constant steady states (if $K(x) = \alpha \exp(-\lambda|x|)$, $\alpha \neq 0$, $\lambda > 0$), trivial steady states (for fast decaying potentials: $|K(x)| + |K'(x)| < ce^{-ax}$, $a, c > 0$), blow-up solutions (if $K(x) = K_0(x) + \bar{K} \cdot (x)_+$, with $\bar{K} > 0$, $(x)_+ = x$ for $x \geq 0$ and $(x)_+ = 0$ for $x < 0$), globally-defined solutions (if $K(x) = K_0(x) + \bar{K} \cdot (x)_+$, with $\bar{K} \leq 0$, $(x)_+ = x$ for $x \geq 0$ and $(x)_+ = 0$ for $x < 0$), and globally-defined Dirac masses. We emphasise that all these pattern formation results in [26] are theoretical.

We will return to these types of nonlocal hyperbolic models in the next chapter, when we will discuss models in two spatial dimensions. As we will see there, the addition of another spatial dimension can allow for the emergence of multiple lanes and vortices.

5.3 Derivation of Nonlocal Hyperbolic Systems

We assume now that individuals/cells can move either left or right on an 1D domain, and thus we consider systems of left-moving and right-moving individuals/cells. Because these individuals can change their movement direction (in response to nearby or more distant conspecifics), we start our investigation of these nonlocal systems by focussing first on the turning behaviour of individuals.

Similar to the stochastic derivation of local hyperbolic systems (4.3) (see Sect. 4.2), one can derive a nonlocal version of these models using a correlated random walk approach [7]. For this, consider a population of size N. As for the local systems, the probability of a *randomly chosen right or left-moving individual* to be found inside the interval $[x - \Delta x/2, x + \Delta x/2)$ at time t is defined by Eq. (4.1). To describe the turning behaviour, we assume that the probability of changing direction has two components: a random component and a directed component. This assumption is consistent with the biological behaviour of organisms, as remarked by Lotka [87, p. 360]: "the type of motion presented by living organisms ...can be regarded as containing both a systematically directed and also a random element". Therefore, the probability of a right-moving (left-moving) individual to turn left/right is given by

$$\lambda^{\pm} = \text{probability of turning randomly} +$$

probability of turning left/right in response to distant neighbors

$$= \frac{\lambda_1 \Delta t}{2} + \frac{\lambda_2 \Delta t}{2} F^{\pm}. \tag{5.11}$$

Here, $F^{\pm}$ are non-dimensional, increasing, uniformly continuous functions of the difference between the right-moving and left-moving neighbours that are located far away. We choose $0 < \lambda_1,\ \lambda_2,\ F^{\pm} < 1$. Moreover, it is biologically realistic to assume that when the numbers of these distant right-moving and left-moving individuals are similar (i.e., their difference is almost zero), then movement is more likely to be random (since individuals do not have any preference in turning towards a particular distant neighbour). Mathematically, this means that $F^{\pm} \approx 0$. To exemplify how we model turning behaviours, let us focus on the repulsive spatial ranges. To avoid collisions with nearby neighbours, an individual will increase its probability of turning around if there are more individuals in front of it moving in the opposite direction, than individuals behind it moving in the same direction (see Fig. 5.4). The turning probability will decrease if there are more individuals behind it moving in the same direction. Hence, $F^{\pm}$ will be defined in terms of

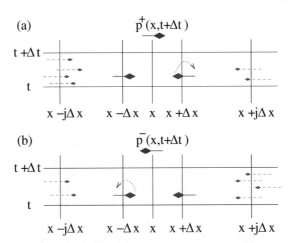

Fig. 5.4 The movement probabilities of (**a**) a right-moving individual, and (**b**) a left-moving individual. A right-moving individual can be positioned at x at time $t + \Delta t$, if at the previous time step t it was at $x - \Delta x$ and kept moving in the same direction, or it was at $x + \Delta x$ and moving left, and changed direction at the end of the time step. A similar explanation holds for a left-moving individual (**b**). The change in direction is either random, or in response to distant individuals positioned at $x \pm j\Delta x$, where j takes values within a certain interaction range (e.g., repulsion, alignment, or attraction ranges)

the difference between right-moving and left-moving neighbours that are within a certain interaction range.

Before giving the expression for $F^\pm$, we make one more assumption. We assume that *all other individuals have the same probabilities* $p^\pm(x,t)$, independent of the location of the right-moving or left-moving individual chosen previously (see Eq. (4.1)). This assumption is made to ensure a rigorous derivation of the nonlocal model using a correlated random walk approach, with the turning being governed by a Poisson process [27, 28]. We note, however, that the assumption might be too restrictive for some biological systems. For example, if individuals tend to avoid being spatially too close to their neighbours, then the observation of an individual at a particular position also implies that the nearby positions might not be occupied [29]. Hence, the probability distributions are correlated. Also, foraging animals might prefer spending more time in some regions than in others [28, 30], and thus spatial positions in those regions are more likely to be occupied. However, as noted by Othmer et al. [28], the macroscopic-level derivation of a transport equation describing the changes in population density does not generally have to be governed by a Poisson process.

Returning to the derivation of the model via a correlated random walk approach, if the chosen individual is at point x, then the expected number of individuals at distance $j\Delta x$ that are moving right or left is $Np^\pm(x + j\Delta x, t)$. Under these assumptions, the probability of turning in response to distant neighbours is described by

$$F^\pm = F\left(\pm N \sum_{j=-\infty}^{\infty} K_w(j\Delta x)\left(p^-(x + j\Delta x, t) - p^+(x - j\Delta x, t)\right)\right),$$

(5.12)

where F is a uniform continuous function of the difference between the left-moving and right-moving neighbours positioned within a certain range described by kernels K_w. Substituting (5.11) and (5.12) into Eq. (4.2) leads to

$$\frac{\partial p^+}{\partial t} + \frac{\Delta x}{\Delta t}\frac{\partial p^+}{\partial x} = -p^+\left(\frac{\lambda_1}{2} + \frac{\lambda_2}{2}F^+\right) + p^-\left(\frac{\lambda_1}{2} + \frac{\lambda_2}{2}F^-\right)$$
$$+ O(\Delta x) \text{ terms},$$

(5.13a)

$$\frac{\partial p^-}{\partial t} - \frac{\Delta x}{\Delta t}\frac{\partial p^-}{\partial x} = p^+\left(\frac{\lambda_1}{2} + \frac{\lambda_2}{2}F^+\right) - p^-\left(\frac{\lambda_1}{2} + \frac{\lambda_2}{2}F^-\right)$$
$$- O(\Delta x) \text{ terms}.$$

(5.13b)

Now let $\Delta x, \Delta t \to 0$, such that $\frac{\Delta x}{\Delta t} \to \gamma$. Since $F^\pm$ are assumed to be uniformly continuous (because F is uniformly continuous), we can interchange the limit and the functions $F^\pm$. Multiplying Eqs. (5.13) by N and using Eq. (4.1), leads to the following nonlocal hyperbolic system

$$\frac{\partial u^+}{\partial t} + \gamma \frac{\partial u^+}{\partial x} = -u^+ \lambda^+[u^+, u^-] + u^- \lambda^-[u^+, u^-], \qquad (5.14a)$$

$$\frac{\partial u^-}{\partial t} - \gamma \frac{\partial u^-}{\partial x} = u^+ \lambda^+[u^+, u^-] - u^- \lambda^-[u^+, u^-]. \qquad (5.14b)$$

Here, $u^\pm$ are probability density functions for the left- and right-moving individuals. The nonlocal turning rates are described by

$$\lambda^+[u^+, u^-] = \frac{\lambda_1}{2} + \frac{\lambda_2}{2} F \left(\int_{-\infty}^{\infty} K_w(s)(u^-(x+s, t) - u^+(x-s, t)) ds \right),$$
$$(5.15a)$$

$$\lambda^-[u^+, u^-] = \frac{\lambda_1}{2} + \frac{\lambda_2}{2} F \left(-\int_{\infty}^{\infty} K_w(s)(u^-(x+s, t) - u^+(x-s, t)) ds \right).$$
$$(5.15b)$$

Remark 5.1 It was shown in [7, 31, 32] that by rescaling the time and space variable ($x = \psi/\epsilon, t = \tau/\epsilon^2, \epsilon \ll 1$), or the speed and the turning rates ($\gamma = \gamma^*/\epsilon, \lambda_{1,2} = \lambda_{1,2}^*/\epsilon^2, \epsilon \to 0$), and then taking the parabolic limit (as $\epsilon \to 0$), model (5.14) can be transformed into a parabolic advection-diffusion equation for the total population density $u = u^+ + u^-$, similar to the one in [8]:

$$\frac{\partial u}{\partial t} = \frac{\partial}{\partial x} \left(D_0 \frac{\partial u}{\partial x} \right) - \frac{\partial}{\partial x} \left(B_0 u \left(F(y^-[u]) - F(y^+[u]) \right) \right), \qquad (5.16)$$

where $y^\pm = \pm \int_{-\infty}^{\infty} K_w(s)(u^-(x+s, t) - u^+(x-s, t)) ds$, $D = \lim_{\epsilon \to 0} \gamma^2/(2\lambda_1)$, and $B = \lim_{\epsilon \to 0} \gamma \lambda_2/(2\lambda_1)$. Note that the nonlocal advection term is generated by the differences in directional changes (i.e., $\lambda^-[u] - \lambda^+[u]$, with $u = u^+ + u^-$ the total density). It was shown in [8] that in the absence of external drift factors, such an advection-diffusion model cannot exhibit persistent travelling pulses. We will return to this particular result in Sect. 5.7, when we will discuss the loss of spatio-temporal patterns in the parabolic limit, as a result of the loss in the bifurcation structure of the models.

5.4 Alignment Models

The most common assumption incorporated in the local as well as some nonlocal hyperbolic models is that individuals/cell change their movement direction to align with their neighbours [33, 34]. One of the first nonlocal models to incorporate long-distance alignment interactions was introduced by Pfistner [33]. Modelling the behaviour of *Myxobacteria* swarms, Pfistner [33] started with Eq. (5.14) and assumed that bacteria turn only as a result of interactions with other bacteria further away (i.e., $\lambda_1 = 0$ in (5.15)), which are detected within a perception interval $[-R, R]$. The turning rates are defined by the following nonlocal terms:

$$\lambda^+[u^+, u^-] = F\left(\int_{-R}^{R} (\alpha(s)u^+(x+s, t) + \beta(s)u^-(x+s, t))ds\right), \qquad (5.17a)$$

$$\lambda^-[u^+, u^-] = F\left(\int_{-R}^{R} (\alpha(s)u^-(x-s, t) + \beta(s)u^+(x-s, t))ds\right), \qquad (5.17b)$$

where α and β are the weight functions for the surrounding densities. The functional F, which describes the turning behaviour, is monotone increasing and positive. The simulation results showed that the model can display stationary swarms [33]. Later, Pfistner modified this model to incorporate moving boundaries for the swarm edges [34]. The dynamics of the swarm was analysed numerically through the retraction and expansion of the boundaries.

5.5 Repulsive, Attractive, and Alignment Models

The assumption that individuals turn only to align with their neighbours might explain the behaviour of some bacteria (e.g., *Myxobacteria* [33]), but it cannot explain the more complex behaviours observed in other groups, such as swarms of insects or flocks of birds. A more realistic assumption would be that individuals turn also to approach other neighbours further away, or to avoid collision with neighbours in the immediate vicinity. A nonlocal model that considers all three social interactions, namely alignment, attraction and repulsion, was introduced by Eftimie et al. [19, 20]. There, the turning rates were described by the following terms:

$$\lambda^\pm[u^+, u^-] = \lambda_1 + \lambda_2 \tanh\left(y^\pm[u^+, u^-] - y_0\right), \qquad (5.18)$$

where

$$y^\pm[u^+, u^-] = y_r^\pm[u^+, u^-] - y_a^\pm[u^+, u^-] + y_{al}^\pm[u^+, u^-]. \qquad (5.19)$$

The terms $y_r^\pm$, $y_{al}^\pm$ and $y_a^\pm$ denote the repulsion, alignment, and attraction interactions that influence the likelihood of turning to the left $(+)$ or to the right $(-)$. To describe these terms, we recall that communication can influence these social interactions. Eftimie et al. [20] started with five communication mechanisms in 1D and correspondingly modelled five different ways (models M1–M5) in which group members can interact with each others (see Fig. 5.5 and Table 5.1). The following assumptions were incorporated into the five models:

- *model M1* assumes that the attractive and repulsive interactions depend on the stimuli received from all neighbours, whereas the alignment depends only on the

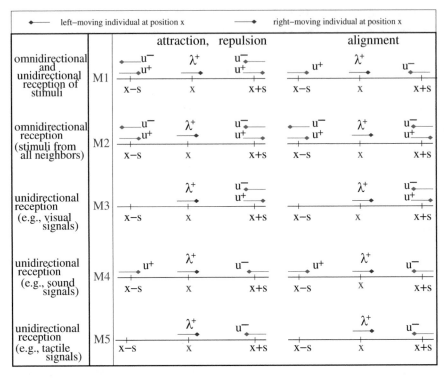

Fig. 5.5 The five sub-models (M1–M5) for signal reception introduced in [20]. A reference individual is positioned at x, while its right-moving (u^+) and left-moving (u^-) neighbours can be positioned at $x+s$ and/or $x-s$. M1: information can be received from all neighbours (for attraction and repulsion), or from neighbours moving towards the reference individual (for alignment); M2: information is received from all neighbours (for attraction, repulsion, alignment); M3: information is received only from ahead with respect to the moving direction of the reference individual (for attraction, repulsion, alignment); M4: information is received only from neighbours moving towards the reference individual; M5: information is received only from ahead, and only from neighbours moving towards the reference individual

stimuli received from those neighbours moving towards the reference individual (this case was investigated in [19]);

- *model M2* assumes that all three social interactions depend on stimuli received from all neighbours;
- *model M3* assumes that the interactions depend only on the information received from ahead (with respect to the moving direction);
- *model M4* assumes that the social interactions depend on the stimuli received from ahead and behind, only from neighbours moving towards the reference individual;
- *model M5* assumes that the social interactions depend on the stimuli received only from ahead and only from neighbours moving towards the reference individual.

Table 5.1 The nonlocal terms used to describe the social interactions in [20]

Model	Social interactions: repulsion $(y_r^\pm)$, attraction $(y_a^\pm)$, alignment $(y_{al}^\pm)$
M1	$y_{r,a}^\pm = q_{r,a} \int_0^\infty K_{r,a}(s)(u(x \pm s) - u(x \mp s))ds$
	$y_{al}^\pm = q_{al} \int_0^\infty K_{al}(s)(u^\mp(x \pm s) - u^\pm(x \mp s))ds$
M2	$y_{r,a}^\pm = q_{r,a} \int_0^\infty K_{r,a}(s)(u(x \pm s) - u(x \mp s))ds$
	$y_{al}^\pm = q_{al} \int_0^\infty K_{al}(s)(u^\mp(x \pm s) + u^\mp(x \mp s) - u^\pm(x \pm s) - u^\pm(x \mp s))ds$
M3	$y_{r,a}^\pm = q_{r,a} \int_0^\infty K_{r,a}(s)(u(x \pm s))ds$
	$y_{al}^\pm = q_{al} \int_0^\infty K_{al}(s)(u^\mp(x \pm s) - u^\pm(x \pm s))ds$
M4	$y_{r,a}^\pm = q_{r,a} \int_0^\infty K_{r,a}(s)(u^\mp(x \pm s) - u^\pm(x \mp s))ds$
	$y_{al}^\pm = q_{al} \int_0^\infty K_{al}(s)(u^\mp(x \pm s) - u^\pm(x \mp s))ds$
M5	$y_{r,a}^\pm = q_{r,a} \int_0^\infty K_{r,a}(s)u^\mp(x \pm s)ds$
	$y_{al}^\pm = q_{al} \int_0^\infty K_{al}(s)u^\mp(x \pm s)ds$

These terms are the translations of the diagrams from Fig. 5.5 into mathematical equations, after summing up the information received from all neighbours ($s \in (0, \infty)$). The terms $y_{r,a}^-$ and y_{al}^- were obtained through a process similar to the one described in Fig. 5.5. For each of the five models, the equations for $y_r^\pm$ and $y_a^\pm$ are almost identical (since the effect of the repulsive and attractive interactions is similar but opposite—hence the negative sign in front of $y_a^\pm$ term in Eq. (5.19)). As mentioned in the main text, q_a, q_r, and q_{al} give the strength of the attraction, repulsion, and alignment interactions, respectively. Kernels $K_a(s)$, $K_r(s)$ and $K_{al}(s)$ give the spatial ranges for the attractive, repulsive and alignment interactions; see also Fig. 5.2. The total density is $u = u^+ + u^-$

All these different assumptions are described by mathematical equations in Table 5.1. We emphasise that not all these models are biologically realistic (see also Fig. 5.5 for the biological interpretation of some of the assumptions). However, this modelling approach proposed a straightforward way to incorporate various communication mechanisms into a mathematical model.

Remark 5.2 For individual-based models, there are two modelling approaches used to explain the empirical observations: the *metric-distance approach* (where interactions among individuals depend only on the spatial distance at which these individuals are located, which leads to short-range repulsion, medium-range alignment/orientation, and long-range attraction interactions) and the *topological-distance approach* (with the assumption is that individuals interact only with 6-7 of their neighbours [35]). By considering these communication mechanisms in model (5.14)–(5.19), we connect the metric and topological approaches through the assumption that individuals can interact only with those neighbours that they can perceive (within some specific spatial ranges). It is very plausible that individuals respond only to those neighbours that "catch their attention"—although experimental evidence supporting this idea is lacking at the moment, due to the difficulty of gathering data regarding group-level communication. Partan [36] showed that at behavioural level, the mechanisms that lead to the integration of signals from neighbours involve not only communication but also perception and attention. Ballerini et al. [35] do suggest that the number of 6–7 neighbours that were shown

to influence the movement of a reference individual in a flock of starlings, is likely
the result of cortical incorporation of the visual input. Therefore, model (5.14)–
(5.19) combines topological-like interactions (via communication with specific
neighbours) with metric-distance interactions (via interaction zones).

To investigate the types of patterns displayed by this class of nonlocal hyperbolic
models, the authors in [19, 20] performed numerical simulations on a finite
domain $[0, L]$ with periodic boundary conditions. To check whether the patterns
were not an artefact of the periodic boundary conditions, the authors performed
simulations also with Neumann and Dirichlet boundary conditions. The results
were consistent for all these different conditions (at least for some small time,
before the aggregations reached the boundaries). The numerical scheme chosen to
discretise the equations was a first-order upwind scheme. However, the simulations
were also checked with a second-order MacCormack scheme. The integrals were
discretised using the Simpson's method, and were wrapped around the domain
(due to the periodic boundary conditions). Finally, the initial conditions for the
simulations were chosen to be small random perturbations of spatially homogeneous
steady states $(u^+, u^-) = (u_*^+, u_*^-)$ (with left-moving and right-moving individuals
uniformly spread over the whole domain).

The numerical results in [19, 20] suggested that alignment is a necessary ingre-
dient for the movement in the absence of any external environmental cues. More
precisely, attraction and repulsion alone lead to stationary groups. However, when
alignment was included, it was possible to obtain moving groups. The movement of
these aggregations was the result of Hopf bifurcations (while the attractive-repulsive
interactions alone lead only to real steady-state bifurcations; we will return to the
discussion of these bifurcations in Sect. 5.7. An alternative to including alignment
was to consider an environmental drift, which induces asymmetry in the reception
of signals [7]. For example, if we focus on attraction/repulsion term corresponding
to the general model M2, we can define the nonlocal terms as follows:

$$y_{r,a}^{\pm} = q_{r,a} \int_0^{\infty} K_{r,a}(s)\big(p_{\pm}u(x \pm s) - p_{\mp}u(x \mp s)\big)ds, \qquad (5.20)$$

where $p_{\pm}$ describe the communication signals received from the right (subscript
"+") or from the left (subscript "−"). Figure 5.6 depicts the asymmetry in the perception of signals, where $p_+ > p_-$. This assumption of asymmetric communication
(which translates into asymmetric equations; see the discussion below), can lead to
moving aggregations of individuals [7].

Incorporating various communication mechanisms (as shown by the terms in
Table 5.1) leads to more complex spatial and spatio-temporal patterns [20]; see
also Fig. 5.7. Some of these patterns, such as stationary pulses, travelling pulses, or
ripples, have been previously obtained with other hyperbolic [37, 38] or parabolic
models [39]. However, the majority of patterns described in [20] are new: breathers,
travelling breathers, feathers, zigzags, and semi-zigzags (see Fig. 5.7). A detailed
description of these patterns is given in Table 1.2. Since the majority of these

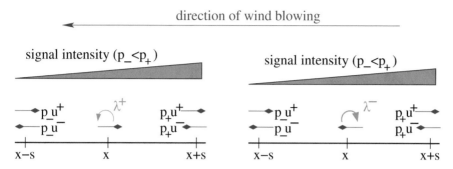

Fig. 5.6 Caricature depiction of asymmetric reception of signals from neighbours, under the effect of an external environmental drift, such as wind blowing from a certain direction (i.e., from right in the above figure)

patterns have not been seen in local models, it suggests that simple local interaction mechanisms might not be sufficient to explain some of the observed biological behaviours, such as the zigzagging or the expanding/contracting flocks of birds [40]. This latter behaviour is described mathematically by breathers or travelling breathers—see Table 1.2. The importance of nonlocal interactions for the emergence of the complex spatio-temporal patterns summarised in Fig. 5.7 is reinforced by the comparison with the patterns shown in Fig. 1.4 (obtained with the nonlocal individual-based model introduced in [41]).

Numerical simulations of model (5.14)+(5.18)+(5.19) also revealed that some of these patterns can be obtained only for specific assumptions regarding the communication signals. For example, breathers were obtained only with model M4, and feathers were obtained only with model M3. There were a few other patterns, such as stationary pulses, travelling trains and travelling pulses, which could be obtained with the majority of nonlocal models. It is interesting to remark that these three patterns are observed in almost every biological aggregation: stationary pulses describe stationary/sleeping aggregations [42], travelling pulses describe travelling/migrating aggregations [43], travelling trains describe waves of activity that pass through aggregations [44]. A summary of some of the patterns exhibited by models M2-M5 is shown in Table 5.2; see also [20]. Given the very large parameter space corresponding to this model, a full investigation of this space with the aim of identifying all possible patterns is very time consuming. (Note that non-dimensionalising the model does not lead to a significant reduction in the number of parameters.)

Because of their complexity, these nonlocal hyperbolic models have not been the subject of a very thorough analytical investigation, with the aim of classifying the numerically-observed patterns. So far, the investigation of some of the patterns has been carried out in [21, 31, 45–47]. These studies focused on the use of weakly-nonlinear analysis and centre manifold reductions to understand the formation of those patterns that arise near codimension-1 and codimension-2 bifurcation points.

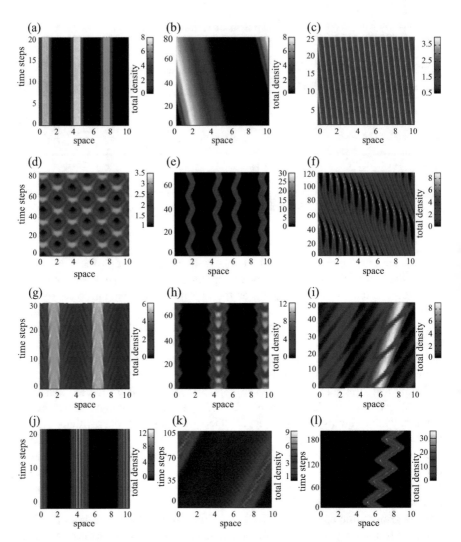

Fig. 5.7 Finite size patterns exhibited by the nonlocal hyperbolic model (5.14) with constant speed [20]. The turning rates are described by Eq. (5.18). The repulsive, alignment, and attractive terms $y_j^{\pm}$, $j = r, al, a$ are given in Table 5.1. The patterns are obtained with the five communication models (M1–M5) described in Table 5.1. The following parameters (and their meaning) are the same as in [20]. (**a**) Stationary pulses—type (1) (for model M2 with $q_{al} = 0, q_a = 4, q_r = 0.5, \tau = 1$); (**b**) travelling pulse (for model M1 with $q_{al} = 2, q_a = 1.6, q_r = 0.5, \tau = 1$); (**c**) travelling trains (for model M3 with $q_{al} = 2, q_a = 0, q_r = 0, \tau = 0.03$); (**d**) ripples (for model M5 with $q_{al} = 2, q_a = 1.5, q_r = 1.1, \tau = 1$); (**e**) zigzag pulses (for model M4: with $q_{al} = 0, q_a = 2, q_r = 2, \tau = 1$); (**f**) semi-zigzag pulses (for model M2 with $q_{al} = 2.2, q_a = q_r = 0, \tau = 0.3$); (**g**) feathers (for model M3 with $q_{al} = 0, q_a = 6, q_r = 6.4, \tau = 1$); (**h**) breathers (for model M4 with $q_{al} = 0, q_a = 2, q_r = 1, \tau = 1$); (**i**) travelling breathers (for model M4 with $q_{al} = 2, q_a = 4, q_r = 4, \tau = 1$); (**j**) Stationary pulses—type (2) (for model M2 with $q_{al} = 0, q_a = 2, q_r = 2.4, \tau = 1.0$); (**k**) Travelling feathers (for model M3 with $q_{al} = 2, q_a = 1.9, q_r = 2.0, \tau = 1.0$); (**l**) Travelling zigzags (for model M2 with $q_{al} = 2.0, q_a = 4.14, q_r = 1.0, \tau = 1.0$). The rest of the parameters are: $\gamma = 0.1, s_r = 0.25, s_{al} = 0.5, s_a = 1.0, m_r = s_r/8, m_{al} = s_{al}/8, m_a = s_a/8$

Table 5.2 Summary of some of the spatial and spatio-temporal patterns exhibited by the nonlocal model (5.14) with communication mechanisms M2–M5

Patterns	M2	M3	M4	M5
Travelling train	✓	✓	✓	
Travelling pulse	✓	✓	✓	
Stationary pulse	✓		✓	✓
Zigzag pulse	✓		✓	
Travelling zigzag	✓		✓	
Semi-zigzag pulse			✓	
Breather			✓	
Travelling breather			✓	
Feather		✓		
Travelling feather		✓		
Ripple			✓	✓
Blow-up	✓			

To identify all patterns displayed by model (5.14) more intensive numerical simulations are necessary, to cover the whole parameter space

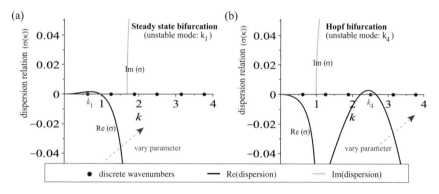

Fig. 5.8 Dispersion relation for model M4, obtained after substituting perturbations of the spatially homogeneous steady states $u^{\pm} = u^* + a_{\pm}e^{\sigma t + ik_j x}$ into the linearised model (5.14) with M4 nonlocal terms. Black curves depict the real part of this dispersion relation: $Re(\sigma(k_j))$. Green curves depict the imaginary part of this dispersion relation: $Im(\sigma(k_j))$. We observe: (**a**) a codimension-1 (k_1) Steady-state bifurcation; (**b**) a codimension-1 (k_4) Hopf bifurcation. The filled black circles on the k-axis represent the discrete wavenumbers $k_j = 2\pi j/L$, with L =domain length

(Note that a codimension-1 bifurcation is the result of variations in one parameter, while a codimension-2 bifurcation is the result of variations in two parameters).

In Fig. 5.8 we graph two examples of dispersion relations near two codimension-1 bifurcation points: (a) a steady-state bifurcation point; (b) a Hopf bifurcation point. A detailed investigation of the patterns arising near these steady-state and Hopf points was performed in [21], where the authors used weakly-nonlinear analysis to show that two patterns, namely stationary pulses and travelling pulses, arise from

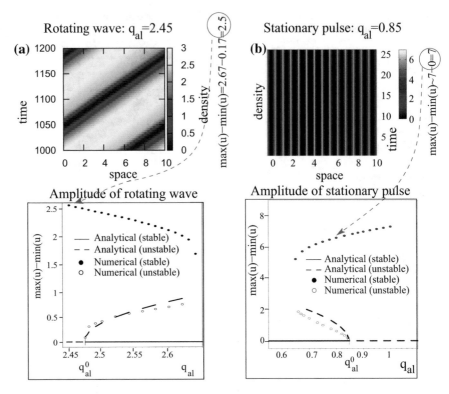

Fig. 5.9 (a) Travelling pulse patterns (top figure), and the subcritical bifurcation of these patterns from spatially homogeneous steady states (bottom figure); (b) Stationary pulses (top figure), and the subcritical bifurcation of these patterns from spatially homogeneous steady states (bottom figure)

subcritical bifurcations from spatially homogeneous steady states; see Fig. 5.9. Note that the stationary pulses are the result of real bifurcations (i.e., at the bifurcation point, the dispersion relation associated with the hyperbolic system (5.14) is real; similar to the case in Fig. 5.8a), while the travelling pulses are the result of complex (Hopf) bifurcations (i.e., the dispersion relation at the bifurcation point is purely imaginary; similar to the case in Fig. 5.8b). The mathematical mechanisms behind the rest of the patterns (e.g., breathers, feathers, zigzags and semi-zigzags) are still to be investigated. We should stress that, similar to the case of local hyperbolic systems, the ripples generated by the nonlocal models (see Fig. 5.7d) are also obtained via Hopf bifurcations.

The subcritical bifurcations which cause the formation of stationary pulses and travelling trains in Fig. 5.9 are associated with the existence of a critical threshold for the density of the group: animal groups with densities below this threshold will disperse, while groups with densities above the threshold will become even more dense and persist for a longer time. Biologically, the existence of such a threshold suggests that the density of the group is very important for the persistence of well coordinated aggregations. This result is consistent with experimental observations of density-dependent transitions between disordered and ordered behaviours in some species of ants [48] or locusts [49]. The bifurcation diagrams at the bottom of Fig. 5.9 offer a better understanding of the role of social interactions (i.e., alignment) on the structure of these patterns. More precisely, in the case of stationary groups with high individual turning rates, alignment has an aggregative effect, with the groups becoming more dense (see Fig. 5.9b). However, in case of moving groups (with relatively low individual turning rates), the effect of alignment is opposite: density decreases as the groups become more elongated (see Fig. 5.9a). When alignment becomes very large, the moving groups disintegrate and the individuals spread over the domain (as the spatially homogeneous steady state—with zero amplitude—is stable).

In addition to these codimension-1 bifurcations, the class of nonlocal hyperbolic models described by (5.14) can exhibit also codimension-2 bifurcations: Hopf/Hopf, Hopf/Steady-state and Steady-state/Steady-state; see Fig. 5.10. Because the seminal studies in [50, 51] classify rigorously the patterns displayed by systems with various symmetries and different bifurcation structures, the starting point for the investigation of patterns arising near the codimension-2 points depicted in Fig. 5.10 was the identification of $O(2)$ symmetry (reflection+translation) for the whole class of models (5.14) defined on a finite domain $[0, L]$:

$$\text{translation:} \quad \theta.u^{\pm}(x, t) = u^{\pm}(x + \theta, t), \quad \theta \in [0, L), \tag{5.21a}$$

$$\text{reflection:} \quad \kappa.(u^+(x, t), u^-(x, t)) = (u^-(L - x, t), u^+(L - x, t)). \tag{5.21b}$$

In [45] the authors used both symmetry theory and weakly nonlinear analysis to identify all types of spatial and spatio-temporal patterns that could arise in the

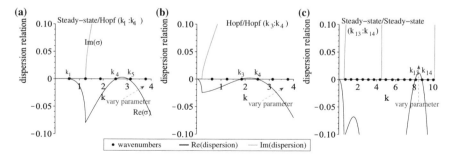

Fig. 5.10 Dispersion relation for model M4, obtained after substituting perturbations of the spatially homogeneous steady states $u^{\pm} = u^* + a_{\pm}e^{\sigma t + ik_j x}$ into the linearised model (5.14) with M4 nonlocal terms. Black curves depict the real part of this dispersion relation: $Re(\sigma(k_j))$. Green curves depict the imaginary part of this dispersion relation: $Im(\sigma(k_j))$. We observe (**a**) a codimension-2 ($k_1 : k_4$) Steady-state/Hopf bifurcation; (**b**) a codimension-2 ($k_3 : k_4$) Hopf/Hopf bifurcation; (**c**) a codimention-2 ($k_{13} : k_{14}$) Steady-state/Steady-state bifurcation

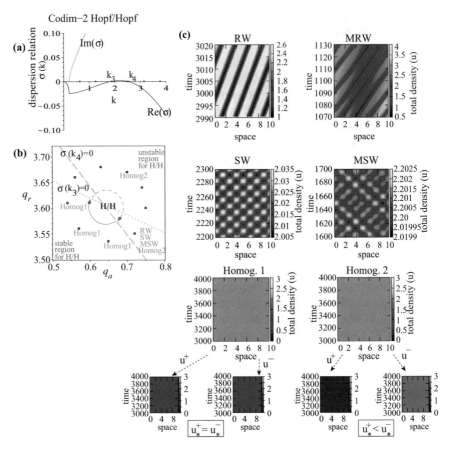

Fig. 5.11 (a) Dispersion relation for a codim-2 Hopf/Hopf bifurcation, as we vary parameters q_r and q_a. (b) Neutral stability curves around a codim-2 Hopf/Hopf (H/H) point, in the (q_r, q_a) plane. (c) Patterns around the Hopf/Hopf point: rotating waves (RW; or travelling pulses) shown here for $q_a = 0.72$ and $q_r = 3.55$; modulated rotating waves (MRW; or travelling breathers) shown here for $q_a = 0.77$ and $q_r = 3.56$; standing waves (SW; or ripples) shown here for $q_a = 0.72$ and $q_r = 3.55$; modulated standing waves (MSW; or modulated ripples) shown here for $q_a = 0.66$ and $q_r = 3.56$; spatially homogeneous solutions: "Homog. 1" shown here for $q_a = 0.54$ and $q_r = 3.61$ (where $u_*^+ = u_*^- = 1$), and "Homog. 2" shown here for $q_a = 0.7$, $q_r = 3.67$ (where $u_*^+ \approx 0.3 \ll u_*^- = 1.7$)

neighbourhood of a Hopf/Hopf bifurcation point; see Fig. 5.11. For the numerical simulations in Fig. 5.11, small perturbations were applied to the spatially homogeneous steady state $u_*^+ = u_*^-$ (here called "Homog.1"), for different parameter values in the (q_a, q_r) space (see the black dots around the H/H bifurcation point in Fig. 5.11b). It was shown analytically (with the help of amplitude equations obtained using the weakly nonlinear analysis; see also Chap. 8) that modulated standing waves (or modulated ripples) bifurcate from standing waves (or ripples), which in turn bifurcate subcritically (i.e., unstable) from the spatially homogeneous steady states. Also the rotating waves (i.e., travelling pulses) bifurcate subcritically from the homogeneous steady states (even if in Fig. 5.11 it is shown that for some values of q_r and q_a these patterns could persist for very long time). We should mention here that since the majority of the patterns around the H/H point are unstable, the dynamics of the system approaches asymptotically another spatially homogeneous steady state point (called here "Homog.2", with $u_*^+ \neq u_*^-$). The results in [45] also identified parameter regions in the neighbourhood of H/H bifurcation points where multiple patterns (i.e., group behaviours) could exist simultaneously. Biologically, this means that for the same parameter values (e.g., same speed and same turning rates), the groups can display different behaviours. The existence of these subcritical bifurcations suggests that the transitions between different behaviours do not necessarily involve changes in the parameter values (as it was shown numerically in previous studies [52]). These transitions could be intrinsic to the group.

A similar classification of patterns can be performed also around the Hopf/Steady-state and Steady-state/Steady-state bifurcation points; see the results in [47]. Figure 5.12 shows (a) the dispersion relation for a particular Hopf/Steady-state (H/Ss) bifurcation, (b) the neutral stability curves for the k_1 and k_4 wavenumbers, (c) the bifurcation diagram around the Hopf/Steady-state point, and (d) the patterns obtained around the Hopf/Steady-state point. As for the Hopf/Hopf case discussed above, the majority of the patterns bifurcate subcritically (i.e., they form unstable branches). Nevertheless, we note that we can have also patterns that bifurcate supercritically (i.e., they form stable branches): rotating waves (i.e., travelling pulses) and stationary pulses.

These two cases presented above (i.e., codimension-1 bifurcations for the M2 model, and codimension-2 bifurcations for the M4 model) illustrate the complex bifurcating dynamics exhibited by the nonlocal models (5.14). As we have seen in Fig. 5.7, other communication mechanisms (e.g., M3 or M5—see Fig. 5.5) could exhibit different patterns, which might also arise around codimension-1/codimension-2 bifurcation points. A detailed bifurcation investigation of these communication models could shed light, for example, on the formation of feather-like patterns (which seem to be specific to model M3). However, such an investigation is still an open problem.

5.6 Multiple Populations

As seen in Chap. 4, the hyperbolic models can be easily generalised to describe the dynamics of interacting populations. However, different populations likely use different communication mechanisms to interact with each other. Eftimie [53] generalised model (5.14) to describe the movement and aggregation of two populations $u^\pm$ and $v^\pm$ that perceive/emit differently the information from/to their

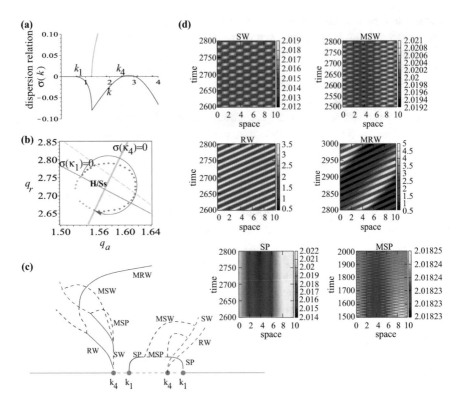

Fig. 5.12 (**a**) Dispersion relation for a codim-2 Hopf/Steady-state bifurcation. (**b**) Neutral stability curves around a codim-2 Hopf/Steady-state point, in the (q_r, q_a) plane. (**c**) Bifurcation diagram created by straightening the path around the Hopf/Steady-state point (in a clockwise direction). (**d**) Patterns around the Hopf/Steady-state point: rotating waves (RW; or travelling pulses); modulated rotating waves (MRW; or travelling breathers); standing waves (SW; or ripples); modulated standing waves (MSW; or modulated ripples); stationary pulses (SP); modulated stationary pulses (MSP)

neighbours:

$$\frac{\partial u^+}{\partial t} + \gamma \frac{\partial u^+}{\partial x} = -\lambda_u^+[u^+, u^-, v^+, v^-]u^+ + \lambda_u^-[u^+, u^-, v^+, v^-]u^-, \quad (5.22a)$$

$$\frac{\partial u^-}{\partial t} - \gamma \frac{\partial u^-}{\partial x} = \lambda_u^+[u^+, u^-, v^+, v^-]u^+ - \lambda_u^-[u^+, u^-, v^+, v^-]u^-, \quad (5.22b)$$

$$\frac{\partial v^+}{\partial t} + \gamma \frac{\partial v^+}{\partial x} = -\lambda_v^+[u^+, u^-, v^+, v^-]v^+ + \lambda_v^-[u^+, u^-, v^+, v^-]v^-, \quad (5.22c)$$

$$\frac{\partial u^-}{\partial t} - \gamma \frac{\partial u^-}{\partial x} = \lambda_v^+[u^+, u^-, v^+, v^-]v^+ - \lambda_v^-[u^+, u^-, v^+, v^-]v^-. \quad (5.22d)$$

Table 5.3 Nonlocal social interaction terms $(y_j^{u,v,\pm}, \; j \in \{a, al, r\})$ for the two-populations model (5.22)

Model	Social interactions: attraction $(y_a^{Mj,\pm})$, repulsion $(y_r^{Mj,\pm})$, alignment $(y_{al}^{Mj,\pm})$
M2	$y_{r,a}^{M2,\pm} = q_{r,a} \int_0^\infty K_{r,a}(s)\,(u(x \pm s) + v(x \pm s) - u(x \mp s) - v(x \mp s))\,ds$ $y_{al}^{M2,\pm} =$ $q_{al} \int_0^\infty K_{al}(s) \big(u^\mp(x \mp s) + u^\mp(x \pm s) + v^\mp(x \mp s) + v^\mp(x \pm s) - u^\pm(x \mp s)$ $\qquad\qquad -u^\pm(x \pm s) - v^\pm(x \mp s) - v^\pm(x \pm s)\big)\,ds$
M3	$y_{r,a}^{M3,\pm} = q_{r,a} \int_0^\infty K_{r,a}(s)\,(u(x \pm s) + v(x \pm s))\,ds$ $y_{al}^{M3,\pm} = q_{al} \int_0^\infty K_{al}(s) \big(u^\mp(x \pm s) + v^\mp(x \pm s) - u^\pm(x \pm s) - v^\pm(x \pm s)\big)\,ds$
M4	$y_{r,a}^{M4,\pm} =$ $q_{r,a} \int_0^\infty K_{r,a}(s) \big(u^\mp(x \pm s) + v^\mp(x \pm s) - u^\pm(x \mp s) - v^\pm(x \mp s)\big)\,ds$ $y_{al}^{M4,\pm} = q_{al} \int_0^\infty K_{al}(s) \big(u^\mp(x \pm s) + v^\mp(x \pm s) - u^\pm(x \mp s) - v^\pm(x \mp s)\big)\,ds$
M2&M3	$y_{r,a,al}^{u,\pm} = y_{r,a,al}^{M2,\pm}, \quad y_{r,a,al}^{v,\pm} = y_{r,a,al}^{M3,\pm}$
M2&M4	$y_{r,a,al}^{u,\pm} = y_{r,a,al}^{M2,\pm}, \quad y_{r,a,al}^{v,\pm} = y_{r,a,al}^{M4,\pm}$
M3&M4	$y_{r,a,al}^{u,\pm} = y_{r,a,al}^{M3,\pm}, \quad y_{r,a,al}^{v,\pm} = y_{r,a,al}^{M4,\pm}$

Here we show only the communication mechanisms M2, M3 and M4. For the model M2&M3, the social interaction terms $y_{r,a,al}^{u,\pm}$ are given by $y_{r,a,al}^{M2,\pm}$, while $y_{r,a,al}^{v,\pm}$ are given by $y_{r,a,al}^{M3,\pm}$; see also Eq. (5.23). Similar explanations hold for models M2&M4 and M3&M4

The turning functions $\lambda_{u,v}^\pm$ now depend on the interactions between the two subpopulations:

$$\lambda_{u,v}^\pm[u^\pm, v^\pm] = \lambda_1 + \lambda_2 f(y_r^{u,v,\pm}[u^\pm, v^\pm] - y_a^{u,v,\pm}[u^\pm, v^\pm] + y_{al}^{u,v,\pm}[u^\pm, v^\pm]). \tag{5.23}$$

The social interaction terms $y_j^{u,v,\pm}, \; j = r, a, al$ are described in Table 5.3. Note that in this table we describe only three communication mechanisms: M2, M3 and M4. Since model M1 was a combination of M2 (for attraction-repulsion interactions) and M4 (for alignment interactions), and model M5 did not seem to add any new results, these two models were ignored in [53].

To identify the types of patterns obtained with these two-population models, Eftimie [53] performed numerical simulations with initial conditions random perturbations of the spatially homogeneous steady states $u^\pm = u_*$ and $v^\pm = v_*$. The numerical approach was the same as for the one-population model: a second-order MacCormack finite difference scheme (see also Chap. 7).

In contrast to the one-population models that displayed deterministic patterns, the two-population models can exhibit also chaotic patterns. Some of these patterns are shown in Fig. 5.13: chaotic feathers (panel (a")) and chaotic zigzags (panels (b"), (c")). For comparison purposes, the parameters and the initial conditions are the same across the horizontal rows. We note here that for the same parameter values, the use of multiple communication mechanisms (panels (a")–(c")) can lead to behaviours (patterns) that are not necessarily predicted by the behaviour of the subpopulations that use only one communication mechanisms (panels (a)–(c),

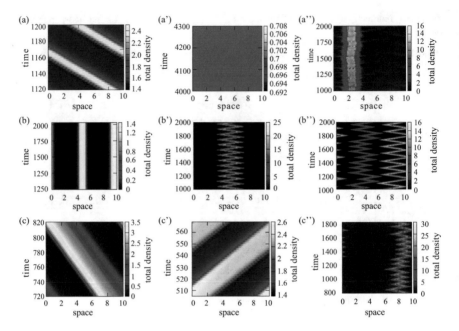

Fig. 5.13 Patterns obtained with one or two communication mechanisms. (**a**) Travelling pulses (M3: $q_a = 2$, $u_* = 0.8$); (**a'**) Spatially homogeneous state (M4: $q_a = 2$, $v_* = 0.3$); (**a''**) Chaotic feathers (M3&M4: $q_a = 2$, $u_* = 0.8$, $v_* = 0.3$); (**b**) Stationary pulses (M2: $q_a = 6.7$, $u_* = 0.1$); (**b'**) Zigzags (M4: $q_a = 6.7$, $v_* = 1.0$); (**b''**) Chaotic zigzags (M2&M4: $q_a = 6.7$, $u_* = 0.1$, $v_* = 1.0$); (**c**) Travelling pulses (M2: $q_a = 2.0$, $u_* = 0.5$); (**c'**) Travelling pulses (M3: $q_a = 2.0$, $v_* = 0.9$); (**c''**) Chaotic zigzags (M2&M3: $q_a = 2.0$, $u_* = 0.5$, $v_* = 0.9$). The rest of parameters are $q_r = q_{al} = 2.0$, $\gamma = 0.1$, $s_a = 1.0$, $s_{al} = 0.5$, $s_r = 0.25$, $\lambda_1 = 0.2$, $\lambda2 = 0.9$, $L = 10$

(a')–(c')). A more detailed investigation of the types of patterns obtained with two communication mechanisms is shown in Fig. 5.14, for different initial sub-population sizes u_* and v_*. The simultaneous use of different communication mechanisms can lead to the emergence of spatial patterns in cases where the use of only one communication mechanism leads to spatially homogeneous steady states (as in Fig. 5.14b, c, for small u_*, v_*). The reverse is also true: the use of multiple communication mechanisms can lead to spatially homogeneous states in cases where the use of one mechanism leads to patterns (as in Fig. 5.14a, for large u_*, v_*).

Finally, for the two-population model, it was shown numerically in [53] that the use of different communication mechanisms can lead to the spatial sorting of individuals inside the groups. For example, Fig. 5.15 shows that the fully aware (M2) individuals that receive full information about the location and the movement direction of all their neighbours are very likely to be positioned at the centre of the aggregation. In contrast, individuals that receive only partial information about the location and the movement direction of their neighbours (e.g., the M4-individuals), are more likely to be positioned at the edges of the aggregation. This self-sorting

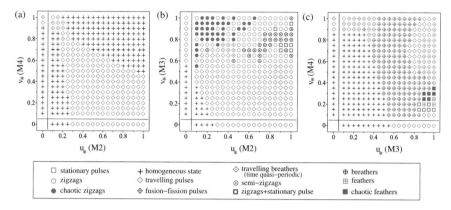

Fig. 5.14 Summary of patterns obtained with two-population nonlocal hyperbolic models (5.22), for different initial populations $u^+ = u^- = u_*$ and $v^+ = v^- = v_*$. (a) Model M2&M4; (b) Model M2&M3; (c) Model M3&M4. The parameters are: $q_a = 2$, $q_r = 2$, $q_{al} = 2$, $\gamma = 0.1$, $L = 10$, $s_a = 1.0$, $s_{al} = 0.5$, $s_r = 0.25$, $\lambda_1 = 0.2$, $\lambda_2 = 0.9$. By fixing the parameter values and changing only the communication mechanisms (i.e., the direction to/from which signals are emitted/perceived), it is possible to obtain different types of spatio-temporal patterns corresponding to different aggregation behaviours at group level.

behaviour characterises all types of spatial patterns, from stationary and travelling pulses to chaotic patterns and fusion-fission patterns.

Avoidance Behaviours The majority of mathematical models for collective behaviour focus on avoidance behaviours in the context of repulsion (to avoid collision with nearby neighbours [54]). While this type of social interaction occurs on short spatial ranges, the more complex avoidance behaviour can occur on both shorter and longer spatial ranges (when neighbours are detected far away). Avoidance behaviours have been observed in animal and human communities in the context of epidemiological infections [55–58], food catching and sharing among conspecifics and heterospecifics [59], interactions with predators [60]. Since communication can influence how animals/humans interact with each other, and whether they avoid or not their conspecifics, Eftimie and Coulier [61] investigated avoidance and learning behaviours in sub-populations that use different communication mechanisms. To this end, the authors focused only on two communication mechanisms, namely M2 and M4, and assumed that population u (which communicates via M2) avoids population v (which communicates via M4); see also Fig. 5.16. The avoidance behaviour is manifested in the alignment and attractive interactions: population u does not align with v, nor is it attracted to v.

As expected, avoidance behaviours can lead to the segregation of individuals inside the aggregations, leading even to different behaviours exhibited by the two populations which stay connected; see Fig. 5.17. Moreover, the introduction of avoidance behaviours in the u population can lead to changes in group-level patterns. Figure 5.18 summarises the patterns that can be obtained for a fixed initial

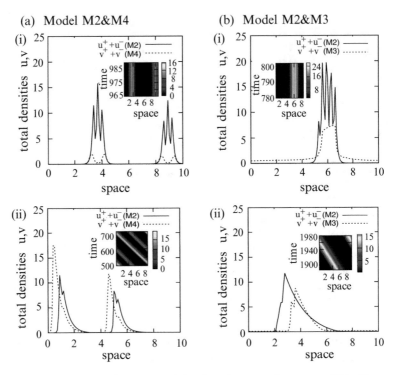

Fig. 5.15 Spatial structure of various spatial and spatio-temporal patterns exhibited by models M2&M3 and M2&M4 (similar patterns are obtained also with M3&M4). The continuous and dotted curves describe the total density of subpopulations $u = u^+ + u^-$ and $v = v^+ + v^-$. The inset shows a "bird's eye view" of the total population density $(u + v)$ as a function of time and space. (**a**) Model M2&M4: (i) Stationary pulses for $q_a = 6.7$, $u^{\pm}_{*,M2} = 0.7$, $v^{\pm}_{*,M4} = 0.1$; snapshot at $t = 990$; (ii) Travelling pulses for $q_a = 6.7$, $u^{\pm}_{*,M2} = 0.6$, $v^{\pm}_{*,M4} = 0.9$; snapshot at $t = 670$; (**b**) Model M2&M3: (i) Stationary pulses for $q_a = 2.0$, $u^{\pm}_{*,M2} = 0.9$, $v^{\pm}_{*,M3} = 0.7$; snapshot at $t = 800$; (ii) Travelling pulses for $q_a = 2.0$, $u^{\pm}_{*,M2} = 0.9$, $v^{\pm}_{*,M3} = 0.5$; snapshot at $t = 1910$; The rest of parameters are $q_r = q_{al} = 2.0$, $\gamma = 0.1$, $s_a = 1.0$, $s_{al} = 0.5$, $s_r = 0.25$, $\lambda_1 = 0.2$, $\lambda_2 = 0.9$, $L = 10$

conditions for population v (i.e., $v^{\pm}(x, 0) = v^* + Rand(0..0.01))$ and various densities for population u (i.e., $u^{\pm}(x, 0) = u^* + Rand(0..0.01)$ with $u^* \in [0.1, 1]$). We note in this case a shift in the parameter values for u^* where travelling pulse patterns can occur: from low u^* densities in the absence of avoidance (see also Fig. 5.14a), to medium-high u^* densities in the presence of avoidance. One needs also to acknowledge that for different parameter values, it is possible to obtain completely different patterns in the presence/absence of avoidance behaviours. However, this is an aspect open to investigation.

Learning Tolerant Behaviours While avoidance behaviours are common in human and animal communities, they are closely related to learning aspects regarding

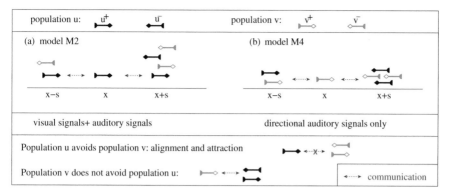

Fig. 5.16 Description of communication models M2 and M4, in the case where population $u^\pm$ (which communicates via M2) avoids population $v^\pm$ (which communicates via M4). It is assumed that the avoidance behaviour affects only the alignment and attraction social interactions (since repulsion is by default an avoidance behaviour)

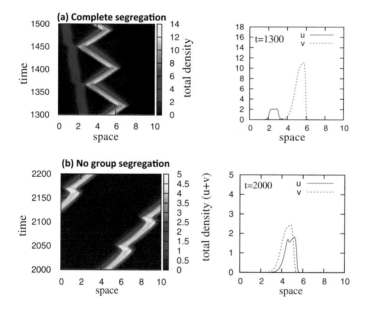

Fig. 5.17 Examples of patterns (**a**) with group separation, and (**b**) without group separation, obtained with the avoidance model (5.22). For more details regarding the parameter values used for numerical simulations see [61]

life in a community. In particular, it is possible for individuals to change their attitudes towards neighbours based on interactions with more tolerant conspecifics (including interactions with conspecifics never exposed to avoidance behaviours, as seen in certain fish populations [62]). To investigate the effect of learning to tolerate neighbours, Eftimie and Coulier [61] generalised model (5.22) by assuming that

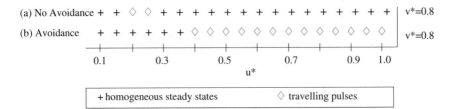

Fig. 5.18 Comparison between patterns obtained (**a**) without avoidance behaviour (as in Fig. 5.14a), and (**b**) with avoidance behaviour in the u-population, for fixed initial v-population size ($v^* = 0.8$) and different initial u-population sizes ($u^* \in (0.1, 1.0)$). The rest of parameters are as in Fig. 5.14a

population $u^{\pm}$ (which uses communication mechanism M2) is composed of a sub-population $u_1^{\pm}$ that avoids population $v^{\pm}$, and a sub-population $u_2^{\pm}$ that tolerates population $v^{\pm}$. It was assumed that $u^{\pm} = u_1^{\pm} + u_2^{\pm}$. The learning behaviour at the level of population $u^{\pm}$ is described by the transitions rates a_{12} (from tolerance to avoidance) and a_{21} (from avoidance to tolerance). As for all reaction-transport models with left/right sub-populations, we assume that a tolerant population which becomes intolerant has the same probability of becoming left-moving or right-moving (and hence the rates $\frac{a_{12}}{2}$ in Eqs. (5.24a)–(5.24b)). In a similar manner, the intolerant population that learns to become tolerant can become either left-moving or right-moving (and hence the rates $\frac{a_{21}}{2}$). Population v uses the communication mechanism M4, and is tolerant with population u. The new model is described by the following equations:

$$\frac{\partial u_1^+}{\partial t} + \gamma \frac{\partial u_1^+}{\partial x} = -\lambda_{u_1}^+[u_{1,2}^+, u_{1,2}^-, v^+, v^-]u_1^+ + \lambda_{u_1}^-[u_{1,2}^+, u_{1,2}^-, v^+, v^-]u_1^-$$
$$- a_{21}u_1^+u_2 + a_{12}\frac{u_1}{2}u_2, \tag{5.24a}$$

$$\frac{\partial u_1^-}{\partial t} - \gamma \frac{\partial u_1^-}{\partial x} = \lambda_{u_1}^+[u_{1,2}^+, u_{1,2}^-, v^+, v^-]u_1^+ - \lambda_{u_1}^-[u_{1,2}^+, u_{1,2}^-, v^+, v^-]u_1^-$$
$$- a_{21}u_1^-u_2 + a_{12}\frac{u_1}{2}u_2, \tag{5.24b}$$

$$\frac{\partial u_2^+}{\partial t} + \gamma \frac{\partial u_2^+}{\partial x} = -\lambda_{u_2}^+[u_{1,2}^+, u_{1,2}^-, v^+, v^-]u_2^+ + \lambda_{u_2}^-[u_{1,2}^+, u_{1,2}^-, v^+, v^-]u_2^-$$
$$+ a_{21}u_1\frac{u_2}{2} - a_{12}u_2^+u_1, \tag{5.24c}$$

$$\frac{\partial u_2^-}{\partial t} - \gamma \frac{\partial u_2^-}{\partial x} = \lambda_{u_2}^+[u_{1,2}^+, u_{1,2}^-, v^+, v^-]u_2^+ - \lambda_{u_2}^-[u_{1,2}^+, u_{1,2}^-, v^+, v^-]u_2^-$$

$$+ a_{21}u_1 \frac{u_2}{2} - a_{12}u_2^- u_1, \tag{5.24d}$$

$$\frac{\partial v^+}{\partial t} + \gamma \frac{\partial v^+}{\partial x} = -\lambda_v^+[u^+, u^-, v^+, v^-]v^+ + \lambda_v^-[u^+, u^-, v^+, v^-]v^-, \tag{5.24e}$$

$$\frac{\partial v^-}{\partial t} - \gamma \frac{\partial v^-}{\partial x} = \lambda_v^+[u^+, u^-, v^+, v^-]v^+ - \lambda_v^-[u^+, u^-, v^+, v^-]v^-. \tag{5.24f}$$

Here $u_{1,2} = u_{1,2}^+ + u_{1,2}^-$, and all individuals in the populations move at the same constant speed γ. The turning rates $\lambda_{u_1,u_2}^\pm$ are described in terms of avoidance and learning behaviours:

$$\lambda_{u_1,u_2}^\pm = \lambda_1 + \lambda_2 f\left(y_r^{u_1,u_2;\pm} - y_a^{u_1,u_2;\pm} + y_{al}^{u_1,u_2;\pm}\right), \tag{5.25}$$

The nonlocal interaction terms $y_j^{u_1,u_2;\pm}$ (for a population assumed, for example, to communicate via mechanisms M2) are given in Table 5.4. The interaction terms for population v (assumed, for example to communicate via mechanism M4) are the same as in Table 5.3.

Numerical simulations performed in [61] with this model showed that avoidance and learning behaviours can change the spatial structure of the aggregations—see Fig. 5.19. In particular, while in the absence of avoidance the M2-individuals can be found at the back of a travelling aggregation (with the M4-individuals leading the group), the introduction of avoidance in the M2 population lead them move

Table 5.4 Nonlocal social interaction terms ($y_j^{u_1,u_2;\pm}$, $j \in \{a, al, r\}$) for the three-populations model (5.24) with avoidance and learning behaviours [61]

Population	Social interactions: attraction $\left(y_a^{u_{1,2};\pm}\right)$, repulsion $\left(y_r^{u_{1,2};\pm}\right)$, alignment $\left(y_{al}^{u_{1,2};\pm}\right)$
u_1	$y_r^{u_1;\pm} = \pm q_r \int_0^\infty K_r(s)\,(u_1(x+s) + u_2(x+s) - u_1(x-s) - u_2(x-s)$
	$\qquad\qquad + v(x+s) - v(x-s))\,ds$
	$y_a^{u_1;\pm} = \pm q_a \int_0^\infty K_r(s)\,(u_1(x+s) + u_2(x+s) - u_1(x-s) - u_2(x-s))\,ds$
	$y_{al}^{u_1;\pm} = \pm q_{al} \int_0^\infty K_{al}(s)\,\left(u_1^-(x+s) + u_1^-(x-s) - u_1^+(x+s) - u_1^+(x-s)\right.$
	$\qquad\qquad \left.+ u_2^-(x+s) + u_2^-(x-s) - u_2^+(x+s) - u_2^+(x-s)\right)\,ds$
u_2	$y_r^{u_2;\pm} = \pm q_r \int_0^\infty K_r(s)\,(u_1(x+s) + u_2(x+s) - u_1(x-s) - u_2(x-s)$
	$\qquad\qquad + v(x+s) - v(x-s))\,ds$
	$y_a^{u_2;\pm} = \pm q_a \int_0^\infty K_a(s)\,(u_1(x+s) + u_2(x+s) - u_1(x-s) - u_2(x-s)$
	$\qquad\qquad + v(x+s) - v(x-s))\,ds$
	$y_{al}^{u_2;\pm} = q_{al} \int_0^\infty \left(u_1^-(x+s) + u_1^-(x-s) - u_1^+(x+s) - u_1^+(x-s) + u_2^-(x+s)\right.$
	$\qquad\qquad + u_2^-(x-s) - u_2^+(x+s) - u_2^+(x-s) + v^-(x+s) + v^-(x-s)$
	$\qquad\qquad \left.- v^+(x+s) - v^+(x-s)\right)\,ds$

Here shown only the communication model M2. (Communication mechanism M4 is similar to the one described in Table 5.3.) Note that both u_1 and u_2 are repelled by each other and by population v within the repulsion range. Moreover, population u_1 is only attracted to and aligns with population u_2 (within attraction and alignment ranges). In contrast, population u_2 is attracted to and aligns with both u_1 and v populations

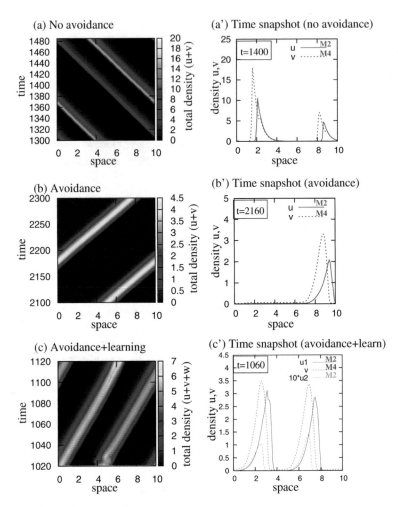

Fig. 5.19 Spatial distribution of individuals inside moving aggregations: (**a**) with no avoidance behaviours; (**b**) with avoidance behaviours displayed by u-population (communicating via M2); (**c**) with avoidance and learning behaviours displayed by u population. For a more clear description of the tolerant u-subpopulation (since this population is very small), we graph here the subpopulation u_2 magnified by 10 times (i.e., $10 \times u_2$). Initial conditions are random perturbations (of magnitude 0.01) of spatially homogeneous steady states: (**a, a'**) $u^* = 0.4, v^* = 0.8$, (**b, b'**) $u^* = 0.1$, $v^* = 0.2$, (**c, c'**) $u_1^* = 0.3, u_2^* = 0.001, v^* = 0.4$. The rest of the parameters are: $a_{12} = a_{21} = 0.2$, $q_a = 6.7, q_{al} = q_r = 2, \gamma = 0.1, \lambda_1 = 0.2, \lambda_2 = 0.9$. The nonlocal interactions are modelled by translated Gaussian kernels, with $s_r = 0.25, s_{al} = 0.5, s_a = 0.5$, on a domain of length $L = 10$ with periodic boundary conditions

towards the front of the moving group (with the M4-individuals now trailing in the back). The assumption that intolerant M2-individuals can learn from tolerant M2-neighbours influenced the spatial segregation of the M2 population: the intolerant M2-individuals were found at the front of the moving group while the tolerant M2-individuals were found in the middle of the moving group (mingling with the tolerant M4-individuals). This result suggests that tolerance.

To conclude this section, we note that this class on nonlocal models which incorporate various inter-individual communication mechanisms can be easily generalised to investigate different aspects of social interactions in human and animal communities. Classical intra-species and inter-species ecological interactions, such as predation, competition or cooperation, can be easily incorporated into these nonlocal hyperbolic models.

5.7 Analytical Approaches for the Investigation of Patterns: Loss of Patterns in the Parabolic Limit

In Remark 5.1 we have mentioned that the limiting parabolic equation (5.16) cannot exhibit persistent travelling pulses in the absence of external drift factors (as it was shown numerically in [8]). This aspect was investigated analytically by Buono and Eftimie [31], who focused on the symmetry and bifurcation structure of the original hyperbolic model (5.14) and the limiting parabolic model (5.16). The authors first showed that both the hyperbolic system (5.14) and the parabolic equation (5.16) are $O(2)$-equivariant for all communication mechanisms (M1, ..., M5), with the $O(2)$ action given by Eq. (5.21). Then they considered the linear operator $\mathcal{L}$ (linearised at the spatially homogeneous steady state $u = u^*$) associated with the parabolic equation (5.16),

$$\mathcal{L}(\xi) = D_0\xi_{xx} + \nabla h(u^*, K^+ \star u^*, K^- \star u^*, 0, 0, 0)$$
$$\times (\xi, K^+ \star \xi, K^- \star \xi, \xi_x, K^+ \star \xi_x, K^- \star \xi_x)^\top \tag{5.26}$$

with

$$h(u, K^+ \star u, K^- \star u, u_x, K^+ \star u_x, K^- \star u_x)$$
$$:= -B_0 \frac{\partial}{\partial x}\left(u\left(F(y^-[u]) - F(y^+[u])\right)\right) \tag{5.27}$$

and kernels $K(s) = q_r K_r(s) - q_a K_a(s)$ and convolutions

$$K^\pm \star u = \int K(s)u(x \pm s, t)ds, \quad \text{and} \quad K^\pm(s) \star u_x = \int K(s)u_x(x \pm s, t)ds. \tag{5.28}$$

Buono and Eftimie [31] showed that the spectrum of the linear operator can have only real eigenvalues, and thus the linear system cannot exhibit $O(2)$-symmetric Hopf bifurcations. We note here that introducing an external environmental forcing as discussed in Fig. 5.6 (or using an odd kernel as in [8]) leads to a forced symmetry breaking from $O(2)$ to $SO(2)$ (where the reflection symmetry is lost).

Because the systems of parabolic equations usually display richer dynamics (in terms of pattern formation) compared to single parabolic equations—see the Turing bifurcations—Buono and Eftimie [31] considered also the parabolic limit of systems of two different populations ($u^\pm$ and $v^\pm$) which interact via different communication mechanisms (Mi and Mj, with $i \neq j \in \{2, \ldots, 5\}$). The hyperbolic model was described by Eq. (5.22), and the limit parabolic system was described by the following two equations:

$$\frac{\partial u}{\partial t} = D_u \frac{\partial^2 u}{\partial x^2} - B_u \frac{\partial}{\partial x}\left(u\left(f(y^{u,-}[u,v]) - f(y^{u,+}[u,v])\right)\right), \qquad (5.29a)$$

$$\frac{\partial v}{\partial t} = D_v \frac{\partial^2 v}{\partial x^2} - B_v \frac{\partial}{\partial x}\left(v\left(f(y^{v,-}[u,v]) - f(y^{v,+}[u,v])\right)\right), \qquad (5.29b)$$

with the diffusion and advection parameters

$$D_{u,v} = \lim_{\epsilon \to 0} \frac{\gamma}{2\lambda_1^{u,v}}, \quad B_{u,v} = \lim_{\epsilon \to 0} \frac{\gamma \lambda_2^{u,v}}{2\lambda_1^{u,v}}. \qquad (5.30)$$

Note that in the limit $\epsilon \to 0$, the nonlocal terms $y^{u,v,\pm}$ are the same for the communication mechanisms M2 and M4, and for the mechanisms M3 and M5:

- For M2 and M4:

$$y^{u,\pm}[u,v] = \pm \int_0^\infty K^u(s)\left((u+v)(x+s) - (u+v)(x-s)\right)ds. \qquad (5.31)$$

- For M3 and M5:

$$y^{v,\pm}[u,v] = \int_0^\infty K^v(s)\left(u(x \pm s) + v(x \pm s)\right)ds. \qquad (5.32)$$

The kernels $K^{u,v}(s)$ are given as follows:

For M2,M3: $K^{u,v}(s) = q_r K_r(s) - q_a K_a(s),$ \qquad (5.33a)

For M4,M5: $K^{u,v}(s) = \frac{1}{2}(q_r K_r(s) - q_a K_a(s) + q_{al} K_{al}(s)).$ \qquad (5.33b)

Even system (5.29) is $O(2)$-symmetric with respect to the action given by Eq. (5.21) [31]. Moreover, if $\lambda_u^\pm = \lambda_v^\pm$ (which implies that $B_u = B_v$ and $D_u = D_v$), then the parabolic system (5.29) is symmetric with respect to the interchange $(u,v) \to (v,u)$, and the symmetry group is $O(2) \times \mathbb{Z}_2$ [31].

Linear stability analysis of a generic spatially homogeneous equilibrium point (u^*, v^*) showed that the parabolic system (5.29) can exhibit $O(2)$-symmetric steady-state bifurcations independent of the communication mechanism used. (Note that these bifurcations give rise to stationary aggregations.) Moreover, system (5.29) can exhibit $O(2)$-symmetric Hopf bifurcations only if the two communication mechanisms used by populations u and v are such that two parameters, $\bar{b}^u$ and $\bar{b}^v$, have opposite signs with $(B)\bar{b}^v + \bar{b}^u > 0$ and $-(B)\bar{b}^v - (D)^2\bar{b}^u > 0$, where $D = D_v/D_u$, $B = B_v/B_u$,

$$\bar{b}^u = -a_{Mj}iu^* f'(y^{u,-}[u^*, v^*]) \int_0^\infty K^u(s)\sin(k_n s)ds, \qquad (5.34a)$$

$$\bar{b}^v = -a_{Mj}iv^* f'(y^{v,-}[u^*, v^*]) \int_0^\infty K^v(s)\sin(k_n s)ds, \qquad (5.34b)$$

and a_{Mj} are different coefficients for the interaction kernels corresponding to different communication mechanisms: $a_{Mj} = 4$ for $Mj =$M2,M4 and $a_{Mj} = 2$ for $Mj =$M3,M5. The upper indices "u, v" for the interaction kernels $K^{u,v}(s)$ differentiate these kernels if we assume that different populations use different communication mechanisms that have different spatial actions (e.g., M2 corresponds to long-range communication via visual and auditory stimuli, while M5 corresponds to short-range communication via tactile stimuli). The dispersion relation that results from the linear stability analysis of system (5.29), and which gives the eigenvalues of the linearised system, is given in the following (and graphed in Fig. 5.20a):

$$\sigma^2 + \sigma\left(D + 1\right)k_n^2 + ik_n(Bb^v + b^u)) + k_n^4 D + ik_n^3(Bb^v + Db^u) = 0. \qquad (5.35)$$

The Hopf bifurcations exhibited by the limiting parabolic system (5.29) lead to rotating waves or modulated rotating waves, as shown in Fig. 5.20 (for more details, see [31]).

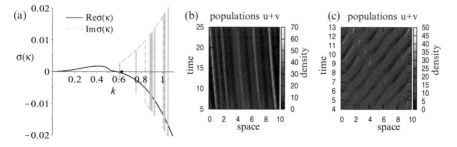

Fig. 5.20 (a) Example of dispersion relation $\sigma(k_1)$ as given by Eq. (5.35), showing complex eigenvalues in the parameter space where $\bar{b}^u > 0$ and $\bar{b}^v < 0$. (b) Examples of rotating waves (i.e., travelling pulses) exhibited by nonlocal parabolic system (5.29). (c) Examples of modulated rotating waves exhibited by nonlocal parabolic system (5.29). For details regarding the parameter values see [31]

Therefore, while both the hyperbolic and (corresponding) parabolic models for self-organised animal behaviours via communication mechanisms are $O(2)$-symmetric, their bifurcation structure is quite different: the hyperbolic models can exhibit all codimension-1 (Hopf and steady-state) and codimension-2 (Hopf/Hopf, Hopf/steady-state, steady-state/steady-state) bifurcations, but the corresponding parabolic models can exhibit only steady-state and steady-state/steady-state bifurcations. Therefore, the parabolic limit caused the loss of one of the mechanisms responsible for obtaining moving aggregations (described by rotating waves), namely the Hopf bifurcations with $O(2)$ symmetry. To obtain Hopf bifurcations in the limiting systems of parabolic equations (5.29), the authors in [31] needed to impose the assumption that different communication mechanisms influence differently the strength and spatial range of social interactions (such that parameters $\bar{b}^u$ and $\bar{b}^v$ acquire opposite signs).

5.8 Systems with Density-Dependent Speeds

Despite the complexity of the patterns exhibited by hyperbolic systems with density-dependent turning rates, these models cannot explain all observed group behaviours. For example, they cannot explain the spontaneous splitting and merging behaviours observed in flocks of birds [40], herds of ungulates [63], or schools of fish [64]. These behaviours have been previously associated with changes in individuals' velocity [63, 64]. To address this issue, one can increase the complexity of the nonlocal models by considering density-dependent speeds.

As mentioned earlier, organisms can speed-up to approach conspecifics further away (attraction), or slow-down to avoid collisions from those near-by (repulsion) [63, 65]. As with the turning behaviour, speed adaptation happens through communication via different signals, such as long-range visual or acoustic signals [64–66], or short-range tactile signals [39, 64, 65]. As an example, pollock fish seem to use vision for attractive interactions that keep the school together, and lateral line for repulsive interactions that prevent collisions during swimming [64].

The previous description of nonlocal 1-equation models with density-dependent speeds (see Eq. (5.1)) can be easily generalised to hyperbolic systems. The general equations describing hyperbolic systems with density-dependent speeds and density-dependent turning rates are

$$\frac{\partial u^+}{\partial t} + \frac{\partial (\Gamma^+[u^+, u^-]u^+)}{\partial x} = -\lambda^+[u^+, u^-]u^+(x, t) + \lambda^-[u^+, u^-]u^-(x, t),$$
$$(5.36a)$$

$$\frac{\partial u^-}{\partial t} - \frac{\partial (\Gamma^-[u^+, u^-]u^-)}{\partial x} = \lambda^+[u^+, u^-]u^+(x, t) - \lambda^-[u^+, u^-]u^-(x, t).$$
$$(5.36b)$$

Here, $\Gamma^{\pm}[u^+, u^-]$ are the density-dependent speeds, and $\lambda^{\pm}[u^+, u^-]$ are the density-dependent turning rates (see Eq. (5.18)). Since individuals' speeds cannot be arbitrarily large, one can choose $\Gamma^{\pm}$ to be bounded functions [20]. (This is in contrast with the assumption of linear function employed by one-equation models.) Moreover, it would make sense biologically to assume that increasing the intensity of signals would lead to an increased speeding-up or slowing-down behaviour. Mathematically, this means that $\Gamma^{\pm}$ are increasing functions of the perceived signals. An example of such a function is tanh [7, 17]:

$$\Gamma^+[u^+, u^-] = \gamma \left(1 + \tanh \left(q_a \int_0^\infty K_a(s) \left(u(x+s, t) - u(x-s, t) \right) ds \right.\right.$$
$$\left.\left. -q_r \int_0^\infty K_r(s) \left(u(x+s, t) - u(x-s, t) \right) ds \right) \right), \qquad (5.37a)$$

$$\Gamma^-[u^+, u^-] = \gamma \left(1 + \tanh \left(-q_a \int_0^\infty K_a(s) \left(u(x+s, t) - u(x-s, t) \right) ds \right.\right.$$
$$\left.\left. +q_r \int_0^\infty K_r(s) \left(u(x+s, t) - u(x-s, t) \right) ds \right) \right). \qquad (5.37b)$$

Analytical and numerical results for model (5.36)–(5.37) showed a variety of new and interesting patterns [7, 17]. In particular, when the kernels K_j, $j = r, a$ are discontinuous at the origin (i.e., Morse interaction kernels), the hyperbolic model exhibits not only finite-density patterns, but also blow-up patterns [17]. In particular, the sign of the jump at the origin ($[K] = K(0+) - K(0-)$, where $K(x) = -q_a K_a(x) + q_r K_r(x)$) determines whether blow-up solutions are possible [17]. When $[K] \geq 0$ (i.e., repulsion is stronger than attraction), the L^∞-norm of the solution is bounded for all times and the solution exists globally in $W^{2,1}(\mathbb{R}) \times W^{2,1}(\mathbb{R})$ [17]. In contrast, when $[K] < 0$ (i.e., attraction is stronger than repulsion), the amplitude of the solution may blow up in finite time provided that the initial data is large enough such that

$$\max_{x \in \mathbb{R}} u_0^{\pm}(x) > \frac{\beta}{\alpha}. \qquad (5.38)$$

Here, β and α are constants which depends on the L^∞ norms of the speed and turning rate functions (see [17] for details). The blow-up time is

$$T = -\frac{1}{\beta} \log(1 - \frac{\beta}{\alpha} \max_{x \in \mathbb{R}} u_0^{\pm}(x)) \qquad (5.39)$$

It is worth mentioning that condition (5.38) is only a sufficient condition for finite time blow-up. Numerical results have shown that initial data which does not satisfy (5.38) may still blow up in finite time (as seen in Fig. 5.22b). This blow-up behaviour can be prevented by large random turning rates (λ_1). In fact, the transition from blow-up patterns to finite-density patterns can be explained by an increase

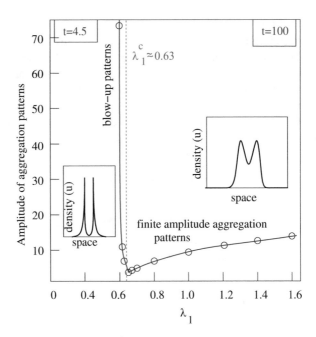

Fig. 5.21 Description of a bifurcation in the amplitude of solutions as the random turning rate λ_1 increases past the threshold $\lambda_1^c \approx 0.63$ (which depends on other model parameters). For graphical purposes, in the blow-up region we show the amplitude of the pattern at $t = 4.5$, while in the finite-amplitude region we show the amplitude of the pattern at $t = 100$, when the pattern has started to stabilise. See also [17]

in λ_1 [17]; see Fig. 5.21. Note that the constant random turning rate λ_1 induces a diffusive effect into the hyperbolic model (as individuals turn randomly either left or right).

There are actually two types of blow-up patterns, stationary or moving, depending on the strength of the repulsive interactions q_r (see Fig. 5.22 and Table 1.2 for a description of these patterns). For weak repulsive interactions the blow-up solutions are stationary (see Fig. 5.22a). However, strong repulsive interactions cause left-moving and right-moving individuals to move away from each other while approaching the spatial location where the blow-up takes place (see Fig. 5.22b). The two-point blow-up patterns (Fig. 5.22) are caused by the densities u^+ and u^- becoming infinite at two different spatial positions, x_1^* and x_2^*:

$$\lim_{t \to T} ||u^+(x_1^*, t)||_\infty = \infty, \quad \lim_{t \to T} ||u^-(x_2^*, t)||_\infty = \infty, \tag{5.40}$$

for stationary blow-ups, and

$$\lim_{t \to T, x \to x_1^*} ||u^+(x, t)||_\infty = \infty, \quad \lim_{t \to T, x \to x_2^*} ||u^-(x, t)||_\infty = \infty, \tag{5.41}$$

for moving blow-ups.

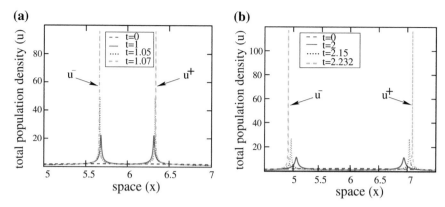

Fig. 5.22 Blow-up patterns exhibited by a nonlocal hyperbolic model with density-dependent speed introduced by Fetecau and Eftimie [17]. Shown is the total density $u = u^+ + u^-$. The initial condition is $u^\pm(x, 0) = e^{-(x/2-L/4)^2}$, where L is the domain length. (**a**) Stationary blow-up patterns. (**b**) Moving blow-up patterns. Using the notation from [17], the parameter values are: (**a**) $q_{al} = 1$, $q_a = 1$, $q_r = 0.5$, $\lambda_1 = 0$, $\lambda_2 = 0$; (**b**) $q_{al} = 1$, $q_r = 0.8$, $q_a = 1$, $\lambda_1 = 0.2$, $\lambda_2 = 0$. Condition (5.38) is satisfied for the parameters in (**a**), but not for the parameters in (**b**). Note that for larger times, the numerical code breaks down (extremely large densities lead to numerical oscillations, and the solution becomes negative)

In general, the assumption of density-dependent speed for a hyperbolic model suggests possible shock solutions. This is true for many of the local models discussed here [67, 68]. However, as shown by Fetecau and Eftimie [17], for the nonlocal hyperbolic systems (5.36)–(5.37) the assumption of having turning rates and speeds that are increasing and bounded does not allow for the formation of shocks. This is consistent with other results obtained with nonlocal traffic flow models which do not exhibit shock formation due to the smoothing effect that the nonlocal terms have on the velocity [69]. (The effect is similar to the effect caused by a viscous term.)

When kernels K_j, $j = r, a$ are continuous at the origin, model (5.36)–(5.37) was shown to exhibit group-splitting behaviours similar to those observed in nature (for example, in schools of fish [70]), as well as more complex chaotic patterns [7]. Some of these chaotic patterns are shown in Fig. 5.23a, b. To emphasise the chaotic nature of the patterns in panels (a),(b), we graph in Fig. 5.23a', a" and b', b" the total population densities $u = u^+ + u^-$ at two different time steps: (a'), (b') $t = 99$, (a"), (b") $t = 100$. We note that an analytical investigation of the formation of these patterns (which is likely the result of the interactions between different unstable states) is still an open problem.

Remark 5.3 Note that taking the formal parabolic limit of the system (5.36) with density-dependent speeds (via the rescaling of time and space variables $x = x^*/\epsilon$, $t = t^*/\epsilon^2$, or the rescaling of speed and turning rates $\gamma = \gamma^*/\epsilon$, $\lambda^\pm = \lambda^\pm_*/\epsilon^2$, with

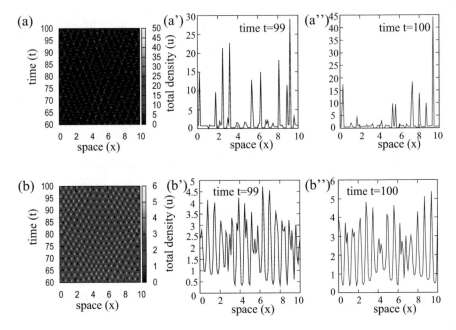

Fig. 5.23 Examples of chaotic spatio-temporal patterns exhibited by model (5.36) with density-dependent speeds (5.37). (**a**) Pattern obtained for $q_{al} = 0$, $q_{al} = 0.5$, $q_r = 0.1$, $\lambda_1 = 0.2$, $\lambda_2 = 0.9$. (**a'**), (**a''**) Time-snapshots of the pattern in (**a**). (**b**) Pattern obtained for $q_{al} = 0.6$, $q_{al} = 0.5$, $q_r = 0.5$, $\lambda_1 = 0.2$, $\lambda_2 = 0.9$. (**b'**), (**b''**) Time-snapshots of the pattern in (**b**)

$\epsilon \ll 1$) we obtain (after dropping the "∗" for simplicity) an advection-diffusion equation of the form [7]:

$$\frac{\partial u}{\partial t} = D \frac{\partial^2 u}{\partial x^2} - B \frac{\partial}{\partial x}\big(\alpha^*[u]u\big) - \gamma^* \frac{\partial}{\partial x}\big(g^*[u]u\big), \qquad (5.42)$$

where we had $\Gamma^{\pm}[u^+, u^-] = \gamma(1 \pm g[u])$, and rescaled $g[u] = \epsilon g^*[u]$. We observe that the nonlocal component of the individuals' speed, $g[u]$, gives rise to a drift which is added to the drift caused by the difference in the turning rates: $\alpha[u] = \lambda_2 \epsilon \alpha^*[u]$, with $\alpha^* = \frac{1}{\epsilon}\big(f^*(y^-[u]) - f^*(y^+[u])\big)$. Therefore, we expect similar qualitative behaviours for the parabolic equation (5.42), irrespective of whether the communication mechanisms influence only the turning rates, only the speed, or both the turning rates and the speed.

We conclude this section by observing that even these nonlocal hyperbolic models with density-dependent speeds can be generalised to describe the interactions between multiple species (which use different communication mechanisms). However, to our knowledge, no such models have been developed yet, to investigate the effect of multiple communication mechanisms on speeding-up/slowing-down behaviours.

5.9   Systems Incorporating Environmental Stochasticity

Even if the nonlocal hyperbolic/kinetic models discussed in the previous sections can exhibit chaotic dynamics in the presence of deterministic inter-individual inter-actions (as shown numerically in Figs. 5.13a", c" and 5.23), we need to emphasise that inter-individual communication can be affected by noise (e.g., abiotic, biotic or anthropogenic; see Fig. 5.24 and the discussion in [71]). Ecological, physiological and psychological studies have shown that anthropogenic (i.e., human-made) noise can decrease the accuracy of animal communication [72–74], can increase the energetic cost of animals (who try to move away from noise and change their feeding patterns) [75], can restrict animal movement (which can impact whole ecosystems) [76], or can impact the well functioning of different physiological systems (e.g., cardiac, auditory, neural or immune systems) in various animal species [77–82]. Therefore noise likely impacts that movement dynamics of the whole population and the interactions among conspecifics [83]. Moreover, many of the species that self-organise (e.g., birds) form relatively small aggregations, and in this case, random fluctuations in population size—caused, for example, by environmental stochasticity—can become important.

A generalisation of nonlocal models (5.14)–(5.19) could include the explicit incorporation of environmental and anthropogenic stochasticity. However, we need to emphasise that models (5.14)–(5.19) do incorporate implicitly a sort of stochastic-ity, as they are derived using a correlated random walk approach (see Sect. 5.3) and the resulting solutions can be seen as probability densities [84]. Thus, the explicit incorporation of environmental and anthropogenic noise into the hyperbolic/kinetic models discussed above adds another layer of stochasticity.

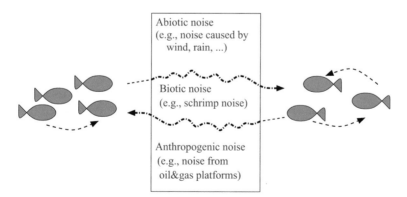

Fig. 5.24 Caricature description of the effect of noise (abiotic, biotic, anthropogenic) on the emission/reception of communication signals from conspecifics

Eftimie [84] has recently investigated the effect of anthropogenic noise on the dynamics of nonlocal models (5.14)–(5.19), under two assumptions:

- As the presence of noise was shown to lead to changes in the emission/reception of communication signals [85, 86], Eftimie [84] assumed that in noisy environments individuals try to increase the detection level of signals emitted by neighbours. Therefore, the term describing the perception of neighbours (see also Eq. (5.20)) is now time-dependent:

$$y_{r,a}^{\pm} = q_{r,a} \int_0^{\infty} K_{r,a}(s)\big(p_{\pm}(t)u(x \pm s) - p_{\mp}(t)u(x \mp s)\big)ds, \qquad (5.43)$$

with $p_{\pm}(t) = p_{\pm}^* + W(t)$, with $W(t) = Rand[0, n_m]$ (i.e., $W(t)$ is a uniform randomly distributed number within the interval $[0, n_m]$, where the maximum noise level is $n_m \leq 1$). This particular choice of $W(t)$ follows the approach taken by the individual-based models for collective movement of particles/cells/bacteria/animals discussed in Chap. 1 (see, for example, [88]). Moreover, the general terms $p_{\pm}$ (with p_+ possibly distinct from p_-) were used to describe also the asymmetric perception of neighbours due to abiotic noise (e.g., wind blowing from a certain direction; see Fig. 5.6);

- In the presence of noise individuals try to increase their turning rates (to detect better their neighbours):

$$\lambda^{\pm} = \lambda_1(t) + \lambda_2(t)\tanh(y^{\pm}[u^+, u^-] - y_0), \qquad (5.44)$$

with $\lambda_{1,2} = \lambda_{1,2}^* + W(t)$ and $W(t) = Rand[0, n_m]$. This assumption is similar to the assumptions made by the majority of individual-based models for collective dynamics of particles/cells/animals/etc.; see also the discussion in Chap. 1.

Since the large majority of individual-based models in the literature for self-organised biological aggregations (e.g., see [88–90]) discuss the effect of noise on the transition between ordered aggregations (where all individuals are moving in one direction) and disordered aggregations (with individuals facing different directions), the simulations in [84] started with parameter values that generated (in the absence of noise) one travelling pulse/aggregation; see the caption of Fig. 5.26.

An investigation into the effect of perturbing the $p_{\pm}$ values showed (see Fig. 5.25) that: (i) higher noise leads to the destruction of travelling aggregation patterns (panels (a),(b)); and (ii) a heterogeneous environment (characterised by asymmetric perception of neighbours: $p_+^* \neq p_-^*$) is more likely to destroy the travelling aggregation patterns compared to a homogeneous environment (i.e., $p_+^* = p_-^*$); see panel (c) in Fig. 5.25.

The effect of noise on the turning rates $\lambda_{1,2}(t)$ is illustrated in Fig. 5.26. Panels (a) show three examples of spatio-temporal patterns when we perturb *deterministically* the turning rates by a constant (in time) value n_m: (i) a stationary pulse for $n_m = 0.8$, (ii) a regular zigzag for $n_m = 0.3$, and (iii) two travelling pulses for $n_m = 0.1$. Panels (b) show three examples of spatio-temporal patterns when

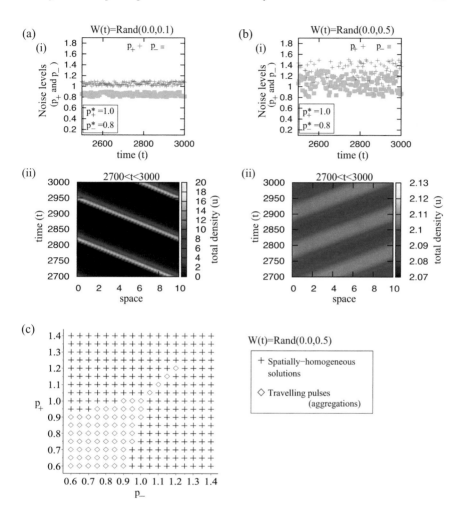

Fig. 5.25 Pattern formation in a heterogeneous environment ($p_+ = 1.0$, $p_- = 0.8$), as we increase the noise level in $p_\pm$: (**a**) $W(t) = Rand[0, 0.1]$, (**b**) $W(t) = Rand[0, 0.5]$. The bifurcation diagram in (**c**) shows that travelling pulses persist mostly when $p_+ \approx p_-$ (and thus the environment is homogeneous, being characterised by symmetric perception of neighbours ahead/behind a reference individual). The parameter values used for the simulations are: $\gamma = 0.1$, $\lambda_1^* = 0.1$, $\lambda_2^* = 0.9$, $q_a = 2.0$, $q_r = 0.5$, $q_{al} = 1.0$, $y_0 = 0.2$

we decrease the maximum noise level n_m: (i) a stationary pulse for $n_m = 0.8$, (ii) a chaotic zigzag for $n_m = 0.3$, and (iii) a travelling pulse for $n_m = 0.1$. Sub-panels (iv) and (iv') show bifurcation diagrams for changes in the maximum amplitude of the patterns as we decrease the n_m values. The red amplitude values are obtained when we start with $n_m = 1$ and we decrease it towards $n_m = 0.1$, while the black amplitude values are obtained when we start with $n_m = 0.1$ and we increase it towards $n_m = 1$. We observe that for the parameters used in these

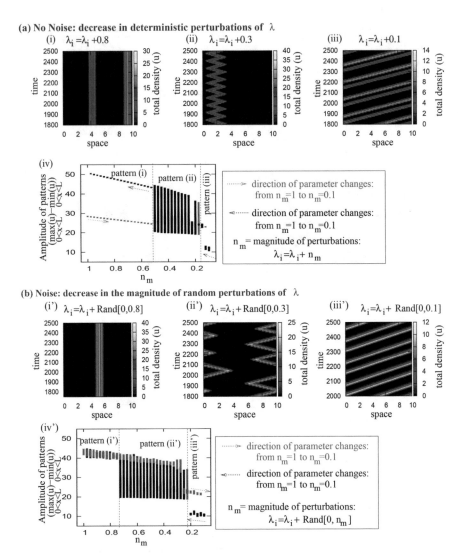

Fig. 5.26 (a) Spatio-temporal patterns (sub-panels (i)–(iii)) and bifurcation diagram (sub-panel (iv)) obtained numerically when we perturb $\lambda_{1,2}$ deterministically with a constant n_m. (b) Spatio-temporal patterns (sub-panels (i)–(iii)) and bifurcation diagram (sub-panel (iv)) obtained numerically when we perturb $\lambda_{1,2}$ randomly with a uniformly distributed number chosen within the interval $[0, n_m]$. The parameter values used for the simulations are: $\gamma = 0.1$, $\lambda_1^* = 0.1$, $\lambda_2^* = 0.9$, $q_a = 2.0$, $q_r = 0.5$, $q_{al} = 1.0$, $p_- = p_+ = 1$. The domain length is $L = 10$, and periodic boundary conditions are considered

numerical simulations (and listed in the caption of Fig. 5.26), either random or deterministic perturbations of the turning rates $\lambda_{1,2}$ lead to similar transitions from stationary pulses, to zigzags (chaotic or deterministic) and eventually to travelling pulses. However, the detailed bifurcation structure (i.e., the parameter values at which these transitions occur, and the number of bifurcating branches arising at the bifurcation points) is different for the deterministic and stochastic cases. Moreover, we notice a sort of hysteresis phenomenon during the transition between different patterns (chaotic or deterministic). For example, if we start with $n_m = 1$ in the deterministic case (Fig. 5.26(iv)), we obtain a stationary pulse (pattern (a)) which persists also as we decrease n_m below the critical value $n_m = 0.54$. Only by decreasing $n_m \in (0.16, 0.2)$ leads to a zigzag pattern (b). Similar observations hold also for the stochastic case (Fig. 5.26(iv')): decreasing n_m below 0.76 does not lead to an immediate transition to a zigzag pattern (i.e., the solution stays on the higher-amplitude red branch). This transition occurs as n_m is decreased even further: $n_m \in (0.22, 0.3)$.

A detailed, rigorous investigation of the bifurcation structure of these stochastic systems is currently an open problem (as it will be discussed later in Sect. 8.8).

References

1. R. Larkin, R. Szafoni, Integr. Comp. Biol. **48**(1), 40 (2008)
2. A. Farnsworth, Auk **122**(3), 733 (2005)
3. M.C. Hager, G.S. Helfman, Behav. Ecol. Sociobiol. **29**, 271 (1991)
4. K. Zuberbühler, R. Noë, R. Seyfarth, Anim. Behav. **53**(3), 589 (1997)
5. S. Wich, C. Nunn, Behav. Ecol. Sociobiol. **52**(6), 474 (2002)
6. T. Das, K. Safferling, S. Rauch, N. Grabe, H. Boehm, J. Spatz, Nat. Cell Biol. **17**, 276 (2015)
7. R. Eftimie, Modelling group formation and activity patterns in self-organising communities of organisms. Ph.D. thesis, University of Alberta (2008)
8. A. Mogilner, L. Edelstein-Keshet, J. Math. Biol. **38**, 534 (1999)
9. L. Edelstein-Keshet, J. Watmough, D. Grünbaum, J. Math. Biol. **36**(6), 515 (1998)
10. N. Armstrong, K. Painter, J. Sherratt, J. Theor. Biol. **243**(1), 98 (2006)
11. R. Seyfarth, D. Cheney, P. Marler, Science **201**, 801 (1980)
12. I. Barber, OIKOS **101**, 331 (2003)
13. J.B.J. rgensen, Trends Ecol. Evol. **25**, 292 (2010)
14. A. Leverentz, C. Topaz, A. Bernoff, SIAM J. Appl. Dyn. Syst. **8**(3), 880 (2009)
15. C.M. Topaz, A.L. Bertozzi, SIAM J. Appl. Math. **65**, 152 (2004)
16. A. Mogilner, L. Edelstein-Keshet, L. Bent, A. Spiros, J. Math. Biol. **47**, 353 (2003)
17. R. Fetecau, R. Eftimie, J. Math. Biol. **61**(4), 545 (2010)
18. R. Eftimie, M. Perez, P.L. Buono, Math. Biosci. **289**, 96 (2017)
19. R. Eftimie, G. de Vries, M.A. Lewis, F. Lutscher, Bull. Math. Biol. **69**(5), 1537 (2007)
20. R. Eftimie, G. de Vries, M.A. Lewis, Proc. Natl. Acad. Sci. USA **104**(17), 6974 (2007)
21. R. Eftimie, G. de Vries, M. Lewis, J. Math. Biol. **59**, 37 (2009)
22. X. Zhang, A. Chen, D.D. Leon, H. Li, E. Noiri, V. Moy, M. Goligorsky, Am. J. Physiol. Heart Circ. Physiol. **286**(1), H359 (2004)
23. D. Morale, V. Capasso, K. Oelschläger, J. Math. Biol. **50**, 49 (2005)
24. K. Fellner, G. Raoul, Math. Comput. Modell. **53**, 1436 (2011)
25. G. Raoul, Differ. Integr. Equ. **25**(5/6), 417 (2012)
26. M. Bodnar, J. Velazquez, J. Differ. Equ. **222**(2), 341 (2006)
27. M. Kac, Rocky Mt. J. Math. **4**, 497 (1974)
28. H.G. Othmer, S.R. Dunbar, W. Alt, J. Math. Biol. **26**, 263 (1988)

29. D. Grünbaum, A. Okubo, in *Frontiers in Mathematical Biology*. Lecture Notes in Biomathematics, vol. 100, ed. by S.A. Levin (Springer, Berlin Heidelberg, 1994), pp. 296–325
30. J. Hutchinson, P. Waser, Biol. Rev. **82**, 335–359 (2007)
31. P.L. Buono, R. Eftimie, J. Math. Biol. **71**(4), 847 (2014)
32. J. Carrillo, R. Eftimie, F. Hoffmann, Kinet. Relat. Models **8**(3), 413 (2015)
33. B. Pfistner, in *Biological Motion*. Lecture Notes on Biomathematics, vol. 89, ed. by W. Alt, G. Hoffmann (Springer, Berlin, 1990), pp. 556–563
34. B. Pfistner, J. Biol. Syst. **3**, 579 (1995)
35. M. Ballerini, N. Cabibbo, R. Candelier, A. Cavagna, E. Cisbani, I. Giardina, V. Lecomte, A. Orlandi, G. Parisi, A. Procaccini, M. Viale, V. Zdravkovic, Proc. Natl. Acad. Sci. USA **105**(5), 1232 (2008)
36. S. Partan, in *Handbook of Multi Sensory Processes*, ed. by C. Calvert, B. Stein (The MIT Press, Cambridge, 2004), pp. 225–240
37. F. Lutscher, A. Stevens, J. Nonlinear Sci. **12**, 619 (2002)
38. K. Kang, A. Scheel, A. Stevens (2018). arXiv
39. O. Igoshin, A. Mogilner, R. Welch, D. Kaiser, G. Oster, Proc. Natl. Acad. Sci. USA **98**, 14913 (2001)
40. H. Pomeroy, F. Heppner, Auk **109**, 256 (1992)
41. C. Zmurchok, G. de Vries, PLoS One **13**(6), e0198550 (2018)
42. W. Allee, Q. Rev. Biol. **2**(3), 367 (1927)
43. G. Harris, S. Thirgood, J. Hopcraft, J. Cromsigt, J. Berger, Endanger. Species Res. **7**, 55 (2009)
44. F. Siegert, C. Weijer, Curr. Biol. **5**(8), 937 (1995)
45. P.L. Buono, R. Eftimie, Math. Models Methods Appl. Sci. **24**(2), 327–357 (2014)
46. P.L. Buono, R. Eftimie, in *Mathematical Sciences with Multidisciplinary Applications*. Springer Proceedings in Mathematics & Statistics, vol. 157 (Springer, Berlin, 2016), pp. 29–59
47. P.L. Buono, R. Eftimie, SIAM J. Appl. Dyn. Syst. **13**(4), 1542 (2014)
48. M. Beekman, D.J.T. Sumpter, F.L.W. Ratnieks, Proc. Natl. Acad. Sci. USA **98**(17), 9703 (2001)
49. J. Buhl, D.J.T. Sumpter, I.D. Couzin, J.J. Hale, E. Despland, E.R. Miller, S.J. Simpson, Science **312**, 1402 (2006)
50. M. Golubitsky, I. Stewart, D.G. Schaeffer, *Singularities and Groups in Bifurcation Theory*, vol. II (Springer, New York, 1988)
51. M. Golubitsky, I. Stewart, *The Symmetry Perspective: From Equilibrium to Chaos in Phase Space and Physical Space* (Birkhäuser, Basel, 2002)
52. I.D. Couzin, J. Krause, R. James, G. Ruxton, N.R. Franks, J. Theor. Biol. **218**, 1 (2002)
53. R. Eftimie, J. Theor. Biol. **337**, 42 (2013)
54. W. Guo, X. Wang, X. Zheng, Physica A **432**, 87 (2015)
55. V. Curtis, Trends Immunol. **35**(10), 457 (2014)
56. C. Sarabian, V. Curtis, R. McMullan, Philos. Trans. R. Soc. Lond. B Biol. Sci. **373**(1751), 20170256 (2018)
57. M. Kavaliers, E. Choleris, Philos. Trans. R. Soc. Lond. B Biol. Sci. **373**(1751), 20170206 (2018)
58. G. Rubin, R. Amlôt, L. Page, S. Wessely, Br. Med. J. **339**, 1 (2009)
59. L. Leaver, L. Hopewell, C. Caldwell, L. Mallarky, Anim. Cogn. **10**(1), 23 (2007)
60. T. Pitcher, C. Wyche, in *Predators and Prey in Fishes. Developments in Environmental Biology of Fishes*, vol. 2, ed. by D. Noakes, D. Lindquist, G. Helfman, J. Ward (Springer, Dordrecht, 1983), pp. 193–204
61. R. Eftimie, A. Coulier, Math. Model Nat. Phenom. **10**(2), 27 (2015)
62. D. Álvarez, A. Nicieza, J. Fish Biol. **63**(6), 1565 (2003)
63. S. Gueron, S.A. Levin, D.I. Rubenstein, J. Theor. Biol. **182**, 85 (1996)
64. B. Partridge, Sci. Am. **246**(6), 114 (1982)
65. D. Helbing, P. Molnar, Phys. Rev. E **51**(5), 4282 (1995)
66. W. Takken, Invertebr. Reprod. Dev. **36**(1–3), 67 (1999)
67. K. Lika, T. Hallam, J. Math. Biol. **38**, 346 (1999)
68. R. Mickens, SIAM Rev. **30**(4), 629 (1988)

69. D. Helbing, A. Hennecke, V. Shvetsov, M. Treiber, Math. Comput. Model. **35**(5–6), 517 (2002)
70. J.K. Parrish, L.E. Keshet, Science **284**, 99 (1999)
71. H. Brumm (ed.), *Animal Communication and Noise* (Springer, Berlin Heidelberg, 2013)
72. L. Rabin, B. McCowan, S. Hooper, D. Owings, Int. J. Comp. Psychol. **16**, 172 (2003)
73. T. Ord, J. Stamps, Proc. Natl. Acad. Sci. USA **105**(48), 18830 (2008)
74. A. Codarin, L. Wysocki, F. Ladich, M. Picciulin, Mar. Pollut. Bull. **58**, 1880 (2009)
75. F. VanDyke, R. Broke, H. Shaw, B. Ackerman, T. Hemker, F. Lindzey, J. Wildl. Manage. **50**(1), 95 (1986)
76. M. Tucker, K. Böhning-Gaese, W. Fagan, J. Fryxell, B.V. Moorter, S. Alberts, A. Ali, A. Allen, N. Attias, T. Avgar, H. Bartlam-Brooks, B. Bayarbaatar, J. Belant, A. Bertassoni, D. Beyer, L. Bidner, F. van Beest, S. Blake, N. Blaum, C. Bracis, D. Brown, P. de Bruyn, F. Cagnacci, J.M. Calabrese, C. Camilo-Alves, S. Chamaillé-Jammes, A. Chiaradia, S. Davidson, T. Dennis, S. DeStefano, D. Diefenbach, I. Douglas-Hamilton, J. Fennessy, C. Fichtel, W. Fiedler, C. Fischer, I. Fischhoff, C. Fleming, A. Ford, S. Fritz, B. Gehr, J. Goheen, E. Gurarie, M. Hebblewhite, M. Heurich, A. Hewison, C. Hof, E. Hurme, L. Isbell, R. Janssen, F. Jeltsch, P. Kaczensky, A. Kane, P. Kappeler, M. Kauffman, R. Kays, D. Kimuyu, F. Koch, B. Kranstauber, S. LaPoint, P. Leimgruber, J. Linnell, P. López-López, A. Markham, J. Mattisson, E.P. Medici, U. Mellone, E. Merrill, G. de Miranda Mourào, R. Morato, N. Morellet, T. Morrison, S. Diaz-Munoz, A. Mysterud, D. Nandintsetseg, R. Nathan, A. Niamir, J. Odden, R. O'Hara, L. Oliveira-Santos, K. Olson, B. Patterson, R. de Paula, L. Pedrotti, B. Reineking, M. Rimmler, T. Rogers, C. Rolandsen, C. Rosenberry, D. Rubenstein, K. Safi, S. Saïd, N. Sapir, H. Sawyer, N. Schmidt, N. Selva, A. Sergiel, E. Shiilegdamba, J. Silva, N. Singh, E. Solberg, O. Spiegel, O. Strand, S. Sundaresan, W. Ullmann, U. Voigt, J. Wall, D. Wattles, M. Wikelski, C. Wilmers, J. Wilson, G. Wittemyer, F. Zieba, T. Zwijacz-Kozica, T. Mueller, Science **358**(6374), 466 (2018)
77. P. Marler, M. Konishi, A. Lutjen, M. Waser, Proc. Natl. Acad. Sci. USA **70**, 1393 (1973)
78. E. Chang, M. Merzenich, Science **300**, 498 (2003)
79. S. Sobrian, V. Vaughn, W. Ashe, B. Markovic, V. Djuric, B. Jankovic, Environ. Res. **73**, 227 (1997)
80. J. Hochel, R. Pirow, M. Nichelmann, Comp. Biochem. Physiol. **131**, 805 (2002)
81. H. Ising, B. Kruppa, Noise Health **6**, 5 (2004)
82. I. Croy, M. Smith, K. Waye, Br. Med. J. Open **3**, e002655 (2013)
83. C. Francis, C. Ortega, A. Cruz, Curr. Biol. **45**, 1415 (2009)
84. R. Eftimie, in *BIOMATH* (2018)
85. D. proppe, C. Sturdy, C.S. Clair, PLoS One **6**, e25413 (2011)
86. F. Ladich, in *Animal Communication and Noise*, ed. by H. Brumm (Springer, Berlin, 2013), pp. 65–90
87. A.J. Lotka, *Elements of Physical Biology* (Williams and Wilkins, Baltimore, 1924)
88. T. Vicsek, A. Czirók, E. Ben-Jacob, I. Cohen, O. Shochet, Phys. Rev. Lett. **75**(6), 1226 (1995)
89. T. Vicsek, A. Czirok, I.J. Farkas, D. Helbing, Physica A **274**, 182 (1999)
90. T. Vicsek, A. Zafeiris, Phys. Rep. **517**(3–4), 71 (2010)

Chapter 6
Multi-Dimensional Transport Equations

6.1 General Description of 2D and 3D Models

The 1D spatial models described in the previous chapters are easy to simulate numerically and to investigate analytically. However, they are not very biological realistic, since the majority of behaviours among individuals in cellular, bacterial or animal communities occur in 2D or 3D. Moreover, while some 2D/3D patterns can be reduced to similar 1D patterns (e.g., stationary pulses corresponding to stationary aggregations, or travelling pulses (rotating waves) corresponding to travelling aggregations), it is expected that many other 2D/3D patterns cannot be reduced to simpler 1D patterns. Therefore, to understand the mechanisms behind the formation of various biological aggregations observed in nature we need to focus on 2D/3D models.

The last two decades have seen the development of a large variety of 2D/3D models describing self-organised phenomena in cells [1–6], animals [7–11] and even humans [12–15]. While the majority of these models are local, many of the recent models include also nonlocal interactions [2, 8, 9]. As mentioned before for the 1D models, many of the 2D/3D models in the literature for self-organised biological aggregations are of parabolic type, thus describing dominant random movement behaviours; see, for example, [1, 2] and the references therein. Nevertheless, the latest developments in experimental cell biology and ecology have led to an accumulation of data regarding individual velocities and turning rates, which in turn led to the further development of kinetic models; see, for example, [3–5, 14, 15] and the references therein. Due to the complexity of these (mesoscale) kinetic models, which renders them difficult to be investigated analytically and numerically, some studies consider hydrodynamic and/or parabolic limits of these models [10, 11, 16] (since these macroscopic models are somewhat easier to be studied). We will discuss this aspect in more detail in Sect. 6.8.

© Springer Nature Switzerland AG 2018
R. Eftimie, *Hyperbolic and Kinetic Models for Self-organised Biological Aggregations*, Lecture Notes in Mathematics 2232,
https://doi.org/10.1007/978-3-030-02586-1_6

(a) particles/cells crossing a
     control volume V

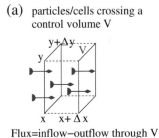

Flux=inflow−outflow through V

(b)

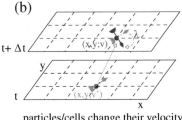

particles/cells change their velocity
(from v' to v) with rate λ (constant or
density−dependent)

Fig. 6.1 Description of (**a**) the "flow" approach (where the arrows describe particles/cells that enter/exit a control volume V), and (**b**) the velocity-jump approach for the derivation of a 2D model (generally a macroscopic model for (**a**) and a mesoscopic/kinetic model for (**b**)). For the case in (**b**), we note that one can have also a spatial movement of particles/cells from some position (x', y') to position (x, y)

In the following we discuss briefly the general structure of the 2D/3D kinetic models for biological aggregations. To this end, we first note that while car traffic models are usually in one spatial dimension (due to the 1D structure of roads and highways), the models for pedestrian dynamics and more generally those for cell and animal movement are usually described in two or three spatial dimensions [17]. These (space) multi-dimensional models can be derived either using a "flow" approach (where one focuses on the density of particles/cells/animals/pedestrians that cross a surface or a volume; see Chapter 2 in [18], and Fig. 6.1a), or using a 2D generalisation of the classical Goldstein-Kac approach or the more general velocity-jump approach (see Fig. 6.1b), or using the binary collision of active-particles approach. These different approaches allow for the incorporation of various reaction terms describing birth/death processes, or inflow/outflow of individuals into the domain.

We start the discussion of 2D kinetic models by focusing in Sect. 6.2 on models with local interactions (which are generalisations of the 1D models discussed in Chaps. 3 and 4). In Sect. 6.3 we focus on a class of nonlocal transport models where the nonlocal interactions impact only the velocity. These models are 2D generalisations of the 1D models discussed in Sect. 5.2. Next, in Sects. 6.4 and 6.5, we discuss two types of nonlocal kinetic models where the turning is influenced by interactions between particles/cells/animals (and the velocity could be constant or density-dependent): (1) models derived from velocity-jump processes, and (2) kinetic models for active particles. We emphasise that throughout this chapter we focus mostly on models that consider direct cell-cell/animal-animal interactions; although in Sect. 6.6 we mention also a few models that consider indirect cell-cell interactions via an external signal (i.e., a chemical either produced by the cells in the environment [19], or that is produced inside the cells [20]). Finally, in Sect. 6.11 we present some kinetic models that incorporate explicitly stochastic terms. To balance the exposition of this long list of various kinetic models in 2D, in

Sects. 6.7, 6.8, and 6.9 we present in more detail some analytical approaches used to manipulate the models in such a way that they become more amenable to analysis: the mean-field approach used to derive mesoscale kinetic models from microscale individual-based models (see the models in Chap. 1), the hydrodynamic limit and the grazing collision limit of kinetic models to obtain macroscale hyperbolic models (with respect to space).

To help the reader, we summarise in Fig. 6.2 some of the aggregation patterns obtained with the 2D kinetic models discussed in this Chapter: (a) stationary pulses, (b) travelling pulses, (c) spatially homogeneous states with individuals aligned in one direction, (d) vortices. In Fig. 6.3 we show a splitting of the aggregation patterns obtained with the kinetic model introduced in [8]. Note that this splitting behaviour is usually accompanied by a merging behaviour (not shown here). As the simulations are performed on a domain with periodic boundary conditions, the two groups will meet again and merge. We will refer to these patterns throughout the rest of the sections.

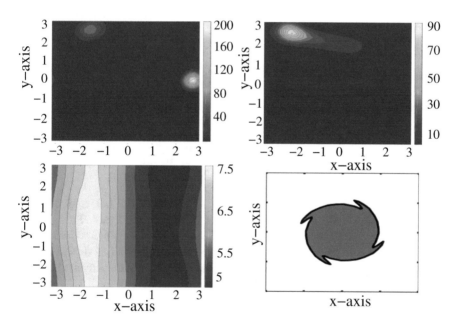

Fig. 6.2 Examples of 2D patterns exhibited by kinetic models. (**a**) Stationary aggregations (pulses) as in [8]; (**b**) Travelling pulses as in [8]; (**c**) Lane patterns as in [8]; (**d**) Vortices [7]. The patterns in panels (**a**), (**b**), (**c**) were simulated using the same parameter values as in [8] (but on a shorter domain $[-3.14, 3.14]$), where we show the total population density (as given by the color bar) averaged over the orientation ϕ: $\rho(x, y; t) = \int u(x, y; t, \phi)d\phi$. Panel (**d**) reproduced from [7]: Copyright ©2004 Society for Industrial and Applied Mathematics. Reprinted with permission. All rights reserved

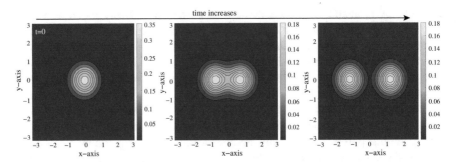

Fig. 6.3 2D splitting of aggregation patterns, exhibited by the kinetic model introduced in [8]; see Eqs. (6.25)–(6.28). The numerical simulations were performed on a square domain $[-L, L]$ with $L = 3.14$, with social interactions having magnitudes $q_a = 5$, $q_{al} = 5$, $q_r = 1$, and speed $\gamma = 1$. Initial conditions are 2D Gaussian functions describing an aggregation (see left panel), and boundary conditions are periodic. A spectral method is used to deal with the integrals in model (6.25)–(6.28); see also Chap. 7. As time increases, the initial aggregation splits into 2 sub-groups that are moving apart. Shown is the total population density averaged over the orientation angle ϕ: $\rho(x, y; t) = \int u(x, y; t, \phi)d\phi$. The color bar describes the range over which the population density varies

6.2   Local Advective Models in 2D

The 2D movement of pedestrians has often been described with the help of first-order transport models [21, 22]

$$\frac{\partial u}{\partial t} + \frac{\partial f(u)}{\partial x} + \frac{\partial g(u)}{\partial y} = F(u), \tag{6.1}$$

where $u(x, y, t) \in \mathbb{R}^+$ is the total population density, and $F(u) \in \mathbb{R}$ describes pedestrian inflow/outflow into the domain or birth/death processes. These models are usually derived using the "flow" approach discussed in the previous section (see also Fig. 6.1a). We emphasise that in addition to these first-order models in 2D, one could have also second-order transport models for pedestrian movement [17]

$$\frac{\partial u}{\partial t} + \frac{\partial (uv_x)}{\partial x} + \frac{\partial (uv_y)}{\partial y} = 0, \tag{6.2a}$$

$$\frac{\partial v_x}{\partial t} + v_x \frac{v_x}{\partial x} + v_y \frac{\partial v_x}{\partial y} = F_1(\mathbf{x}, u, \mathbf{v}, \nabla_{v_0}u), \tag{6.2b}$$

$$\frac{\partial v_y}{\partial t} + v_x \frac{v_y}{\partial y} + v_y \frac{\partial v_y}{\partial y} = F_2(\mathbf{x}, u, \mathbf{v}, \nabla_{v_0}u), \tag{6.2c}$$

where $\mathbf{v} = (v_x, v_y) \in \mathbb{R}^2$ is the speed, $v_0 \in \mathbb{R}$ represents some movement direction, and $F_{1,2} \in \mathbb{R}$ describe terms that influence the changes in the velocity in x and y directions.

As an example of a first-order transport model for pedestrian dynamics, Hughes [21] considered the following hypotheses regarding pedestrian motion:

(1) pedestrians' speed is determined only by the density of their surrounding neighbours;
(2) pedestrians move in direction perpendicular to a potential Φ that describes the task they face to reach their common destination;
(3) pedestrians try to minimise their travel time, while avoiding at the same time high densities;
(4) there is no pedestrian inflow/outflow into the domain.

These assumptions translate into the following $f(u)$ and $g(u)$ functions in Eq. (6.1):

$$f(u) = ug_0(u)f_0^2(u)\frac{\partial\Phi}{\partial x}, \quad g(u) = ug_0(u)f_0^2(u)\frac{\partial\Phi}{\partial y}. \tag{6.3}$$

Here $f_0(u) = a - bu$ describes the speed (with a, b some constants), and $g_0(u)$ is a factor that models discomfort at very high densities, with

$$g_0(u)f_0(u) = \frac{1}{\sqrt{\left(\frac{\partial\Phi}{\partial x}\right)^2 + \left(\frac{\partial\Phi}{\partial y}\right)^2}}. \tag{6.4}$$

For model (6.1)–(6.3), Hughes [21] discussed the existence of two possible types of pedestrian flows (similar to the one-dimensional traffic flow behaviours in [23]): a fast-moving, low-density flow called "supercritical flow", and a slow-moving high-density flow called "subcritical flow". The author then used this first-order transport model to describe numerically the movement and distribution of pedestrians around a pillar, with a focus on the two types of pedestrian flows (subcritical and supercritical).

In the context of second-order models, Bellomo and Dogbé [17] discussed various pedestrian movement strategies, such as movement along straight lines towards the target, or movement towards paths with small population density gradients. Each strategy corresponds to different functions F_1 and F_2. For example, the case of pedestrian moving along straight lines was described by $F_1 = F_2 = f_1 + f_2$, with

$$f_1 = \alpha(v_e(u)v_0 - v), \quad f_2 = -\frac{K^2(u)}{u}\nabla_{v_0}u. \tag{6.5}$$

Here f_1 models the adaptation of pedestrians velocity to an equilibrium velocity v_e, and f_2 models the influence of local population density gradients. It is worth mentioning that the results in [17] are mainly analytical, the authors showing that the various models they derived are actually of hyperbolic type. No numerical simulations are performed to show how the strategies incorporated into the models affect the groups-level behaviour of pedestrians.

The classical Goldstein-Kac derivation [24, 25] of 1D hyperbolic systems for moving particles (which was discussed in Chap. 3) can be carried out also in two spatial dimensions. In this case, the derivation of the model gives rise to a system of four equations for particles moving East, West, North and South (or right, left, up and down) [26]:

$$\frac{\partial u_E}{\partial t} + \gamma \frac{\partial u_E}{\partial x} = -\lambda u_E + \lambda u_M, \tag{6.6a}$$

$$\frac{\partial u_W}{\partial t} - \gamma \frac{\partial u_W}{\partial x} = -\lambda u_W + \lambda u_M, \tag{6.6b}$$

$$\frac{\partial u_N}{\partial t} + \gamma \frac{\partial u_N}{\partial y} = -\lambda u_N + \lambda u_M, \tag{6.6c}$$

$$\frac{\partial u_S}{\partial t} - \gamma \frac{\partial u_S}{\partial y} = -\lambda u_S + \lambda u_M, \tag{6.6d}$$

where u_j, $j \in E, W, N, S$, describe the density of organisms moving East, West, North and South, and u_M is given by $u_M = \frac{1}{4}(u_E + u_W + u_N + u_S)$. As in 1D, the turning events are the result of Poisson processes.

Unfortunately, manipulating Eqs. (6.6) to obtain a telegraph equation in 2D (similar to Eq. (4.5), which has a finite speed of propagation [27]) cannot lead to a closed form equation for the total population density u_M [28]. (Note that the 2D telegraph equation does exist, but it cannot be obtained as a limit from Eqs. (6.6)). Even if some work to solve this problem has been done in [27], this negative result prompted scientists to focus on 2D kinetic transport models as generalisations of local and nonlocal 1D hyperbolic models [26, 28–32]. We will discuss some of these kinetic models in the following sections.

6.3   Nonlocal Advective Models in 2D

The most straightforward generalisation of the 1D nonlocal model (5.1) to two spatial dimensions (2D) is given by

$$\frac{\partial u}{\partial t} + \nabla(\Gamma[u]u) = 0, \tag{6.7}$$

where the velocity $\Gamma[u]$ incorporates also the prescribed trajectories $\mathbf{v}(\mathbf{x})$ followed by the particles:

$$\Gamma[u] = \gamma(K \star u)\mathbf{v}(\mathbf{x}), \quad \mathbf{x} \in \mathbb{R}^2. \tag{6.8}$$

An example of such as model was introduced in [33] to investigate pedestrian movement. There the authors chose the vector $\mathbf{v}(\mathbf{x})$ to be the unit tangent at $\mathbf{x}$ to the geodesic curve connecting point $\mathbf{x}$ with the final destination of the individual positioned at $\mathbf{x}$. To add more biological realism, the velocity can be changed slightly, as in [33, 34]:

$$\Gamma[u] = \gamma(u)\big(\mathbf{v}(\mathbf{x}) + I(u)\big), \quad \text{with } I(u) = -\epsilon \frac{\nabla(K \star u)}{\sqrt{1 + ||\nabla(K \star u)||^2}}. \tag{6.9}$$

Here, the term $I(u)$ describes pedestrian deviation from the prescribed path, in an attempt to avoid entering regions with high-density neighbours (which are perceived through long-distance interactions) [34]. Models of the type (6.7)–(6.9) have been shown to admit weak solutions [33, 35] that depend continuously on the initial data [33, 34]. Moreover, numerical simulations performed with these models (using a Lax-Friedrichs method with dimensional splitting) showed the emergence of lane formation for pedestrians, as in Fig. 6.2d [33, 34]. Moreover, in [34], the authors noted that the number of pedestrian lanes was correlated to the size of the support of the convolution kernel K. Finally, the authors studied the effect of the domain shape (and in particular the presence of columns inside the domain) on the merging/splitting of lanes. We remark that in [34], the lane patterns were obtained under the assumption that pedestrians receive information from their neighbours positioned both ahead and behind them (i.e., isotropic environment). This is equivalent to the communication mechanisms M2 described in Chap. 5.

To generalise model (5.1) to two spatial dimensions, Topaz and Bertozzi [7] used the Hodge decomposition theorem [36] (which states that a vector field in the plane can be decomposed uniquely into a divergence-free component and a gradient component). Thus, the authors decomposed the velocity vector field into a divergence-free component Γ_t (also known as transverse component, or incompressible component; with $\nabla \cdot \Gamma_t \equiv 0$) and a gradient component Γ_l (also knows as longitudinal component; with $\nabla \times \Gamma_l \equiv 0$):

$$\Gamma = \Gamma_t + \Gamma_l := \nabla^\perp \Psi + \nabla \Phi. \tag{6.10}$$

This decomposition can be re-applied directly to the interaction kernel K (since the functions considered in [7] had integrable gradients, and the convolution commuted with the derivatives):

$$K = \nabla^\perp N + \nabla P, \tag{6.11}$$

where P describes the interaction pressure and N describes some additional motion that allows for rotation. The authors then investigated two cases:

1. *Incompressible motion:* $K = \nabla^\perp N$, with $\nabla \cdot \Gamma = 0$. For a Gaussian kernel $N = G_d(x) = (1/d^2)exp(-|x|^2/d^2)$, the authors showed that the model can exhibit vortex solutions. The equation for the boundary $z(\alpha, t)$ of the solution (with α a parameter on the boundary) is

$$\frac{dz(\alpha, t)}{dt} = \int_0^{2\pi} N(z(\alpha, t) - z(\alpha', t))z_\alpha(\alpha', t)d\alpha'. \tag{6.12}$$

The analytical solutions are backed up by numerical simulations; see also Fig. 6.2d. Here, the authors used the Simpson's rule to discretise the integral, and a combination of fourth-order Runge-Kutta method and fourth-order Adams-Bashforth method to discretise Eq. (6.12).

2. *Potential motion:* $K = \nabla P$. For $P = -G_d$, the motion is convective and the population is transported through space while it spreads away. For $P = G_d$, the motion will lead to spatial aggregations (stationary pulses).

Bertozzi et al. [37] focused on model (6.7) (which was derived as the continuum limit of an individual-based model), with nonlocal velocity

$$\Gamma[u] = \int_{\mathbb{R}^3} K(x - y)|(x - y)|u(y, t)dy, \qquad (6.13)$$

and investigated the stability and bifurcation dynamics of a single ring pattern. The authors showed that a low-mode instability can lead to the deformation of the ring pattern, while a high-mode instability can lead to the disintegration of the pattern. They also noted that an analysis of the stability of double-mill patterns for these transport models is still an open question.

Models (6.7) can be easily generalised to describe the interactions between multiple populations in case of panic and repulsive situations, as well as follower-leader dynamics and predator-prey interactions [35]. We will return to the discussion of these models in Sect. 6.10.

6.4 Velocity-Jump Processes

Experimental results showed that some bacteria (e.g., *Escherichia coli*) move with a constant velocity in a straight line, stop, and then choose a new direction with a certain probability [38]. This behaviour, called "run and tumble" (which was discussed in more detail at the beginning of Chap. 4; see also Fig. 4.1), can be modelled by a velocity-jump process [30, 39]. This modelling approach is not restricted only to bacteria, but can be used to model the behaviour of other organisms (e.g., amoeboid cells [40] or birds [41]) that move following similar rules. The general equation describing the movement of these organisms is

$$\frac{\partial}{\partial t}u(\mathbf{x}, t, \mathbf{v}) + \nabla_{\mathbf{x}} \cdot \mathbf{v}u(\mathbf{x}, t, \mathbf{v}) = -\lambda u(\mathbf{x}, t, \mathbf{v}) + \lambda \int_V T(\mathbf{v}, \mathbf{v}')u(\mathbf{x}, t, \mathbf{v}')d\mathbf{v}',$$

$$(6.14)$$

where $u(\mathbf{x}, t, \mathbf{v})$ is the density distribution function for organisms at time $t \geq 0$ and spatial position $\mathbf{x} \in R^n$, which move with velocity $\mathbf{v} \in R^n$. Parameter λ describes the random turning rate (with λ^{-1} the mean run length time between the random choices of direction [30]). Generally, it is assumed that either $\lambda = \lambda_0 =$constant, or

$\lambda = \lambda_0 + \lambda_1(u)$ (in which case the waiting time between jumps increases with density) [30]. Finally, the turning kernel $T(\mathbf{v}, \mathbf{v}')$ describes the probability of a velocity change from $\mathbf{v}'$ to $\mathbf{v}$, if a jump in velocity occurs. To ensure the conservation of organisms, the kernel must satisfy

$$\int T(\mathbf{v}, \mathbf{v}')d\mathbf{v} = 1. \qquad (6.15)$$

Note that the set of velocities $V \subset R^n$ is compact and symmetric (which implies that if $\mathbf{v} \in V$ then $-\mathbf{v} \in V$). This allows the multi-dimensional models (6.14) to be reduced to the previous one-dimensional hyperbolic models (4.3) by considering $T(\mathbf{v}, \mathbf{v}') = \delta(\mathbf{v} + \mathbf{v}')$ and only two possible values for the velocity: $\mathbf{v} = \gamma$ and $\mathbf{v} = -\gamma$ [30]. If we ignore the interactions between particles (cells), we obtain the transport equation (6.7) with constant velocity $\Gamma[u] = \mathbf{v}$. For a derivation of Eq. (6.14) in a biological context, see [30].

Equation (6.14) is similar to the classical Boltzmann equation, derived in the context of kinetic theory of gas molecules [42, 43]. The right-hand-side of Eq. (6.14) can be interpreted as a collision operator

$$Q(u, u) = -\lambda u(\mathbf{x}, t, \mathbf{v}) + \lambda \int_V T(\mathbf{v}, \mathbf{v}')u(\mathbf{x}, t, \mathbf{v}')d\mathbf{v}', \qquad (6.16)$$

which, for Boltzmann equation, has to satisfy the three conservation properties of the collision process:

1. conservation of mass: $\int_V Q(u, u)dv = 0$;
2. conservation of momentum: $\int_V Q(u, u)vdv = 0$;
3. conservation of energy: $\int_V Q(u, u)|v|^2dv = 0$.

At equilibrium ($Q(u, u) = 0$), the distribution $u(x, v, t)$ of the Boltzmann equation is a local Maxwellian distribution:

$$u(x, v, t) = \frac{\bar{u}(x, t)}{(2\pi)^{3/2}}e^{-\frac{|v-p_0|^2}{2}}, \qquad (6.17)$$

with $\bar{u}(x, t) = \int_V u(x, v, t)dv$ the mass density and $p_0(x, t)$ the mean velocity $p_0 = \int_V u(x, v, t)vdv$. We need to emphasise here that in the context of the Boltzmann equation, the collision operator incorporates some assumptions that might not be considered biologically realistic: binary collision between pairs of identical molecules, negligible influence of any possible external force, and uncorrelated velocities and positions of a molecule. Nevertheless, as we will see next, these types of models are being often applied to describe collective movement in biology.

In general, Eq. (6.14) can incorporate reaction terms describing population dynamics as a result of individuals/cells/bacteria entering or leaving the domain, or as a result of birth and death events [26, 32, 44]:

$$\frac{\partial}{\partial t}u + \mathbf{v} \cdot \nabla_{\mathbf{x}}u = -\lambda u + \lambda \int_V T(\mathbf{v}, \mathbf{v}')u(\mathbf{x}, t, \mathbf{v}')d\mathbf{v}' + \frac{1}{|V|}m(\bar{u}) - g(\bar{u})u, \quad (6.18)$$

with $u = u(\mathbf{x}, t, \mathbf{v})$, $|V| = \int_V d\sigma_v$, and $\bar{u} = \int_V u(\mathbf{x}, t, \mathbf{v})d\sigma_v$. Here, $m(\bar{u})$ describes the proliferation of the population, and $g(\bar{u})$ describes the decay (death) of the population. In [45], the changes in population size (as given by the reaction terms) are caused by pedestrians entering or leaving the system.

The complexity of models (6.14) is generated by the form of the turning kernel $T(\mathbf{v}, \mathbf{v}')$ (assuming that λ is constant or might depend on the local density u). One simple choice of the turning kernel in one space dimension was discussed in [30]: $T(\mathbf{v}, \mathbf{v}') = \delta(\mathbf{v} + \mathbf{v}')$. This choice implies that individuals change direction each time a choice is made [30]. More complex models can be obtained by assuming density-dependent turning kernels T and/or density-dependent rates λ. The majority of models that describe the collective movement of organisms focus on the role of one type of social interaction (which can usually be thought as *alignment*) in determining the new movement direction [30, 45–50]. A few kinetic models focus on the *attractive and repulsive* interactions that influence the velocity of organisms [41]. Even fewer kinetic models incorporate all three social interactions: *repulsion, attraction and alignment* [8]. In the following we discuss three examples of kinetic models that incorporate different social interactions, and implicitly incorporate different communication mechanisms.

- *Turning depends on alignment.* An example of a kinetic model where individual turning depends on alignment was introduced by Boissard et al. [50] to describe ant-trail formation through the deposition of pheromones and interaction with these pheromones through alignment interactions. The model assumes nonlocal sensing of pheromones deposited by the ants at different positions in space (and this implies a sort of nonlocal communication among ants, through the pheromones they deposit). The equations describing the evolution of the ant distribution function $(F(\mathbf{x}, \omega, t))$ and the pheromone distribution function $(G(\mathbf{x}, \omega, t))$ are:

$$\frac{\partial G(\mathbf{x}, \omega, t)}{\partial t} = v_d F(\mathbf{x}, \omega, t) - v_e G(\mathbf{x}, \omega, t), \quad (6.19a)$$

$$\frac{\partial F(\mathbf{x}, \omega, t)}{\partial t} + c\omega \cdot \nabla_{\mathbf{x}}F(\mathbf{x}, \omega, t) = Q_r(F) + Q_p(F), \quad (6.19b)$$

$$Q_p(F)(\mathbf{x}, \omega, t) = \lambda_p \gamma(T_R(\mathbf{x}, t)) \int_{S^1} \Phi_P(\omega, \omega')\big(h_p(\omega)F(\mathbf{x}, \omega', t)$$
$$- h_p(\omega')F(\mathbf{x}, \omega, t)\big)d\omega', \quad (6.19c)$$

$$Q_r(F)(\mathbf{x}, \omega, t) = \lambda_r \int_{S^1} \Phi_r(\omega, \omega')\big(F(\mathbf{x}, \omega', t) - F(\mathbf{x}, \omega, t)\big)d\omega', \quad (6.19d)$$

$$h_p(\omega) = (g_R^{sym})^{|k|}(\omega), \quad g_R^{sym}(\mathbf{x}, \omega, t) = \frac{S_R(\mathbf{x}, \omega, t) + S_r(\mathbf{x}, \omega, t)}{2T_R(\mathbf{x}, t)},$$

$$(6.19e)$$

$$S_R(\mathbf{x}, \omega, t) = \frac{1}{\pi R^2} \int_{|\mathbf{x}-\mathbf{y}|<R} G(\mathbf{y}, \omega, t) d\mathbf{y}, \quad T_R(\mathbf{x}, t) = \int_{\mathbf{S}^1} S_R(\mathbf{x}, \omega, t) d\omega.$$

$$(6.19f)$$

Here $\mathbf{x} \in \mathbb{R}^2$, $\omega, \omega' \in \mathbf{S}^1$ (with ω = post-jump velocity and ω' = pre-jump velocity), and $t \geq 0$. Equation (6.19a) describes the deposition of pheromones at a rate v_d, and their degradation at a rate v_e. Equation (6.19b) describes the movement of ants at a constant speed c in direction ω, as a results of the random velocity jumps (given by operator Q_r) and trail recruitment jumps (given by operator Q_p). For the jump probabilities, it is assumed that they depend on some equilibrium probabilities: $h_r(\omega) = 1/2\pi$ and h_p depending on the density of pheromones on the trails in direction ω. Moreover R is the maximum distance within which an ant can perceive the pheromone molecules (i.e., their perception radius) and $S_R(\mathbf{x}, \omega, t)$ describes the density of pheromones pointing in direction ω that can be perceived by an ant positioned at $\mathbf{x}$. The term $T_R(\mathbf{x}, t)$ describes the total pheromone density within the perception radius of an ant. Parameter λ_p is the trail-recruitment frequency, parameter λ_r is the random velocity jump frequency, and γ is a dimensionless increasing function of T describing the increase in the trail recruitment as the pheromone density increases.

The authors did not show any numerical simulations for this kinetic model (but they did ran simulations with an individual-based model that formed the basis of the derivation for this particular kinetic model). Instead, they focused on analytical approaches (i.e., hydrodynamic limit) to reduce—under some approximations regarding the perception radius R and the transition probabilities Φ_r, Φ_p—this kinetic model to a simpler macroscopic model for the evolution of the ant density $\rho(\mathbf{x}, t) = \int_{\mathbf{S}^1} F(\mathbf{x}, \omega, t) d\omega$ and the pheromone density $T(\mathbf{x}, t) = \int_{\mathbf{S}^1} G(\mathbf{x}, \omega, t) d\omega$. For the details behind this hydrodynamic limit approach, we refer the reader to [50]. And for a more in depth exposition of the hydrodynamic limit for the general Boltzmann equation, we refer the reader to [51].

- *Turning depends on attraction-alignment.* Carrillo et al. [41] derived a Boltzmann-type model corresponding to the Cucker-Smale individual-based model for bird flocking (see also Eq. (1.3)). Recall that the Cucker-Smale model assumed that particles/birds adjusted their velocities to a weighted average of the relative velocity, with the weights being decreasing functions of the distance between birds. The kinetic version of the Cucker-Smale model is given by the following Boltzmann-type equation

$$\frac{\partial u(\mathbf{x}, \mathbf{v}, t)}{\partial t} + \mathbf{v} \cdot \nabla_{\mathbf{x}} u(\mathbf{x}, \mathbf{v}, t) = Q(u, u)(\mathbf{x}, \mathbf{v}, t), \quad \mathbf{x} \in \mathbb{R}^d, \ \mathbf{v} \in \mathbb{R}^d, \ d \geq 1,$$

$$(6.20)$$

with the collision operator

$$Q(u, u)(\mathbf{x}, \mathbf{v}) = \int_{\mathbb{R}^d} \int_{\mathbb{R}^d} \left(\frac{1}{J} u(\mathbf{x}, \mathbf{v}_*) u(\mathbf{y}, \mathbf{w}_*) - u(\mathbf{x}, \mathbf{v}) u(\mathbf{y}, \mathbf{w}) \right) d\mathbf{w} d\mathbf{y}.$$
(6.21)

Here (v_*, w_*) are the pre-collision velocities and (v, w) are the post-collision velocities for two birds with positions and velocities $(\mathbf{x}, \mathbf{v})$ and $(\mathbf{y}, \mathbf{w})$. These velocities are related through the following formulas:

$$v = (1 - \eta a(\mathbf{x} - \mathbf{y}))v^* + \eta a(\mathbf{x} - \mathbf{y})w^*, \tag{6.22a}$$

$$w = \eta a(\mathbf{x} - \mathbf{y})v^* + (1 - \eta a(\mathbf{x} - \mathbf{y}))w^*, \tag{6.22b}$$

where η measures the intensity of the velocity change in the binary interactions, and the weighting term models the communication distance:

$$a(\mathbf{x}) = \frac{1}{(1 + |\mathbf{x}|^2)^\beta}. \tag{6.23}$$

Finally, the term J in (6.21) is the Jacobian $J = (1 - 2\gamma a)^d$ of the transformation of (v^*, w^*) into (v, w). Note that the velocity adaptation term has an alignment effect, while the weight function a induces an attraction effect. We emphasise that no simulations have been performed in [41], and thus there was no investigation of the possible patterns displayed by this kinetic model.

We conclude the discussion of these attraction-alignment models, by mentioning that a slightly different version of a Cucker-Smale kinetic model was derived in [52], where the changes in velocity were described in terms of the velocity gradient for the collision operator $Q(u, u)$;

$$\frac{\partial u(\mathbf{x}, \mathbf{v}, t)}{\partial t} + \mathbf{v} \cdot \nabla_{\mathbf{x}} u(\mathbf{x}, \mathbf{v}, t) + \lambda \nabla_{\mathbf{v}} \cdot Q(u, u) = 0, \tag{6.24a}$$

$$Q(u, u)(\mathbf{x}, \mathbf{v}, t) = \int_{\mathbf{R}^2} a(\mathbf{x}, \mathbf{y})(\mathbf{v}_* - \mathbf{v}) u(\mathbf{x}, \mathbf{v}, t) u(\mathbf{y}, \mathbf{v}_* t) d\mathbf{v}_* d\mathbf{y}, \tag{6.24b}$$

where the interaction kernel is similar to the one described above in Eq. (6.23): $a(\mathbf{x}, \mathbf{y}) = A/(1 + |\mathbf{x} - \mathbf{y}|^2)^\beta$. Ha and Tadmor [52] showed the global existence of classical solutions and time-asymptotic flocking behaviour (where particles' velocities approach the velocity of the centre of mass) for compactly supported initial data. We will return to this kinetic model in Sect. 6.7, in the context of its derivation from an individual-based model (the Cucker-Smale model) via the mean-field approach..

- *Turning depends on attraction-repulsion-alignment.* In an attempt to generalise the 1D hyperbolic models introduced in [53, 54] (see also Eqs. (5.14) + (5.18) + (5.19)), Fetecau [8] derived a new 2D kinetic model that incorporated all three types of social interactions (repulsion, attraction and alignment), which were assumed to influence (in an additive manner similar to Eq. (5.19)) the turning rate $\lambda(\mathbf{x}, \phi, t)$ and the turning kernel $T(\phi', \phi)$:

$$\frac{\partial u}{\partial t} + e_\phi \cdot \nabla_{\mathbf{x}} = -\lambda(\mathbf{x}, \phi)u + \int_{-\pi}^{\pi} T(\mathbf{x}, \phi', \phi)u(\mathbf{x}, \phi', t)d\phi'. \qquad (6.25)$$

Note that Eq. (6.25) considers changes in orientation and not velocity—hence the description of T in terms of the turning angles ϕ' and ϕ. The turning rate λ (which models turning from direction ϕ into any other direction) and the turning kernel T (which describes turning from any direction ϕ' into direction ϕ) are related through the following formula

$$\lambda(\mathbf{x}, \phi) = \int_{-\pi}^{\pi} T(\mathbf{x}, \phi, \phi')d\phi'. \qquad (6.26)$$

As mentioned above, the three social interactions (repulsion, attraction and alignment) influence the turning behaviour of particles, and thus

$$T(\phi', \phi) = T_{al}(\phi', \phi) + T_a(\phi', \phi) + T_r(\phi', \phi). \qquad (6.27)$$

Here T_j, $j \in \{al, a, r\}$, are turning kernels modelling reorientation as a result of alignment (T_{al}), attractive (T_a), and repulsive (T_r) interactions. These kernels are described by the following equations [8]:

$$T_{al}(\phi', \phi) = q_{al} \int_{\mathbb{R}^2} \int_{-\pi}^{\pi} K_{al}^d(\mathbf{x}\text{-}\mathbf{s})K_{al}^o(\theta; \phi')w_{al}(\phi' - \phi, \phi' - \theta)u(\mathbf{s}, \theta, t)d\theta d\mathbf{s},$$
$$(6.28a)$$

$$T_a(\phi', \phi) = q_a \int_{\mathbb{R}^2} \int_{-\pi}^{\pi} K_a^d(\mathbf{x}\text{-}\mathbf{s})K_a^o(\mathbf{s}; \mathbf{x}, \phi')w_a(\phi' - \phi, \phi' - \psi)u(\mathbf{s}, \theta, t)d\theta d\mathbf{s},$$
$$(6.28b)$$

$$T_r(\phi', \phi) = q_r \int_{\mathbb{R}^2} \int_{-\pi}^{\pi} K_r^d(\mathbf{x}\text{-}\mathbf{s})K_r^o(\mathbf{s}; \mathbf{x}, \phi')w_r(\phi' - \phi, \phi' - \psi)u(\mathbf{s}, \theta, t)d\theta d\mathbf{s}.$$
$$(6.28c)$$

Here, kernels K_j^d, $j \in \{al, a, r\}$, model the distance of a reference individual (at x) from its neighbours positioned (at s) inside the alignment, attractive, and repulsive ranges. Kernels K_j^o, $j \in \{al, a, r\}$, describe the re-orientation of the individuals facing in direction ϕ' following interactions with neighbours facing in direction θ, and which are inside the alignment, attractive and repulsive ranges. Note that the alignment orientation kernel K_{al}^o depends on the direction θ of neighbours at position $\mathbf{s}$, while the attractive/repulsive orientation kernels $K_{r,a}^o$ depend on the spatial distance between the individuals at $\mathbf{x}$ and neighbours at $\mathbf{s}$.

Finally, w_j, $j \in \{al, a, r\}$, gives the probability of turning from ϕ' to ϕ, as a result of interactions with neighbours located at or moving in direction θ (or ψ=angle between the x-axis and the vector connecting the individuals with positions $\mathbf{x}$ and $\mathbf{s}$). Since λ and T are connected through Eq. (6.26), we note that also the turning rates λ are described in terms of the three social interactions:

$$\lambda(\mathbf{x}, \phi) = \lambda_{al}(\mathbf{x}, \phi) + \lambda_a(\mathbf{x}, \phi) + \lambda_r(\mathbf{x}, \phi), \tag{6.29}$$

with

$$\lambda_{al}(\mathbf{x}, \phi) = q_{al} \int_{\mathbb{R}^2} \int_{-\pi}^{\pi} K_{al}^d(\mathbf{x}\text{-}\mathbf{s}) K_{al}^o(\theta; \phi) u(\mathbf{s}, \theta, t) d\theta d\mathbf{s}, \tag{6.30a}$$

$$\lambda_j(\mathbf{x}, \phi) = q_j \int_{\mathbb{R}^2} \int_{-\pi}^{\pi} K_j^d(\mathbf{x}\text{-}\mathbf{s}) K_j^o(\mathbf{s}; \mathbf{x}, \phi) u(\mathbf{s}, \theta, t) d\theta d\mathbf{s}, \quad j = r, a. \tag{6.30b}$$

For detailed descriptions of the kernels K_j^d and K_j^o, and the re-orientation probabilities w_j, $j = r, al, a$, we refer the reader to [8]. Note the similarity between Eqs. (6.28) and the one-dimensional alignment, attractive, and repulsive terms described in Table 5.1. Because the model in [8] incorporates the repulsive assumption (i.e., individuals require a minimal space around them), it is more realistic than the classical Boltzmann models that assume compressible, infinitely small molecules.

This particular model was shown to produce stationary and moving aggregations (equivalent to the stationary and travelling pulses in 1D; see also Fig. 6.2b) and translating solutions (i.e., spatially homogeneous solutions that are aligned in a preferred direction - the equivalent of 1D case of individuals aligned either left or right, while being homogeneously spread over the domain [53]).

Many of these kinetic models have been investigated numerically. For example, Vauchelet [55] showed that a kinetic model for chemotaxis can exhibit finite size aggregations (stationary pulses), as well as finite-time blow-up patterns. These blow-up patterns can be point-wise or along some lines, depending on the shape of the domain. As mentioned above, Fetecau [8] showed that the kinetic model (6.25)–(6.30) with repulsive-attractive-alignment interactions can display stationary and moving aggregating solutions, as well as spatially-translating solutions.

However, due to the complexity of these kinetic models, an analytical investigation of the resulting patterns is challenging. Perthame [56] summarised some general analytical methods that are used to investigate the kinetic equations (6.14). These methods range from existence and regularity results, to asymptotic problems and the derivation of macroscopic models. In terms of understanding the patterns, the analytical results go as far as proving the existence of bounded or blow-up solutions [55, 57–59]. For simpler models (e.g., models that assume an uniform distribution of velocities after the turning event, or an uniform distribution through space) it is still possible to investigate analytically the structure of solutions.

For example, Schwetlick [32] used a kinetic model with population dynamics to investigate travelling front solutions for three different reaction terms $F(u)$, with $F(0) = F(1) = 0$:

1. a positive source term (e.g., logistic growth): $F > 0$ on $(0, 1)$;
2. a combustion-law-type term: there exists a $\theta \in (0, 1)$ such that $F \equiv 0$ on $[0, \theta]$ and $F > 0$ on $(\theta, 1)$;
3. a bistable growth term (Allee term): there exists a $\theta \in (0, 1)$ such that $F < 0$ on $(0, \theta)$ and $F > 0$ on $(\theta, 1)$.

The results showed that for positive source terms there is an entire interval of possible speeds. For the other two reaction terms, there are single front solutions that move with unique speeds.

Another simple kinetic model that was investigated analytically, was considered in [46, 48] to describe the angular alignment of interacting bundles of cells or filaments:

$$\frac{\partial u}{\partial t} = -u(\theta, t) \int_{-\pi}^{\pi} T[u](\theta, \phi)d\phi + \int_{-\pi}^{\pi} T[u](\phi, \theta)u(\phi, t)d\phi, \qquad (6.31)$$

with T the turning operator. Note that this particular model did not consider any spatial distribution of cells/filaments (i.e., cells/filaments were assumed to be equally spread over the whole domain). While the study in [46] showed analytically and numerically that the model could exhibit stable peak-like solutions (in the orientation space), the study by Kang et al. [48] showed analytically that uni-directional or multi-directional alignment behaviours could result from different initial distributions for the turning angle of cells/filaments (e.g., symmetric or non-symmetric initial data). Moreover, the study in [46] argued against the possibility of having time-periodic solutions bifurcating from stationary solutions (since there were no complex eigenvalues near the constant solutions). We emphasise that the uni-directional alignment behaviour exhibited by the model (6.31) corresponds numerically to the case depicted in Fig. 6.2c.

Remark 6.1 While this monograph is dedicated to the modelling of spatial and spatio-temporal phenomena occurring in various biological aggregations, it is important to emphasise the great applicability of the class of models discussed here. To this end, we note that in recent years Boltzmann-type equations have been used to describe opinion formation and wealth distribution [60–66]. As an example, Düring et al. [65] used a kinetic model to describe the irrationality and herding behaviour of agents in economic markets. The authors defined $u(x, w, t)$ to be the distribution of agents at time t, with rational ($x > 0$) or irrational ($x < 0$) behaviours, which hold an asset of an estimated value $w > 0$ that can change in response to new public information (e.g., financial reports, balance sheet numbers, etc.) [65]. The evolution of this distribution is governed by the following Boltzmann-type equation

$$\frac{\partial u}{\partial t} + \frac{\partial}{\partial x}\big(\Phi(x, w)u\big) = Q_I(u) + Q_H(u, u), \quad (x, w) \in \mathbb{R} \times \mathbb{R}^+, \qquad (6.32)$$

where the drift $\Phi(x, w)$ describes the effect that changes in the asset values have on the rationality/irrationality of trading agents, the operator Q_I models public information, and the operator Q_H models the herding behaviour. (For detailed descriptions of Q_I and Q_H, see [65].) We can interpret this herding behaviour of financial agents as a "collective behaviour" not in the physical space but in a socio-economic space. In addition to taking a grazing collision limit (see the approach described in Sect. 6.9), the authors in [65] also performed numerical simulations for different parameter values measuring, for example, the influence level of public information on the agent's decision.

6.5   Kinetic Models for Active Particles

The kinetic models described previously incorporate detailed information (i.e., position, velocity and movement direction) regarding the individual behaviour of organisms (or particles in general). Also, they generally ignore population dynamics (i.e., death and birth processes). However, complex biological systems have specific attributes that characterise them, such as the dominance behaviour in a group of animals [67], the activation of a population of cells, or gene expression in a population of genes [68]. To account for these complex attributes, new kinetic models have been derived and used to investigate a large variety of biological processes, ranging from the social behaviour of colonies of insects [67], to the evolution of bacterial populations [69], or the evolution of cancer-immune interactions (see the reviews in [68, 70], and the references therein). In all these models the particles/cells are characterised (in addition to position and velocity) by internal biological variables called "activities" [68, 70]. Examples of such "activities" are the degree of recognition of cancer cells by the antigen-presenting cells [71], or the degree of activation of immune cells [72]. In general, the turning kernel T and the random turning rates λ might depend on cells' activity. Usually, these kinetic models also incorporate population dynamics: cell proliferation and death, and in some cases genetic mutations. Thus, the most general equation describing the dynamics of cell populations is

$$\frac{\partial}{\partial t}u(\mathbf{x}, t, \mathbf{v}) + H(t)\cdot\nabla_{\mathbf{v}}u(\mathbf{x}, t, \mathbf{v}) + \mathbf{v}\cdot\nabla_{\mathbf{x}}u(\mathbf{x}, t, \mathbf{v}) = L[u] + G[u] + I[u]. \quad (6.33)$$

It is clear that Eq. (6.33) is the generalisation of (6.14). Here, u is usually a vector ($u = (u_1, \ldots u_n)$) which describes various cell populations (e.g., immune cells, normal endothelial cells, malignant cells [68]). Hence, (6.33) is a system of kinetic equations. Variable $\mathbf{v}$, which previously denoted the "velocity", can now describe also the "activity" state of the cell: $\mathbf{v} = (s, a)$, where $s = velocity$ and $a = (a_1, \ldots a_m)$ denotes m "activity" states. Note that $\mathbf{v}$ can be a continuous variable (if $m = 1$, as in [68]) or a discrete variable (if $m > 1$, as in [73]). The term $H(t)\cdot\nabla_{\mathbf{v}}u$ describes the change in the density u over the activity space. Note that many of the kinetic models which investigate the evolution of cell populations do not consider how these populations change over space (i.e., $\nabla_{\mathbf{x}} u = 0$; see, for example, [68]).

In Eq. (6.33), $L[u]$ is the classical transport operator (similar to the operator describing the right-hand-side of Eq. (6.14)). This term incorporates a constant λ which models the frequency of interactions (or turning rates) (see also (6.14)). $G[u]$ is a conservative operator describing the gain and loss of cells due to conservative encounters (i.e., encounters that modify the activity state a of cells, without generating or destroying cells). Finally, $I[u]$ is an operator describing the proliferation/destruction of cells. For a detailed description of $L[u]$, $G[u]$ and $I[u]$ see, for example, the model in [70], or the model in [16] which will be discussed in more detail in Sect. 6.6.

Remark 6.2 Note that the operators above incorporate the assumption of binary cell-cell interactions [74]. This assumption, which is not very realistic biologically, is being made to simplify the modelling of complex interactions among cells and/or among animals (and to be consistent with the kinetic theory of gasses). However, one should not forget that these cell-cell interactions occur not only via direct cell-cell contact but also indirectly via cytokines and growth factors produced by the cells themselves or by other cells, which leads to interactions among multiple cells at the same time (see [75] for a discussion on modelling tumour-immune interactions). Similarly, animal interactions are not always direct and one-to-one: in many cases animal-animal interactions are the result of different communication mechanisms (e.g., using hearing, olfactive or visual signals), which can be perceived/emitted by various and multiple neighbours (e.g., following eavesdropping [76]); see also Fig. 1.6a. Hence, a cell/animal does not interact with only one other cell/animal but with a number of other cells/animals (within a certain distance).

In the following we present an example of a kinetic model for tumour-immune interactions described in [77], which even if it is in 1D (for the activity variable a) can be easily generalised to 2D (i.e., $a = (a_1, a_2)$) or even to higher dimensions. Starting with Eq. (6.33) for $u = (u_1, u_2)$ (with $u_1 \in \mathbb{R}^+$ describing tumour cells and $u_2 \in \mathbb{R}^+$ describing immune cells), the authors derived the following model:

$$\frac{\partial u_1(t, a)}{\partial t} + \frac{\partial}{\partial a}\left(-\alpha_{12} a A_2(t) u_1(t, a) + \alpha_{13} a A_3^0 u_1(t, a) \right) = -\beta_{12} A_2(t) u_1(t, a)$$
$$+ \beta_{13} A_3^0 a u_1(t, a), \tag{6.34a}$$

$$\frac{\partial u_2(t, a)}{\partial t} + \frac{\partial}{\partial t}\left(-\alpha_{21} a A_1(t) u_2(t, a) \right) = \beta_{21} a A_1(t) u_2(t, a)$$
$$- \lambda\left(u_2(t, u) - u_2^*(a) \right). \tag{6.34b}$$

Here $t \in \mathbb{R}^+$ represents time, and $a \in \mathbb{R}^+$ is an internal state which describes the mutation level of tumour cells (u_1) and the activation level of immune cells (u_2). The nonlocal terms $A_i(t) = \int_0^\infty a u_1(t, a) da$, $i = 1, 2$ describe the activation level of the two populations, while A_3^0 is a constant describing the (healthy) environmental cells. Parameters α_{ij} describe conservative encounters that lead to the weakening of the immune response in the presence of tumour cells (α_{21}), the reduced progression (mutation) of tumour cells in the presence of immune cells (α_{12}), and the increased progression of tumour cells in the presence of healthy environment that has nutrients (α_{13}). Parameters β_{ij} describe proliferative/destructive interactions that increase/decrease cell numbers: elimination (at a rate β_{12}) of tumour cells by the activated immune cells, growth of tumour cells (at a rate β_{13}) in the presence of a healthy environments which feeds them, proliferation of immune cells (at a rate β_{21}) in the presence of mutated tumour cells. Finally, the last term in Eq. (6.34b) describes the tendency of the immune response to relax (at a rate λ) towards a healthy immune state $u_2^*(a)$.

The authors have proven the existence of continuous solutions $u_{1,2}$, and showed analytically that the immune response can go to the healthy state u_2^* provided that tumour mutation level can be controlled, or the tumour grows to infinity if the activation level of immune cells is not very high. No numerical simulations were shown in this study, to confirm the analytical results. However, simulations for a relatively similar model were shown in [78], which depicted travelling pulses (for tumour and immune cells) through the activity space $a \in (0, 1)$ (thus describing the progression of tumour cells and the activation of immune cells).

Remark 6.3 As noted above, the kinetic models for active particles usually focus on multiple interacting populations (i.e., $u = (u_1, \ldots, u_n)$). Since in many cases these models do not consider any spatial heterogeneity, the interactions are usually local in space (but can be non-local in the activity variable). In Sect. 6.10 we will return to this aspect of interacting populations, and consider nonlocal spatial interactions between different populations.

Remark 6.4 Some of the kinetic models for particle swarming and pedestrian movement can be derived from individual-based models, in the case when the number of particles becomes very large (see, Sect. 6.7). In contrast to this approach, the kinetic models which investigate tumour growth as a result of interactions with the immune cells (and the environment) are usually derived using the active particles approach where the flow at time t into the elementary volume $[a, a + da]$ of the state space (with a = activity variable) is the result of cell transport, and conservative, proliferative and destructive cell-cell interactions [74]. These two approaches, namely the macroscopic limit of individual-based models and the active particles approach, mirror the derivation of one-dimensional hyperbolic models, which can be obtained using either a correlated random walk approach [24, 25], or a traffic flow approach (see, for example, Chapter 2 in [18]).

6.6 Multiscale Models with Explicit Dynamics for the Microscale Variables

The kinetic equations discussed in this chapter can be considered multiscale models, since they combine macroscopic-level dynamics related to cell/animal movement with microscopic-level dynamics related to variables describing the internal state of cells/animals (e.g., age, activation state of cells, number of receptors on cell surface, etc.). However, the models presented in the previous sections usually assumed a static microscopic variable (e.g., $\mathbf{v} = (s, a)$), which did not evolve in time (and hence the multiscale aspect could be considered implicit). Recent studies on collective cell dynamics in the context of cancer invasion and metastasis have started to pay particular attention to the temporal dynamics of cells and molecules inside/on cells, and formulate detailed complex models that connect explicitly the kinetic transport equations for cell movement, with ordinary differential equations or integro-differential equations for the dynamics of molecules (i.e., the internal variables that characterise the cells), and even reaction-diffusion equations for the dynamics of various diffusive chemical compounds that control the movement of cells and/or various molecules; see [3, 5, 6, 16, 20, 79] and references therein.

To illustrate these multiscale models with explicit microscale dynamics, we start with a simple model introduced by Dolak and Schmeiser [20] to describe the evolution of the density $u(\mathbf{x}, t, \mathbf{v}, \xi)$ of a cell population (that depends on space $\mathbf{x} \in \mathbb{R}^n$, time $t \in \mathbb{R}^+$, velocity $\mathbf{v} \in V \subset \mathbb{R}^n$ and an internal variable $\xi \in \mathbb{R}^k$). The components of the internal variable ξ are chemical concentrations inside the cells, which can depend on the concentration of an external chemoattractant $S(\mathbf{x}, t)$:

$$\frac{d\xi}{dt} = \eta\big(\xi, S(t, \mathbf{x}(t))\big). \tag{6.35}$$

The evolution of the external chemoattractant is described by the following reaction-diffusion equation, where the production/degradation of the chemical depends on cell densities and the inner states of the cells:

$$\frac{\partial S}{\partial t} = D_s \Delta_{\mathbf{x}} S + v(S, \xi, u). \tag{6.36}$$

Under the assumption that cell movement is not influenced by the internal variables (i.e., the turning operator T is independent of ξ), Dolak and Schmeiser [20] proposed the following equation for the evolution of $u(\mathbf{x}, t, v, \xi)$:

$$\frac{\partial u}{\partial t} + \mathbf{v} \cdot \nabla_{\mathbf{x}} u + \nabla_\xi \cdot (\eta u) = \int_V \Big[T(\mathbf{v}', \mathbf{v}) u(\mathbf{x}, t, \mathbf{v}', \xi) - T(\mathbf{v}, \mathbf{v}') u(\mathbf{x}, t, \mathbf{v}, \xi) \Big] d\mathbf{v}'. \tag{6.37}$$

The authors then derived a macroscopic limit of this kinetic model (where cell movement would be described by a drift-diffusion equation), and applied it to

investigate the dynamics of *Dictyostelium discoideum* cells under the influence of
an external cAMP concentration (modelled by $S(\mathbf{x}, t)$) and two internal chemicals:
a chemical w activated by cAMP and inhibited by a second internal chemical ξ.
Numerical simulations with this macroscopic model for Dictyostelium dynamics
showed the formation of cellular aggregations (i.e., stationary pulses in 2D) in
response to cAMP waves. Note that for the simulations, the authors used an operator
splitting approach (see also Chap. 7), where the reaction terms and the diffusion
terms were discretised separately.

The last few years have seen the development of more complex multiscale
models that incorporate explicitly the evolution of the microscale (internal) variable [3, 16, 19]. To illustrate these complex modelling approaches we focus on a
study by Kelkel and Surulescu [19]. In this article, the authors proposed a multiscale
kinetic model for the dynamics of a cancer cell population $u(t, \mathbf{x}, \mathbf{v}, \mathbf{y})$ that depends
on time $t \in \mathbb{R}^+$, space $\mathbf{x} \in \mathbb{R}^n$, velocity $\mathbf{v} \in V \subset \mathbb{R}^n$, and activity $\mathbf{y} \in \mathbb{R}^2$—
which is related to the concentration of some chemoattractant L that results from
the degradation of extracellular matrix, and to the density of some protein fibres Q
oriented towards $\theta \in \mathbb{S}^{n-1}$ (with $\mathbb{S}^{n-1}$ denoting the unit sphere in $\mathbb{R}^n$):

$$\frac{\partial u}{\partial t} + \mathbf{v} \cdot \nabla_{\mathbf{x}} u + \nabla_{\mathbf{y}} \cdot \left(G(y, Q, L) u \right) = H(u, Q) + C(L, u). \qquad (6.38)$$

Here $H(u, Q)$ is a haptotaxis operator describing the changes in velocity orientation
due to the encounter between cells and fibres of the extracellular matrix:

$$H(u, Q)(t, \mathbf{x}, \mathbf{v}, \mathbf{y})$$

$$= \int_{\mathbf{v} \in V} \int_{\theta \in \mathbb{S}^{n-1}} p_h(t, \mathbf{x}, \mathbf{v}', \mathbf{y}) \psi(\mathbf{v}; \mathbf{v}', \theta) u(t, \mathbf{x}, \mathbf{v}', \mathbf{y}) Q(t, \mathbf{x}, \theta) d\theta d\mathbf{v}'$$

$$- p_h(t, \mathbf{x}, \mathbf{v}, \mathbf{y}) u(t, \mathbf{x}, \mathbf{v}, \mathbf{y}) \int_{\mathbf{v} \in V} \int_{\theta \in \mathbb{S}^{n-1}} \psi(\mathbf{v}'; \mathbf{v}, \theta) Q(t, \mathbf{x}, \theta d\theta d\mathbf{v}'), \quad (6.39)$$

where the velocity space is $V = [s_1, s_2] \times \mathbb{S}^{n-1}$ with $0 \leq s_1 < s_2 < \infty$, p_h
denotes the interaction frequency between cells and fibres, and $\psi(\mathbf{v}; \mathbf{v}', \theta)$ denotes
the probability of a cell that has an initial velocity $\mathbf{v}'$ to change its velocity to $\mathbf{v}$ after
encountering a fibre with orientation θ.

The chemotactic operator $C(u, L)$, describes the changes in velocity orientation
due to cells moving towards the gradient of chemical L:

$$C(u, L)(t, \mathbf{x}, \mathbf{v}, \mathbf{y}) = \int_{\mathbf{v} \in V} p_c(t, \mathbf{x}, \mathbf{v}', \mathbf{y}) K[\nabla L](\mathbf{v}, \mathbf{v}') u(t, \mathbf{x}, \mathbf{v}', \mathbf{y}) d\mathbf{v}'$$

$$- p_c(t, \mathbf{x}, \mathbf{v}, \mathbf{y}) u(t, \mathbf{x}, \mathbf{v}, \mathbf{y}). \qquad (6.40)$$

Here p_c denotes the interaction frequency between cells and the gradient of L, while
turning kernel K depends on ∇L.

Finally, the **y**-divergence term is related to cell membrane reactions, where the vector $\mathbf{y} = (y_1, y_2) \in (0, R_0) \times (0, R_0)$ is such that $y_1 + y_2 < R_0$ (with $R_0 > 0$ representing the maximum concentration of receptors on cell surface). This condition leads to the definition of the space $Y = \{(y_1, y_2) | y_1 + y_2 < R_0\}$, which will be used to define Q and L (see below). Function G that appears in the **y**-divergence term in (6.38) was derived by applying the law of mass action to describe the receptor-fibre and receptor-chemical interactions occurring at cell membrane [19]:

$$\frac{d\mathbf{y}}{dt} = G(\mathbf{y}, Q, L), \quad \text{with } G(\mathbf{y}, Q, L) = \begin{pmatrix} k_1(R_0 - y_1 - y_2)Q - k_{-1}y_1 \\ k_2(R_0 - y_1 - y_2)L - k_{-2}y_2 \end{pmatrix}.$$

$$(6.41)$$

Here k_1 and k_{-1} are the binding and unbinding rates of the complexes formed between the number of receptors on a cell surface and the total density of protein fibres (Q). Similarly, k_2 and k_{-2} are the binding and unbinding rates of the complexes formed between cell receptors and the chemical L.

Finally, the evolution of the variables Q and L is given by the following two equations:

$$\frac{\partial Q}{\partial t} = -\kappa \left(\int_{v \in V} \int_{y \in Y} \left(1 - |\theta \cdot \frac{v}{|v|}|\right) u \, dv \, dy \right) Q \qquad (6.42)$$

$$\frac{\partial L}{\partial t} = \kappa \int_{\mathbb{S}^{n-1}} \left(\int_{v \in V} \int_{y \in Y} \left(1 - |\theta \cdot \frac{v}{|v|}|\right) u \, dv \, dy \right) Q \, d\theta - r_L L + D_L \Delta_x L, \quad (6.43)$$

with κ the degradation rate of matrix fibres.

Kelkel and Surulescu [19] then showed the existence of a unique weak solution for the above system. We emphasise that no numerical simulations were presented in [19] to exemplify the patterns that could be exhibited by this complex kinetic model.

The kinetic (mesoscale) models discussed throughout this chapter incorporate microscopic features into the state variables. However, in the limit to macroscopic models, these microscopic features are lost through averaging. Picolli and Tosin [80] and Cristiani et al. [81] introduced a novel approach to modelling crowd dynamics by deriving a new class of multiscale models that preserve the microscopic dynamics. These models are given in terms of a time-evolving measure μ_t defined on the Borel σ-algebra $B(\mathbb{R}^d)$ (for a definition of a Borel σ-algebra, see any classical textbook on measure theory; e.g., [82]). For $E \in B(\mathbb{R}^d)$, $\mu_t(E) \geq 0$ represents the mass of pedestrians contained in E, for $t \geq 0$. Using this measure, the authors derived two microscopic and macroscopic models, which were then combined in a multiscale model:

- **Microscopic model:** For a population of N pedestrians with positions $P_j(t)$, $j = 1, \dots, N$, the measure μ_t is defined as the sum of Dirac masses centred at each position P_j:

$$\mu_t = \sum_{j=1}^{N} \delta_{P_j(t)}, \quad j = 1, .., N. \tag{6.44}$$

The microscopic model is a system of N coupled equations for the evolution of the positions P_j:

$$\frac{dP_j}{dt} = v[\mu_t](P_j(t)), \quad \text{with } v[\mu_t](x) = v_{desired}(x) + v[\mu_t](x). \tag{6.45}$$

Here, $v_{desired} : \mathbb{R}^d \to \mathbb{R}^d$ is the desired velocity for pedestrians in the absence of mutual interactions, and $v[\mu_t] : \mathbb{R}^d \to \mathbb{R}^d$ is the interaction velocity that results from inter-individual interactions. This interaction velocity incorporates non-local interactions with neighbours:

$$v[\mu_t](P_j) = \sum_{k=1,...,N, P_k \neq P_j} f(|P_k - P_j|)g(\alpha_{kj})\frac{P_k - P_j}{|P_k - P_j|}, \tag{6.46}$$

with function $f : \mathbb{R}^+ \to \mathbb{R}$, $f(x) = -(q_r/x)\chi_{[0,s_r]}(x) + q_a x \chi_{[0,s_a]}(x)$ describing distance-related inter-individual interactions (where $q_{r,a}$ are the magnitudes of repulsive and attractive interactions, and $s_{r,a}$ are the radii for repulsive and attractive interactions). Also, $\alpha_{kj} \in [-\pi, \pi]$ is the angle between vectors $P_k - P_j$ and $v_{desired}(P_j)$, and the function $g : [-\pi, \pi] \to [0, 1]$, $g(s) = \chi_{|s| \leq \alpha}(s)$, describes the angular focus of the pedestrian at P_j.

- **Macroscopic model:** Since the matter is continuous, the measure μ_t is absolutely continuous with respect to the d-dimensional Lebesgue measure $\mathcal{L}^d$. By Radon-Nikodym's theorem, there exists a function $\rho(x, t)$ called the density of μ_t with respect to the measure $\mathcal{L}^d$, given by $d\mu_t = \rho(\cdot, t)d\mathcal{L}^d$. The strong-form equation for the conservation of the density ρ is

$$\frac{\partial \rho}{\partial t} + \nabla \cdot (\rho v) = 0. \tag{6.47}$$

The weak-form equivalent of this conservation equation is

$$\frac{d}{dt} \int_{\mathbb{R}^d} \rho(x, t)\phi(x)dx = \int_{\mathbb{R}^d} \rho(x, t)v(x, t) \cdot \nabla\phi(x)dx, \quad \text{for } \phi \in C_0^1(\mathbb{R}^d), \tag{6.48}$$

with the interaction velocity

$$v[\mu_t](x) = \int_{\mathbb{R}^d} f(|y - x|)g(\alpha_{xy})\frac{y - x}{|y - x|}\rho(y, t)dy. \tag{6.49}$$

- **Multiscale model:** The multiscale model is a linear combination of the microscopic and macroscopic models:

$$
\frac{d}{dt}\left(\theta\sum_{j=1}^{N}\phi(P_j(t))+(1-\theta)\int_{\mathbb{R}^d}\rho(x,t)\phi(x)dx\right)=
$$

$$
\theta\sum_{j=1}^{N}v(t,P_j(t))\cdot\nabla\phi(P_j(t))+(1-\theta)\int_{\mathbb{R}^d}\rho(x,t)v(x,t)\cdot\nabla\phi(x)ds, \quad (6.50)
$$

with $\phi\in C_0^1(\mathbb{R}^d)$ and the interaction velocity

$$
v[\mu_t](x)=\theta\sum_{k=1,\dots,N,P_k(t)\neq x}f(|P_k(t)x|)g(\alpha_{xP_k(t)})\frac{P_k(t)-x}{|P_k(t)-x|}
$$

$$
+(1-\theta)\int_{\mathbb{R}^d}f(|y-x|)g(\alpha_{xy})\frac{y-x}{|y-x|}\rho(y,t)dy. \quad (6.51)
$$

The parameter $\theta\in[0,1]$ weights the coupling between the microscopic and macroscopic scales: $\theta=0$ corresponds to a macroscopic model, while $\theta=1$ corresponds to a microscopic model. Therefore, the first term in (6.51) gives the microscopic contribution to the macroscopic dynamics [81]. For the numerical simulations of this model, the authors used an algorithm that discretised both the microscopic part (by updating the pedestrian positions P_j) and the macroscopic part (by updating the discrete values of the density ρ_i, with i the index of a grid cell E_i). The numerical results showed macroscopic aggregations that have also microscopic structures (e.g., small areas around microscopic individuals, which are caused by inter-individual repulsion; Fig. 6.4a), and macroscopic travelling pulses that move towards a microscopic leader (which has a pre-assigned velocity; Fig. 6.4b) [81]. The model can also exhibit lane formation before and after a bottleneck (Fig. 6.4c), as well as clogging at the bottleneck (Fig. 6.4d).

6.7 Derivation of Mean-Field Models

Over the past few years, more and more researchers focused on the derivation of kinetic models from individual-based models for swarming, while attempting to connect the microscale dynamics of individual-based models to the macroscale group-level dynamics of continuum models [41, 51, 52, 83–87]. The transition from microscale to macroscale models can occur through the derivation of mean-field models, which describe the averaged effect of all other particles on any given particle. This approach is taken when the interaction potential between particles is not sensitive to the position of particles, an assumption that is sometimes called

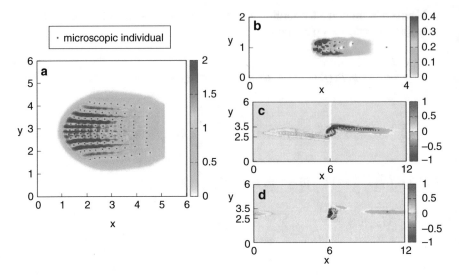

Fig. 6.4 2D spatio-temporal patterns exhibited by the multi-scale model (6.50) (**a**) Macroscopic aggregations containing microscopic individuals; (**b**) Travelling pulses following a microscopic leader; (**c**) Lane formation near a bottleneck; (**d**) Bottleneck clogging. For (**c**), (**d**): blue describes the density of right-moving pedestrians, while yellow describes the density of left-moving pedestrians. Reproduced from [81]. Copyright ©2011 Society for Industrial and Applied Mathematics. Reprinted with permission. All rights reserved

"long-range interaction". Ha and Tadmor [52] started with a N-particle Cucker-Smale model (1.3) and, for large numbers of particles, they derived a mean-field kinetic model similar to the Vlasov model. To this end, they considered the N-particle distribution function

$$u^N = u^N(x_1, v_1, \ldots x_i, v_i, \ldots, x_N, v_N, t), \quad (x_i, v_i) \in \mathbb{R} \times \mathbb{R}^d, \qquad (6.52)$$

and the marginal distribution

$$u^N(x_1, v_1, t) = \int_{\mathbb{R}^{2d(N-1)}} u^N(x_1, v_1, x_-, v_-, t) dx_- dv_-,$$

$$(x_-, v_-) = (x_2, v_2, \ldots, x_N, v_N). \qquad (6.53)$$

The derivation of the kinetic analogue of the Cucker-Smale model was based on the Liouville equation

$$\frac{\partial u^N}{\partial t} + \sum_{i=1}^{N} v_i \cdot \nabla_{x_i} u^N + \frac{\lambda}{N} \sum_{i=1}^{N} \nabla_{v_i} \cdot \left(\sum_{j=1}^{N} a(x_i, x_j)(v_j - v_i) u^N \right) = 0. \qquad (6.54)$$

Integrating (6.54) with respect to $dx_- dv_- = dx_2 dv_2 \ldots dx_N dv_N$ leads to

$$\int_{\mathbb{R}^{2d(N-1)}} \sum_{i=1}^{N} v_i \nabla_{x_1} u^N dx_- dv_- = v_1 \cdot \nabla_{x_1} u^N(x_1, v_1), \qquad (6.55)$$

and

$$\frac{\lambda}{N} \sum_{i=1}^{N} \int_{\mathbb{R}^{2d(N-1)}} \sum_{j=1}^{N} \nabla_{v_i} \cdot \left(a(x_i, x_j)(v_j - v_i) u^N \right) dx_- dv_-$$

$$= \frac{\lambda}{N} \int_{\mathbb{R}^{2d(N-1)}} \sum_{j=2}^{N} \nabla_{v_1} \cdot \left(a(x_1, x_j)(v_j - v_1) u^N \right) dx_- dv_-. \qquad (6.56)$$

Because all particles are identical, the probability density function u^N is symmetrical

$$u^N(\ldots, x_i, v_i, \ldots, x_j, v_j, \ldots) = u^N(\ldots, x_j, v_j, \ldots, x_i, v_i, \ldots), \qquad (6.57)$$

and thus Ha and Tadmor [52] considered only the case $x_j = x_2$:

$$\frac{\lambda}{N} \int_{\mathbb{R}^{2d(N-1)}} \sum_{j=2}^{N} \nabla_{v_1} \cdot \left(a(x_1, x_j)(v_j - v_1) u^N \right) dx_- dv_-$$

$$= \frac{\lambda}{N}(N-1) \int_{\mathbb{R}^{2d(N-1)}} a(x_1, x_2) \nabla_{v_1} \cdot \left((v_2 - v_1) u^N \right) dx_2 dv_2 \ldots dx_N dv_N$$

$$= (\lambda - \frac{\lambda}{N}) \nabla_{v_1} \cdot \left(\int_{\mathbb{R}^{2d}} a(x_1, x_2)(v_2 - v_1) w^N(x_1, v_1, x_2, v_2, t) dx_2 dv_2 \right), \qquad (6.58a)$$

with

$$w^N(x_1, v_1, x_2, v_2, t) = \int_{\mathbb{R}^{2d(N-2)}} u^N dx_3 dv_3 \ldots dx_N dv_N. \qquad (6.59)$$

Therefore, the marginal distribution $u^N(x_1, v_2, t)$ satisfies

$$\frac{\partial u^N}{\partial t} + v_1 \cdot \nabla_{x_1} u^N + \left(\lambda - \frac{\lambda}{N} \right) \nabla_{v_1} \cdot \left(\int_{\mathbb{R}^{2d}} a(x_1, x_2)(v_2 - v_1) w^N dx_2 dv_2 \right) = 0, \qquad (6.60)$$

which in the limit $N \to \infty$ becomes

$$\frac{\partial u}{\partial t} + v_1 \cdot \nabla_{x_1} u + \lambda \nabla_{v_1} \cdot \left(\int_{\mathbb{R}^{2d}} a(x_1, x_2)(v_2 - v_1)w dx_2 dv_2 \right) = 0, \qquad (6.61)$$

with $u = \lim_{N \to \infty} u^N(x_1, v_1)$, $w = \lim_{N \to \infty} w^N(x_1, v_1, x_2, v_2)$. To close this equation, the authors made the assumption that

$$w(x_1, v_1, x_2, v_2, t) = u(x_1, v_1, t)u(x_2, v_2, t). \qquad (6.62)$$

After relabelling $(x_1, v_1) \to (x, v)$ and $(x_2, v_2) \to (x^*, v^*)$, it can be seen that the distribution function $u(x, v, t)$ satisfies the Vlasov-type mean field model

$$\frac{\partial u}{\partial t} + v \cdot \nabla_x u + \lambda \nabla_v \cdot Q(u, u) = 0, \quad \text{with} \qquad (6.63a)$$

$$Q(u, u)(x, v, t) = \int_{\mathbb{R}^2} a(x, x^*)(v^* - v)u(x, v, t)u(x^*, v^*, t)dv^* dx^*. \qquad (6.63b)$$

It has been shown in [52] that as long as the initial data u_0 has finite mass, then (6.63) cannot exhibit finite-time blow-up solutions. In fact, the only possible solutions are the bounded, classical solutions $u \in C^1(\mathbb{R}^{2d} \times [0, T))$ for any $T \in (0, \infty)$.

Chuang et al.[85] started with the individual-based model (1.9) introduced in [88], and further derived a macroscopic model in the limit of large numbers of particles N. To this end, the authors used the approach in [89] and defined a probability density function $u = u(x_1, \ldots, x_N; p_1, \ldots, p_N; t)$ that depends on the positions x_i and momentums p_i, $i = 1, \ldots, N$. They also defined the mass density $\rho(x, t)$, the velocity field $v(x, t)$ and the continuum interaction force $F_U(x, t)$ as follows

$$\rho(x, t) = m \sum_{i=1}^{N} \langle \delta(x_i - x); u \rangle, \qquad (6.64a)$$

$$v(x, t) = \frac{p(x, t)}{\rho(x, t)} = \frac{\sum_{i=1}^{N} \langle p_i \delta(x_i - x); u \rangle}{\rho(x, t)}, \qquad (6.64b)$$

$$F_U(x, t) = \sum_{i=1}^{N} \langle -\nabla_{x_i} U(x_i) \delta(x_i - x); u \rangle. \qquad (6.64c)$$

Here δ is the Dirac delta function, $m = m_i$ is the identical mass of particles and $U(x_i)$ is the attractive-repulsive interaction potential introduced in (1.9), which acts on the particle at x_i. Substituting these terms into the Liouville equation, leads to the following continuum system:

$$\frac{\partial \rho}{\partial t} + \nabla \cdot (\rho v) = 0, \tag{6.65a}$$

$$\frac{\partial (\rho v)}{\partial t} + \nabla \cdot (\rho v v) = \alpha \rho v - 2\beta E_K v + F_U, \tag{6.65b}$$

with

$$F_U(x, t) = \int_{\mathbb{R}^d} -\nabla_x U(x - y) \frac{1}{m^2} \rho(x, t) \rho(y, t) dy. \tag{6.66}$$

Here, $E_K = \rho |v|^2 / 2$ describes the kinetic energy, and U is the interaction potential (see Eq. (1.10)). To compare this continuum model with the original discrete model (1.9), Chuang et al. [85] discretised it using the Lax-Friedrichs scheme. The simulations showed similar profiles for the averaged densities, averaged momentum and averaged tangential velocities, for both the continuum and discrete models. Moreover, both models displayed clump patterns (see also Fig. 6.2f) and lane formation patterns (see also Fig. 6.2d). However, these result hold only in the catastrophic regime. In the parameter regime of H-stability, the individual-based model displays compactly supported solutions, while the continuum model displays a uniform density distribution spread over the entire domain.

Bolley et al. [86] derived a mean-field limit for the following stochastic individual-based model:

$$dx_i = v_i dt, \tag{6.67a}$$

$$dv_i = \sqrt{2} dB_i - F(x_i, v_i) dt - (H \star u)(x_i, v_i) dt, \tag{6.67b}$$

where B_i are N independent Brownian motions in $\mathbb{R}^d$ for particles i ($i = 1, \ldots, N$), and the functions $F, H : \mathbb{R}^{2d} \to \mathbb{R}$ are chosen such that in the absence of noise ($B_i \equiv 0$) the individual-based model (6.67) can be reduced either to the D'Orsogna model or to the Cucker-Smale model discussed in Chap. 1.

- in D'Orsogna model (1.9): $F(x, v) = (\beta |v|^2 - \alpha) v$ and $H(x, v) = -\nabla_x U(x)$;
- in Cucker-Smale model (1.3): $F(x, v) = 0$ and $H(x, v) = a(x)$.

The mean-field continuum model corresponding to (6.67) is given by (in the distributional sense)

$$\frac{\partial u}{\partial t} + v \cdot \nabla_x u - \nabla_v \cdot (u F_0[u]) = \Delta_v u, \quad \text{with } x, v \in \mathbb{R}^d. \tag{6.68}$$

The left-hand side of this equation is a transport operator, which describes the movement of individuals that have velocity v and acceleration $F_0[u] = F + H \star u$ (that results from nonlocal interactions with neighbours). The right-hand side is a velocity diffusion term, which results from the noise in the velocity in Eqs. (6.67). (Note that this term leads to solutions with smooth velocity profiles.) For these stochastic and limit continuum models, Bolley et al. [86] investigated only the existence and uniqueness of solutions; no numerical simulations were performed to investigate the types of patterns exhibited by this mean-field continuum model.

To conclude the discussion of mean-field models derived as limits of individual-based models, we note that Bolley et al. [87] focused on a stochastic version of the classical Vicsek model (1.1) and derived a continuous version of it (defined in terms of the gradient with respect to space, and in terms of the gradient, divergence and Laplace operator with respect to the velocity variable; relatively similar to Eq. (6.68)). A continuum limit of a different version of the Vicsek model (in 3D with Gaussian noise added to the velocity, in contrast to the classical 2D Vicsek model with uniformly distributed noise), as well as the limit of a stochastic version of this new version of the Vicsek model, have been derived in [90]. Again, all these models are investigated only analytically (in terms of the existence of solutions). The authors do not perform numerical simulations to check whether the patterns obtained with the discrete and continuum models are similar, or to investigate the types of patterns exhibited by these models.

6.8   Analytical Approaches for the Investigation of Patterns: Hyperbolic (and Parabolic) Scaling

In general, the kinetic (mesoscopic) models are difficult to be investigated analytically and numerically, due to their complexity generated by the presence of the microscopic variable. One way of simplifying these models (and their analysis) is to reduce them to macroscopic models of hyperbolic and parabolic types. The kinetic (mesoscopic) models (6.14) and (6.33) can be reduced to hyperbolic (macroscopic) models by taking the velocity moments of the distribution function u:

$$\bar{u}(x, t) = \int_{v \in V} u(x, v, t)dv, \qquad (6.69a)$$

$$(p_0 \bar{u})(x, t) = \int_{v \in V} u(x, v, t)v dv, \qquad (6.69b)$$

where $\bar{u}(x, t)$ is the density of particles/individuals/cells and p_0 is the mean velocity. Because of higher order moments, this approach does not lead to a closed model for $\bar{u}$ and p_0. This requires the use of moment closure methods, which connect the distribution u to functions $\bar{u}$ and p_0 [69, 91, 92]. For example, Degond et al. [92] considered two closure methods, namely the monokinetic closure $(u(x, v, t) = \bar{u}(x, t)\delta_{p_0(x,t)}(v))$ and the von Misses-Fisher closure $(u(x, v, t) = \bar{u}(x, t)M_{p_0(x,t)}(v)$, with $M_{p_0} = (1/C)exp(\beta(v \cdot p_0/|p_0|)))$, to transform a Fokker-Planck type equation of the form (6.68) into a macroscopic equation. Another approach involves the use of hyperbolic (hydrodynamic) scaling methods [20, 49, 56, 92, 93]. In particular, by rescaling the variables x and t such that $\xi = \epsilon x$ and $\tau = \epsilon t$ (the "hyperbolic scaling"), one can transforms the kinetic equations (6.14) and (6.33) into hyperbolic ones [68, 93–95] (following expansions in powers of ϵ for the variables in the model):

$$\frac{\partial}{\partial t}\bar{u} + \nabla_x(p_0\bar{u}) = 0, \tag{6.70a}$$

$$\frac{\partial}{\partial t}(p_0\bar{u}) + \nabla_x(p_0\bar{u} \otimes p_0 + P) = F(\bar{u}). \tag{6.70b}$$

where P is a pressure tensor which measures the statistical variation in velocity,

$$P = \int_V (w - p_0) \otimes (w - p_0)U(w)dw, \tag{6.71}$$

and $U(v)$ is the equilibrium distribution of u. The term $F(\bar{u})$ which appears in Eq. (6.70b) is a nonlinear operator described in terms of the turning operator T [93]. If the turning operator models only turning in the absence of any stimuli (e.g., a chemical substance), then $F(\bar{u}) = 0$.

In the context of kinetic models that describe the self-organised dynamics of tumour and immune cell populations, this hyperbolic scaling can lead to hyperbolic systems without source terms [70],

$$\frac{\partial}{\partial t}\bar{u} + \nabla_x(p_0\bar{u}) = 0, \tag{6.72a}$$

$$\frac{\partial}{\partial t}(p_0\bar{u}) + \nabla_x(p_0\bar{u} \otimes p_0 + P) = 0, \tag{6.72b}$$

or hyperbolic systems with source terms related to the conservative and proliferative cell interactions [70]

$$\frac{\partial}{\partial t}\bar{u} + \nabla_x(p_0\bar{u}) = \int G(U)(x, t, \mathbf{v})d\mathbf{v} + \int I[U](x, t, \mathbf{v})d\mathbf{v}, \tag{6.73a}$$

$$\frac{\partial}{\partial t}(p_0 u) + \nabla_x(p_0 u \otimes p_o + P) = \int \mathbf{v}G[U](x, t, \mathbf{v})d\mathbf{v} + \int \mathbf{v}I[U](x, t, \mathbf{v})d\mathbf{v}. \tag{6.73b}$$

Operators $G[U]$ and $I[U]$ describe the conservative and the proliferative/destructive cell interactions, respectively (see Eq. (6.33)). The absence or presence of source terms depends on the scaling of the interaction frequency (λ) and the scaling of the proliferative/destructive interactions ($I[U]$) [70].

The kinetic models (6.14) and (6.33) have been investigated both numerically and analytically, using stability [85], bifurcation [85, 96, 97] and asymptotic techniques

[70, 96]. The results have suggested, for example, conditions on various parameters (e.g., parameters describing conservative interactions among cells [97]) that could lead to the control and elimination of aggregations of tumour cells. They have also suggested conditions that are necessary for the formation of certain group patterns. As an example, [85] showed that the hydrodynamic framework does not allow for double milling patterns (groups formed of individuals rotating in both directions), due to velocity averaging inside a mesh cell. These hydrodynamic models can exhibit only single milling patterns (i.e., groups formed of individuals rotating in one direction; also called vortices). However, double milling patterns are possible with the kinetic framework [84]. Moreover, the hydrodynamic models can exhibit stationary pulses [93], or blow-up patterns for spherically symmetric solutions [98]. They can also exhibit patches (i.e., stationary pulses) and lanes of aggregated individuals [85].

Remark 6.5 Note that a different scaling for the variables x and t (that is, $\xi = \epsilon x$ and $\tau = \epsilon^2 t$, also called the "parabolic scaling") leads to parabolic equations [91, 99].

Remark 6.6 The parameter $\epsilon \ll 1$ describes the ratio between the microscopic length scale of interactions, and the macroscopic scale of the observation domain. For the hydrodynamic scaling, ϵ is also equal to the ratio of the microscopic time scale and the macroscopic observation time [100]. For more detailed discussions of the hydrodynamic limits see, for example, [51, 101] (for hydrodynamic limits of the Boltzmann equation), or [102] (for limits of particle models and more varied kinetic models).

The two asymptotic methods mentioned here (involving hyperbolic and parabolic scalings), have been derived to connect the *mesoscopic*-level kinetic models (which describe the evolution of the distribution function for particles identified by position, velocity and/or activity) to the *macroscopic*-level equations (either parabolic or hyperbolic) for the density of particles. The choice of hydrodynamic vs. parabolic limit could be decided by the phenomenological behaviour of the model. For example, one should choose a hydrodynamic limit if velocity and direction are important features of the phenomenon that has to be modelled. Also, the hydrodynamic limit is a natural choice when one wants to preserve the interactions among particles [70]. The parabolic limit, on the other hand, is preferred when there are collisions among particles [70].

6.9   Analytical Approaches for the Investigation of Patterns: Grazing Collision Limit

The *grazing collisions*, that is collisions with very small deviation angles, were first investigated for Boltzmann operators in the context of physics problems [103–105]. The grazing collision limit transforms the Boltzmann operator into a Fokker-Planck operator, the latter one being more easy to investigate analytically. In the context of kinetic equations for swarming, the grazing collision limit was applied, for example, by Carrillo et al. [41] to a Boltzmann-type model that incorporates the collisional rules of the Cucker-Smale individual-based model (1.3):

$$\frac{\partial u(x, v, t)}{\partial t} + v \cdot \nabla_x u(x, v, t) = Q(u, u)(x, v, t), \tag{6.74a}$$

$$Q(u, u)(x, v, t) = \int_{\mathbb{R}^d} \int_{\mathbb{R}^d} \left(\frac{1}{J} u(x, v_*) u(y, w_*) - u(x, v) u(y, w) \right) dw \, dy, \tag{6.74b}$$

with (v_*, w_*) the pre-collisional velocities and (v, w) the post-collisional velocities, which are related through

$$v^* = (1 - \gamma a(x - y)) v + \gamma a(x - y) w, \tag{6.75a}$$

$$w^* = \gamma a(x - y) v + (1 - \gamma a(x - y)), \qquad a(x) = \frac{1}{(1 + |x|^2)^\beta}, \quad x \in \mathbb{R}^d. \tag{6.75b}$$

Here, γ measures the intensity of the velocity change in the particle-particle interactions. Finally, $J = (1 - 2\gamma a)^d$ is the Jacobian matrix associated with the transformation of (v, w) into (v^*, w^*). For $\gamma < 1/2$ this Jacobian is positive. To avoid the presence of J, Carrillo et al. [41] considered the weak formulation of (6.74):

$$\frac{\partial}{\partial t} \int_{\mathbb{R}^{2d}} \phi(x, v) u(x, v, t) dv dx + \int_{\mathbb{R}^{2d}} (v \cdot \nabla_x \phi(x, v)) u(x, v, t) dv dx = \tag{6.76}$$

$$\int_{\mathbb{R}^{4d}} (\phi(x, v^*) - \phi(x, v)) u(x, v, t) u(y, w, t) dv dx dw dy, \tag{6.77}$$

for all smooth functions ϕ with compact support that satisfy

$$\lim_{t \to 0^+} \int_{\mathbb{R}^{2d}} \phi(x, v) u(x, v, t) = \int_{\mathbb{R}^{2d}} \phi(x, v) u_0(x, v). \tag{6.78}$$

By assuming that $\gamma \ll 1$, one can expand $\phi(x, v^*)$ in Taylor series of $v^* - v$. Keeping only the second-order terms in the collision integral leads to

$$\int_{\mathbb{R}^{4d}} (\phi(x, v^*) - \phi(x, v)) u(x, v, t) u(y, w, t) ds dv dy dw =$$

$$\gamma \int_{\mathbb{R}^{4d}} (\nabla_v \phi(x, v) \cdot (w - v)) a(x - y) u(x, v, t) u(y, w, t) dx dv dy dw$$

$$+ \frac{\gamma^2}{2} \int_{\mathbb{R}^{4d}} \left(\sum_{i,j=1}^d \frac{\partial^2 \phi(x, \bar{v})}{\partial v_i^2} (w_j - v_j)^2 \right) a(x - y)^2 u(x, v) u(y, w) dx dv dy dw. \tag{6.79}$$

Here, $\bar{v} = \theta v + (1 - \theta)v^*$, for $0 \leq \theta \leq 1$. For small γ, the Boltzmann operator $Q(u, u)$ (6.74b) can be approximated by the dissipative operator

$$\nabla_v \cdot \Big(u(x, v, t)\big(H(x)\nabla_v W(v) \star u\big)(x, v, t)\Big). \qquad (6.80)$$

McNamara and Young [105] showed that this operator preserves the same properties as the initial Boltzmann operator. Therefore, the approximated equation in the strong formulation now reads

$$\frac{\partial u}{\partial t} + v \cdot \nabla_x u = \nabla_v \cdot \big(\psi(u)(x, v, t)u(x, v, t)\big), \qquad (6.81)$$

with

$$\psi(u)(x, v, t) = \Big(\big(H(x)\nabla_v W(v)\big) \star u\Big)(x, v, t)$$

$$= \int_{\mathbb{R}^{2d}} \frac{v - w}{(1 + |x - y|^2)^\beta} u(y, w, t)dydw. \qquad (6.82)$$

This is similar to the Vlasov-type equation (6.63) that was derived by Ha and Tadmor [52] via the mean-field limit of the Cucker-Smale model.

In the context of nonlocal kinetic models for animal aggregations, Carrillo et al. [106] focused on a slightly different version of the model introduced by Fetecau [8]:

$$\frac{\partial u}{\partial t} + \gamma e_\phi \nabla_{\mathbf{x}} u = -Q^-[u] + Q^+[u, u], \qquad (6.83)$$

$$\text{with } \ Q^-[u] = Q_r^-[u] + Q_a^-[u] + Q_{al}^-[u], \ \ Q^+[u, u]$$

$$= Q_r^+[u, u] + Q_a^+[u, u] + Q_{al}^+[u, u],$$

$$\text{and } \ Q_j^-[u] = \lambda_j(x, \phi)u, \ \ Q_j^+[u, u]$$

$$= \int_{-\pi}^{\pi} T_j(x, \phi', \phi)u(x, \phi', t)d\phi'), \ \ j = r, a, al,$$

where the nonlocal turning operators contain also a constant (density-independent) term:

$$T_{al}(\mathbf{x}, \phi', \phi) = \frac{\eta_{al}}{2\pi}$$

$$+ \lambda_2 q_{al} \int_{-\pi}^{\pi} \int_{\mathbb{R}^2} K_{al}^o(\mathbf{x} - \mathbf{s}) K_{al}^o(\theta, \phi')\omega_{al}(\phi' - \phi, \phi' - \theta)u(s, \theta, t)dsd\theta,$$

$$T_{r,a}(\mathbf{x}, \phi', \phi) = \frac{\eta_{r,a}}{2\pi}$$

$$+ \lambda_2 q_{r,a} \int_{-\pi}^{\pi} \int_{\mathbb{R}^2} K_{r,a}^d(\mathbf{x} - \mathbf{s}) K_{r,a}^o(\mathbf{s}, \mathbf{x}, \phi')\omega_{r,a}(\phi' - \phi, \phi' - \psi)u(\mathbf{s}, \theta, t)dsd\theta.$$

Since the grazing collision assumes that individuals turn only a small angle upon interactions with neighbours (e.g., moving in direction θ), this leads to a re-scaling of the probability of re-orientation term:

$$\omega_j^\epsilon(\phi - \phi', \phi - \theta) = \frac{1}{\epsilon}g_\epsilon\Big(\frac{\phi - \phi' - \epsilon R(\phi - \theta)}{\epsilon}\Big), \quad j = r, al, a. \tag{6.84}$$

Substituting $\phi' = \phi - \epsilon\beta - \epsilon R(\phi - \theta)$ into the last term (i.e., $\psi(\phi')$) of the weak formulation of the collision operator (shown here only for alignment):

$$\int_{-\phi}^{\phi} Q_{al}[u]\psi(\phi)d\phi = \eta_{al}\int_{-\pi}^{\pi}\Big(\frac{\rho(x,t)}{2\pi} - u(x,\phi,t)\Big)\psi(\phi)d\phi$$

$$+ \int_{-\pi}^{\pi}\int_{-\pi}^{\pi}\int_{\mathbb{R}^2}\Big[\lambda_2 q_{al}K_{al}^d(x-s)K_{al}^o(\theta,\phi)u(x,\phi,t)u(s,\theta,t)$$

$$\Big[\int_{-\pi}^{\pi}\omega_{al}^\epsilon(\phi-\phi')[\psi(\phi')-\psi(\phi)d\phi']\Big]\Big]dsd\theta d\phi, \tag{6.85}$$

and expanding the term in Taylor series about ϕ, leads to the following approximation:

$$\int_{-\pi}^{\pi} Q_{al}[u]\psi(\phi)d\phi = \eta_{al}\int_{-\pi}^{\pi}\Big(\frac{\rho(x,t)}{2\pi} - u(x,\phi,t)\Big)\psi(\phi)d\phi$$

$$- \int_{-\pi}^{\pi}\frac{\partial}{\partial\phi}\big(u(x,\phi,t)C_{al}^\epsilon[u,x,\phi]\big)\psi(\phi)d\phi$$

$$+ \int_{-\pi}^{\pi}\big(u(x,\phi,t)D_{al}^\epsilon[u,x,\phi]\big)\psi(\phi)d\phi. \tag{6.86}$$

Here,

$$C_{al}^\epsilon[u,x,\phi] = \int_{-\pi}^{\pi}\int_{\mathbb{R}^2}\lambda_2 q_{al}K_{al}^d(x-s)K_{al}^o(\theta,\phi)A_{al}^\epsilon(\phi-\theta)u(s,\theta,t)d\theta ds,$$

$$D_{al}^\epsilon[u,x,\phi]$$

$$= \int_{-\pi}^{\pi}\int_{\mathbb{R}^2}\lambda_2 q_{al}K_{al}^d(x-s)K_{al}^o(\theta,\phi)K_{al}^o(\theta,\phi)B_{al}^\epsilon(\phi-\theta)u(s,\theta,t)d\theta ds,$$

$$A_{al}^\epsilon(\phi-\theta) = -\epsilon\big(M_1(\epsilon) + M_0(\epsilon)R(\phi-\theta)\big),$$

$$B_{al}^\epsilon(\phi-\theta) = \frac{\epsilon^2}{2}\big(M_2(\epsilon) + 2M_1(\epsilon)R(\phi-\theta) + M_0(\epsilon)R(\phi-\theta)^2\big), \tag{6.87}$$

with $M_n(\epsilon) = \int_{-\pi}^{\pi}\beta^n g_\epsilon(\beta)d\beta$, $n = 0, 1, 2$, the moment generating functions for $g_\epsilon(\beta)$. Similar expressions can be derived also for the attractive and repulsive collision terms [106]:

$$\int_{-\pi}^{\pi} Q_{r,a}[u]\psi(\phi)d\phi = \eta_{r,a}\int_{-\pi}^{\pi}\left(\frac{\rho(x,t)}{2\pi} - u(x,\phi,t)\right)\psi(\phi)d\phi$$

$$-\int_{-\pi}^{\pi}\frac{\partial}{\partial}\left(u(x,\phi,t)C_{r,a}^{\epsilon}[u,x,\phi]\right)\psi(\phi)d\phi$$

$$+\int_{-\pi}^{\pi}\frac{\partial^2}{\partial\phi^2}\left(u(x,\phi,t)D_{r,a}^{\epsilon}[u,x,\phi]\right)\psi(\phi)d\phi,$$

with

$$C_{r,a}^{\epsilon}[u,x,\phi] = \int_{-\pi}^{\pi}\int_{\mathbb{R}^2}\lambda_2 q_{r,a}K_{r,a}^d(x-s)K_{r,a}^o(s,x,\phi)A_{r,a}^{\epsilon}(s,x,\phi)u(s,\theta,t)dsd\theta,$$

$$D_{r,a}^{\epsilon}[u,x,\phi] = \int_{-\pi}^{\pi}\int_{\mathbb{R}^2}\lambda_2 q_{r,a}K_{r,a}^d(x-s)K_{r,a}^o(s,x,\phi)B_{r,a}^{\epsilon}(s,x,\phi)u(s,\theta,t)dsd\theta,$$

$$A_{r,a}^{\epsilon}(s,x,\phi) = -\epsilon\left(M_1(\epsilon) + M_0(\epsilon)R(\phi-\psi_s)\right),$$

$$B_{r,a}^{\epsilon}(s,x,\phi) = \frac{\epsilon^2}{2}\left(M_2(\epsilon) + 2M_1(\epsilon)R(\phi-\psi_s) + M_0(\epsilon)R(\phi-\psi_s)^2\right). \tag{6.88}$$

Carrillo et al. [106] then showed that the grazing collision limit of model (6.83) with the nonlocal turning operators defined above leads to the following Fokker-Planck equation with non-local advective and diffusive terms in the orientation space:

$$\frac{\partial u}{\partial t} + \gamma e_\phi \cdot \nabla_x u = \lambda_1\left(\frac{1}{2\pi}\rho(x,t) - u(x,\phi,t)\right)$$

$$+\frac{\partial}{\partial\phi}\left(-uC^{\epsilon}[u,x,\phi] + \frac{\partial}{\partial\phi}(uD^{\epsilon}[u,x,\phi])\right), \tag{6.89}$$

with $\lambda_1 = \eta_{al} + \eta_r + \eta_a$, $C^{\epsilon} = C_{al}^{\epsilon} + C_r^{\epsilon} + C_a^{\epsilon}$ and $D^{\epsilon} = D_{al}^{\epsilon} + D_r^{\epsilon} + D_a^{\epsilon}$. Note that this nonlocal model (6.89), with nonlocal diffusion, is in contrast with the majority of the Fokker-Planck models in the literature for self-organised aggregations that consider local diffusion [86, 87, 107]. While these Fokker-Planck models (with local/nonlocal diffusion) are more easy to investigate analytically, there are not many studies that focus also on their numerical investigation (see also the discussion in Sect. 6.7).

6.10 Multiple Population Models with Nonlocal Interactions

As seen in the previous chapters, focusing on interactions between multiple populations is a natural step in the modelling of population dynamics. In particular, local (in space) interactions between different populations are usually considered by the kinetic models for active particles (see Sect. 6.5). However, for kinetic multi-dimensional models that vary in space, modelling the interactions between different populations is complicated by the particular form of the Boltzmann operator (which can depend on both space and orientation/velocity). For example, Fetecau and Meskas [9] derived a system of kinetic equations modelling the predator-prey interactions between two populations, $u^P(\mathbf{x}, \phi, t)$ (the prey) and $u^h(\mathbf{x}, \phi, t)$ (the hunter/predator):

$$\frac{\partial u^P}{\partial t} + \gamma^P e_\phi \cdot \nabla_{\mathbf{x}} u^P = -\lambda^P(\mathbf{x}, \phi) u^P + \int_{-\pi}^{\pi} T^P(\mathbf{x}, \phi', \phi) u^P(\mathbf{x}, \phi', t) d\phi',$$

(6.90a)

$$\frac{\partial u^h}{\partial t} + \gamma^h e_\phi \cdot \nabla_{\mathbf{x}} u^h = -\lambda^h(\mathbf{x}, \phi) u^h + \int_{-\pi}^{\pi} T^h(\mathbf{x}, \phi', \phi) u^h(\mathbf{x}, \phi', t) d\phi',$$

(6.90b)

where the angle ϕ gives the orientation of the populations, $\gamma^{p,h}$ are the speeds of the two populations, $\lambda^{p,h}$ are the turning rates from direction ϕ to any other direction, and $T^{p,h}(\mathbf{x}, \phi', \phi)$ are the turning rates from any direction ϕ' to direction ϕ. While this model includes the same nonlocal attractive-repulsive-alignment interactions introduced in [8] (and described by (6.27)–(6.28)), the interactions in (6.90) are slightly different since: (i) they assume that individuals have a restricted field of vision and cannot receive information from behind (see Fig. 6.5b), and (ii) they model turning in response to predator-prey dynamics. The restricted field of vision is modelled by the multiplicative introduction of another kernel into $\lambda_j, T_j,$

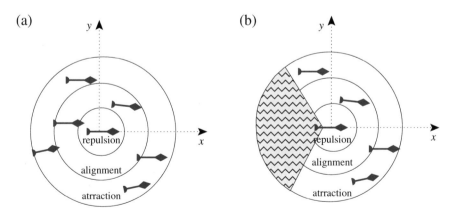

Fig. 6.5 Examples of 2D reception mechanisms. (**a**) The individual at the centre of $x - y$ axes can receive information from all around it (i.e., from ahead, lateral and behind). (**b**) The individual at the centre of $x - y$ axes cannot receive information from behind it (they have a restricted field of vision depicted by the pink region)

$j = r, al, a$, which restricts the contribution of the interaction kernels [9]:

$$K^{bz}(\phi - \psi) = \frac{1}{B}\left(\frac{1}{2}\tanh\left(a(cos(\phi - \psi) + (1 - \frac{b}{\pi}))\right)\right),\tag{6.91}$$

where a determines the "steepness" of the field of vision (i.e., how fast does vision decrease near the edges of the field of vision), b determines the width of the field of vision, and constant B is chosen to normalise the kernel. For numerical simulations the authors represented the space and angle integrals as convolutions, and used Fourier spectral methods to evaluate them. The simulations with prey alone showed [108]: (i) stationary (circular/oval) aggregations (similar to those in Fig. 1.2e); (ii) transient milling aggregations (similar to those in Fig. 1.2a), and (iii) travelling aggregations (similar to those in Fig. 1.2c). The numerical simulations with prey and predators showed: (i') prey engulfment by a stationary predator ring (similar to the pattern in Fig. 1.5a); (ii') splitting and merging of prey aggregations (when prey tries to escape predators), as well as of predator aggregations (when predators split after the prey split); (iii') separated travelling pulse aggregations, where the predator aggregation follows the prey aggregation (similar to the pattern in Fig. 1.5a).

Returning to the advection-only models, we mentioned in Sect. 6.3 that they can be easily generalised to include the interactions between multiple populations. For example, Lécureux-Mercier [35] modelled panic situations by considering two sub-populations moving in two different directions $\mathbf{v}_1$ and $\mathbf{v}_2$:

$$\frac{\partial u_1}{\partial t} + \nabla\left(\Gamma(K_1 \star u_1 + K_2 \star u_2)u_1\mathbf{v}_1(\mathbf{x})\right) = 0,\tag{6.92a}$$

$$\frac{\partial u_2}{\partial t} + \nabla\left(\Gamma(K_1 \star u_1 + K_2 \star u_2)u_2\mathbf{v}_2(\mathbf{x})\right) = 0.\tag{6.92b}$$

Using Kružkov's theory (see Chap. 2), it was shown in [35] that this model has a unique weak entropy solution. Moreover, numerical simulations have shown the formation of lanes (in a 2D domain) formed of left-moving and right-moving sub-populations. In the long term, these lanes have been shown to display fingering behaviours, where left-moving and right-moving sub-populations segregate while they move next to each other.

Another type of interaction modelled by Lécureux-Mercier [35] was the leader-follower dynamics, where the author specified the vector $\mathbf{u}_l \in \mathbb{R}^2$ for the position of the leader, and the density u of the followers:

$$\frac{\partial u}{\partial t} + \nabla\left(\Gamma(u)u(\mathbf{u}_l(t) - \mathbf{x})e^{-||\mathbf{u}_l - \mathbf{x}||}\right) = 0,\tag{6.93a}$$

$$\frac{d\mathbf{u}_l}{dt} = (1 + (K \star u)(\mathbf{u}_l))\boldsymbol{\psi}(t),\tag{6.93b}$$

where $\Gamma(u)$ describes followers' speed, $\mathbf{u}_l - \mathbf{x}$ describes the direction of a follower located at $\mathbf{x}$ and directed towards the leader, and $\boldsymbol{\psi}(t)$ is the direction of the leader.

Numerical simulations for these coupled PDE-ODE models (where the PDEs are discretised with the Lax-Friedrichs scheme and the ODEs are discretised with the Euler scheme) showed splitting groups and travelling groups [109].

6.11 Explicit Stochastic Models

Stochasticity is implicit in higher dimensional kinetic models, as the collisions between particles/cells/bacteria follow the classical rules in the kinetic theory of gases. As discussed at the beginning of this chapter, the derivation of these models using the velocity jump process approach [30] assumes that changes in the orientation/velocity of particles are usually random, being the result of Poisson processes of intensities λ (with $\lambda^{-1} =$ the mean run length between the random choices of direction [30]). Moreover, all the turning kernels in the models discussed above described the probabilities of a velocity/orientation change between two different velocities/directions, given that a re-orientation occurs. On the other hand, as we discussed in Chap. 1, the kinetic models for collective movement are valid for relatively large numbers of particles (in many situations they are being obtained from individual-based models when the number of particles $N \to \infty$). However, in biology, the realistic number of particles/cells/bacteria in a system could be relatively small (e.g., a few thousands of cells/bacteria/animals), and thus environmental fluctuations likely have a strong impact. In this case, the limit $N \to \infty$ might not lead to a biologically realistic approximation of system dynamics. One way to address this issue is to incorporate stochasticity in an explicit manner in these kinetic/hydrodynamic models. To this end, Chavanis [110–113] started with the Euler equations and the Keller-Segel equations, and proposed a class of kinetic and hyperbolic (hydrodynamic) stochastic models that describe bacterial chemotaxis:

$$\frac{\partial u}{\partial t} + \nabla \cdot (u\mathbf{v}) = 0, \tag{6.94a}$$

$$\frac{\partial u\mathbf{v}}{\partial t} + \nabla(u\mathbf{v} \otimes \mathbf{v}) = -D\xi\nabla u + u\nabla c - \xi u\mathbf{v} - \xi\sqrt{2Du}R(\mathbf{x}, t), \tag{6.94b}$$

$$\frac{\partial c(\mathbf{x}, t)}{\partial t} = -kc(\mathbf{x}, t) + D_c\Delta c(\mathbf{x}, t) + hu(\mathbf{x}, t). \tag{6.94c}$$

Here, u describes the density of bacteria, $\mathbf{v}$ is the local velocity, c is the concentration of a chemotactic signal, ξ is a friction coefficient (for the bacteria moving on a fixed matrigel), D is a diffusion coefficient and $R(\mathbf{x}, t)$ is a Gaussian random field satisfying $\langle R(\mathbf{x}, t)\rangle = 0$ and $\langle R^\alpha(\mathbf{x}, t)R^\beta(\mathbf{x}, t)\rangle = \delta_{\alpha\beta}\delta(\mathbf{x} - \mathbf{x}')\delta(t - t')$, where α and β refer to space coordinates and $\langle \cdot \rangle$ denotes the correlation function (i.e., the correlation between random variables at two different points in space and time). In the equation for the dynamics of the chemical, D_c is the diffusion coefficient, k is the degradation rate of the chemical, while h is the production rate of this chemical by the cells themselves. The author focused only on the derivation of the model as a limit of an individual-based stochastic Langevin model, and pattern formation was not investigated for the hyperbolic model. However, pattern formation was investigated for a simplified version of the stochastic Langevin model, and the simulations showed: (i) the formation of a network pattern similar to the pattern in Fig. 1.2i, and (ii) point-wise blow-up aggregations (eventually leading to Dirac peaks, similar to the pattern in Fig. 1.2f) [110].

A kinetic version of the stochastic model for bacterial chemotaxis, derived as a limit of an individual-based stochastic Langevin model, describes the evolution of a distribution function $f(\mathbf{x}, \mathbf{v}, t)$ and a chemical density $c(\mathbf{x}, t)$ [111]:

$$\frac{\partial f}{\partial t} + \mathbf{w} \cdot \frac{\partial f}{\partial \mathbf{x}} + \nabla c \cdot \frac{\partial f}{\partial \mathbf{w}} = \frac{\partial}{\partial \mathbf{w}} \cdot \left(D \frac{\partial f}{\partial \mathbf{w}} + \xi f \mathbf{w} \right) + \frac{\partial}{\partial \mathbf{w}} \cdot \left(\sqrt{2Df} \, Q(\mathbf{x}, \mathbf{w}, t) \right),$$

$$(6.95a)$$

$$\frac{\partial c}{\partial t} = D_c \Delta c - kc + h \int f(\mathbf{x}, \mathbf{w}, t) d\mathbf{w}, \qquad (6.95b)$$

In the absence of noise, this model reduces to a generalised Fokker-Planck equation. Moreover, this kinetic model can be reduced to the hyperbolic model (6.94) by taking the hydrodynamic moments of (6.95a) (see the approach in [110]), where one defines the macroscopic variables u and $\mathbf{v}$ in terms of the above kinetic variables as follows:

$$u(\mathbf{x}, t) = \int f(\mathbf{x}, \mathbf{w}, t) d\mathbf{w}, \quad \mathbf{v}(\mathbf{x}, t) = (1/u) \int f \mathbf{w} d\mathbf{w}, \qquad (6.96)$$

and the Gaussian noise $g(\mathbf{x}, t) = \int \sqrt{2Df} \, Q d\mathbf{w}$.

This class of hydrodynamic models was further generalised to include also nonlocal interaction, as shown in [112, 113]:

$$\frac{\partial u}{\partial t} + \nabla \cdot (u\mathbf{v}) = 0, \qquad (6.97a)$$

$$u\left(\frac{\partial \mathbf{v}}{\partial t} + (\mathbf{v} \cdot \nabla)\mathbf{v} \right) = -d\nabla u - u\nabla \Phi - \xi u \mathbf{v} - \sqrt{2Bu} \, R(\mathbf{x}, t), \qquad (6.97b)$$

$$\text{with } \Phi(\mathbf{x}, t) = \int \mathbf{v}(\mathbf{x} - \mathbf{x}')u(\mathbf{x}', t) d\mathbf{x}' \qquad (6.97c)$$

The authors investigated the evolution of small perturbations around a uniform distribution with $u(\mathbf{x}) = u$, $\Phi(\mathbf{x}) = \Phi$, and $\mathbf{v} = 0$, and identified conditions under which these perturbations grow or decay. For the parameter values that ensure stability, the authors further investigated the correlations of the density fluctuations and correlations of the velocity fluctuations in the presence of noise. However, no numerical simulations were performed to identify the patterns exhibited by this model [112].

For a more in-depth discussion (from a statistical mechanics perspective) of stochastic transport models with applications to vehicular, pedestrian or ant trail traffic, we refer the reader to [114].

References

1. D. Trucu, P. Lin, M. Chaplain, Y. Wang, Multiscale Model. Simul. SIAM Interdisciplinary J. **11**(1), 309 (2013)
2. P. Domschke, D. Trucu, A. Gerisch, M. Chaplain, J. Theor. Biol. **361**, 41 (2014)
3. C. Engwer, T. Hillen, M. Knappitsch, C. Surulescu, J. Math. Biol. **71**, 551 (2015)
4. B. Perthame, M. tang, N. Vauchelet, J. Math. Biol. **73**(5), 1161 (2016)
5. C. Engwer, C. Stinner, C. Surulescu, Math. Models Meth. Appl. Sci. **27**, 1355 (2017)
6. A. Hunt, C. Surulescu, Vietnam J. Math. **45**, 221 (2017)
7. C.M. Topaz, A.L. Bertozzi, SIAM J. Appl. Math. **65**, 152 (2004)
8. R. Fetecau, Math. Model. Method. Appl. Sci **21**(7), 1539 (2011)
9. R. Fetecau, J. Meskas, Swarm Intell. https://doi.org/10.1007/s11721-013-0084-9 (2013)
10. P. Degond, J.G. Liu, S. Motsch, V. Panferov, Methods Appl. Anal. **20**, 89 (2013)
11. P. Degond, A. Frouvelle, S. Merino-Aceituno, Math. Models Methods Appl. Sci. **27**(6), 1005 (2017)
12. Y. Xia, S. Wong, C.W. Shu, Phys. Rev. E **79**, 066113 (2009)
13. N. Bellomo, C. Dogbé, SIAM Rev. **53**, 409 (2011)
14. N. Bellomo, L. Gibelli, Math. Model Meth. Appl. Sci. **25**, 2417 (2015)
15. D. Burini, S.D. Lillo, L. Gibelli, Phys. Life Rev. **16**, 123 (2016)
16. J. Nieto, L. Urrutia, J. Math. Anal. Appl. **433**(2), 1055 (2017)
17. N. Bellomo, C. Dogbé, Math. Models Methods Appl. Sci. **18**, 1317 (2008)
18. R. LeVeque, *Numerical Methods for Conservation Laws* (Birkhäuser, Basel, 1992)
19. J. Kelkel, C. Surulescu, Math. Models Methods Appl. Sci. **22**(3), 1150017 (2012)
20. Y. Dolak, C. Schmeiser, J. Math. Biol. **51**, 595 (2005)
21. R. Hughes, Transport. Res. B **36**, 507 (2002)
22. V. Coscia, C. Canavesio, Math. Models Methods Appl. Sci. **18**, 1217 (2008)
23. M. Lighthill, G. Whitham, Proc. R. Soc. Lond. Ser. A. **229**(1178), 317 (1955)
24. M. Kac, Rocky Mountain J. Math. **4**, 497 (1974)
25. S. Goldstein, Quart. J. Mech. Appl. Math. **4**, 129 (1951)
26. K. Hadeler, *Mathematics Inspired by Biology. Lecture Notes in Mathematics* (Springer Verlag, 1999), chap. Reaction transport systems in biological modelling, pp. 95–150
27. J. Keller, Proc. Natl. Acad. Sci. USA **101**(5), 1120 (2004)
28. E. Codling, M. Plank, S. Benhamou, J. Roy. Soc. Interface **5**(25), 813 (2008)
29. D. Stroock, Probab. Theory Rel. Fields **28**, 305 (1974)
30. H.G. Othmer, S.R. Dunbar, W. Alt, J. Math. Biol. **26**, 263 (1988)
31. W. Alt, J. Math. Biol. **9**, 147 (1980)
32. H. Schwetlick, Annales de l'Institut Henri Poincare **17**(4), 523 (2000)
33. R. Colombo, M. Lécureux-Mercier, Acta Math. Sci. **32**(1), 177 (2012)
34. R. Colombo, M. Garavello, M. Lécureux-Mercier, Math. Models Methods Appl. Sci. **22**(4), 1150023 (2012)
35. M. Lécureux-Mercier, ESAIM Proc. **38**, 409 (2012)
36. A. Majda, A. Bertozzi, *Vorticity and Incompressible Flow* (Cambridge University Press, Cambridge, 2002)
37. A. Bertozzi, J. von Brecht, H. Sun, T. Kolokolnikov, D. Uminsky, Comm. Math. Sci. **13**(4), 955–985 (2015)
38. H. Berg, D. Brown, Nature **239**, 500 (1972)

39. T. Hillen, K. Hadeler, *Analysis and Numerics for conservation laws* (Springer, Berlin, Heidelberg, 2005), chap. Hyperbolic systems and transport equations in mathematical biology, pp. 257–279
40. D. Soll, D. Wessels, *Motion Analysis of Living Cells* (Wiley, New York, Toronto, 1998)
41. J. Carrillo, M. Fornasier, J. Rosado, G. Toscani, SIAM J. Math. Anal. **42**, 218 (2010)
42. C. Cercignani, *The Boltzmann Equation and Its Applications* (Springer, 1987)
43. S. Harris, *An Introduction to the Theory of the Boltzmann Equation* (Courier Corporation, 2012)
44. N. Bellomo, C. Bianca, M. Delitala, Phys. Life Rev. **6**, 144 (2009)
45. D. Helbing, Complex Syst. **6**, 391 (1992)
46. E. Geigant, K. Ladizhansky, A. Mogilner, SIAM J. Appl. Math. **59**(3), 787 (1998)
47. E. Geigant, M. Stoll, J. Math. Biol. **46**, 537 (2003)
48. K. Kang, B. Perthame, A. Stevens, J. Velázquez, J. Differ. Equ. **246**, 1387 (2009)
49. T. Hillen, J. Math. Biol. **53**(4), 585 (2006)
50. E. Boissard, P. Degond, S. Motsch, J. Math. Biol. **66**(6), 1267 (2013)
51. L. Saint-Raymond, *Hydrodynamic Limits of the Boltzmann Equation* (Springer, Berlin, Heidelberg, 2009)
52. S.Y. Ha, E. Tadmor, Kinet. Relat. Models **1**(3), 415 (2008)
53. R. Eftimie, G. de Vries, M.A. Lewis, F. Lutscher, Bull. Math. Biol. **69**(5), 1537 (2007)
54. R. Eftimie, G. de Vries, M.A. Lewis, Proc. Natl. Acad. Sci. USA **104**(17), 6974 (2007)
55. N. Vauchelet, Kinet. Relat. Models **3**(3), 501 (2010)
56. B. Perthame, Bull. Am. Math. Soc. (New Series) **41**(2), 205 (2004)
57. N. Bournaveas, V. Calvez, S. Gutiérrez, B. Perthame, Commun. Part. Differ. Equ. **33**(1), 79 (2008)
58. M. Lachowicz, H. Leszczyński, M. Parisot, Math. Models Methods Appl. Sci. **27**(6), 1153 (2017)
59. N. Bournaveas, V. Calvez, Can. Appl. Math. Q. **18**(3), 253 (2010)
60. G. Toscani, Commun. Math. Sci. **4**(3), 481 (2006)
61. B. Düring, P. Markowich, J.F. Pietschmann, M.T. Wolfram, Proc. R. Soc. A **465**(2112), 3687 (2009)
62. G. Naldi, L. Pareschi, G. Toscani (eds.), *Mathematical Modelling of Collective Behaviour in Socio-Economic and Life Sciences* (Birkhäuser, 2010)
63. D. Maldarella, L. Pareschi, Phys. A Stat. Mech. Appl. **391**(3), 715 (2012)
64. M. Delitala, T. Lorenzi, Kinet. Relat. Models **7**, 29 (2014)
65. B. Düring, A. Jüngel, L. Trussardi, Kinet. Relat. Models **10**(1), 239 (2017)
66. B. Boghosian, Phys. Rev. E **89**, 042804 (2014)
67. E. Jäger, L. Segel, SIAM J. Appl. Math. **52**(5), 1442 (1992)
68. N. Bellomo, M. Delitala, Phys. Life Rev. **5**, 183 (2008)
69. R. Erban, H. Othmer, Multiscale Model. Simul. **3**(2), 362 (2005)
70. N. Bellomo, A. Bellouquid, J. Nieto, J. Soler, Math. Models Methods Appl. Sci. **17**, 1675 (2007)
71. M. Kolev, Int. J. Math. Comput. Sci. **13**(3), 289 (2003)
72. I. Brazzoli, E. Angelis, P.E. Jabin, Math. Models Methods Appl. Sci. **33**, 733 (2010)
73. A. Chauviere, I. Brazzoli, Math. Comput. Model. **43**, 933 (2006)
74. N. Bellomo, N. Li, P. Maini, Math. Models Methods Appl. Sci. **18**(4), 593 (2008)
75. R. Eftimie, J. Bramson, D. Earn, Bull. Math. Biol. **73**(1), 2 (2011)
76. R. Magrath, B. Pitcher, J. Gardner, Behav. Ecol. **20**(4), 745 (2009)
77. E.D. Angelis, B. Lods, Math. Comput. Model. **47**, 196 (2008)
78. N. Bellomo, E.D. Angelis, L. Preziosi, J. Theor. Med. **5**(2), 111 (2003)
79. T. Lorenz, C. Surulescu, Math. Models Methods Appl. Sci. **24**(12), 2383 (2014)
80. B. Piccoli, A. Tosin, Arch. Ration. Mech. Anal. **199**, 707 (2011)
81. E. Cristiani, B. Piccoli, A. Tosin, Multiscale Model. Simul. **9**(1), 155 (2011)
82. T. Tao, *An Introduction to Measure Theory* (American Mathematical Society, 2011)
83. S.Y. Ha, J.G. Liu, Commun. Math. Sci. **7**(2), 297 (2009)

84. J. Carrillo, M. D'Orsogna, V. Panferov, Kinet. Relat. Models **2**, 363 (2009)
85. Y.L. Chuang, M. D'Orsogna, D. Marthaler, A. Bertozzi, L. Chayes, Physica D **232**, 33 (2007)
86. F. Bolley, J. Canizo, J. Carillo, Math. Models Methods Appl. Sci. **21**, 2179 (2011)
87. F. Bolley, J. Canizo, J. Carillo, Appl. Math. Lett. **25**, 339 (2012)
88. M. D'Orsogna, Y. Chuang, A. Bertozzi, L. Chayes, Phys. Rev. Lett. **96**(10), 104302 (2006)
89. J. Irving, J. Kirkwood, J. Chem. Phys. **18**, 817 (1950)
90. P. Degond, S. Motsch, Math. Models Methods Appl. Sci. **20**, 1459 (2008)
91. T. Hillen, Math. Models Methods Appl. Sci. **12**(7), 1 (2002)
92. P. Degond, A. Appert-Rolland, M. Moussaïd, J. Pettré, G. Theraulaz, J. Stat. Phys. **152**, 1033 (2013)
93. F. Filbet, P. Laurencot, B. Perthame, J. Math. Biol. **50**(2), 189 (2005)
94. L. Bonilla, J. Soler, Math. Models Methods Appl. Sci. **11**, 1457 (2001)
95. N. Bellomo, A. Bellouquid, J. Nieto, J. Soler, Math. Comput. Model. **51**, 441 (2010)
96. E.D. Angelis, M. Delitala, A. Marasco, A. Romano, Math. Comput. Model. **37**, 1131 (2003)
97. N. Bellomo, B. Firmani, L. Guerri, Appl. Math. Lett. **12**, 39 (1999)
98. T. Makino, B. Perthame, Jpn. J. Appl. Math. **7**, 165 (1990)
99. T. Hillen, H.G. Othmer, SIAM J. Appl. Math. **61**, 751 (2000)
100. P. Degond, T. Yang, Math. Models Methods Appl. Sci. **20**, 1459 (2010)
101. F. Golse, in *Handbook of Differential Equations. Evolutionary Equations*, Vol. 2, ed. by C. Dafermos, E. Feireisl (Elsevier B.V., 2005)
102. A.D. Masi, E. Presutti, *Mathematical Methods for Hydrodynamic Limits* (Springer, Berlin, Heidelberg, 1991)
103. P. Degond, B. Lucquin-Descreux, Math. Model Methods Appl. Sci. **2**(2), 167 (1992)
104. L. Desvillettes, Transp. Theor. Stat. Phys. **21**(3), 259 (1992)
105. S. McNamara, W. Young, Phys. Fluids A **5**, 34 (1993)
106. J. Carrillo, R. Eftimie, F. Hoffmann, Kinet. Relat. Models **8**(3), 413 (2015)
107. P. Degond, S. Motsch, C. R. Acad. Sci. Paris Ser. I **345**, 555 (2007)
108. J. Meskas. A nonlocal kinetic model for predator-prey interactions in two dimensions. MSc Thesis, Simon Fraser University, Canada (2012)
109. R. Colombo, M. Lécureux-Mercier, J. Nonlinear Sci. **22**, 39 (2012)
110. P.H. Chavanis, C. Sire, Phys. A Stat. Mech. Appl. **384**, 199 (2007)
111. P.H. Chavanis, Commun. Nonlinear. Sci. Numer. Simul. **15**, 60 (2010)
112. P.H. Chavanis, Phys. A Stat. Mech. Appl. **387**, 5716 (2008)
113. P.H. Chavanis, Phys. A Stat. Mech. Appl. **390**(9), 1546 (2011)
114. A. Schadschneider, D. Chowdhury, K. Nishinari, *Stochastic Transport in Complex Systems. From molecules to vehicles* (Elsevier, Amsterdam Oxford, 2011)

Chapter 7
Numerical Approaches for Kinetic and Hyperbolic Models

To understand the spatial and spatio-temporal patterns exhibited by the hyperbolic and kinetic models discussed in the previous chapters, one needs to combine analytical results with numerical simulations. However, the complexity of these local and nonlocal models often leads to numerical difficulties, ranging from the high dimensionality of some equations, to maintaining the conservation properties of other equations, dealing with increased computational cost when discretising the reaction operator (modelling the birth/death of populations, or the transition between different populations), or dealing with the presence of multiple scales [1].

Numerical methods for transport hyperbolic and kinetic equations is a very broad and active research field, which developed very fast over the last few decades. While this evolution was mainly triggered by problems in physics [2], the development of complex kinetic and hyperbolic models for collective behaviours in cell biology and ecology also required the use of new fast and accurate numerical schemes. Among the numerous numerical studies in this area, we mention [1, 3–7].

Due to the large variety of numerical approaches (from finite difference methods, to finite element methods, finite volume methods, Monte Carlo methods and even hybrid methods) is impossible to discuss in detail all contributions to this field. In consequence, in this chapter we aim to give the reader only an brief overview of some of the most common numerical methods used to discretise and simulate the hyperbolic and kinetic models described in the previous chapters. Thus, we start in Sect. 7.1 by discussing some basic finite difference methods used for simple 1D and 2D hyperbolic models. We also mention some simple finite volume methods for linear hyperbolic equations. Then, in Sect. 7.2 we focus on numerical methods developed to deal with the large dimension of kinetic models. In addition, we discuss some numerical techniques developed to deal with stiff problems. We conclude in Sect. 7.3 with a brief discussion of some biologically-realistic boundary conditions.

© Springer Nature Switzerland AG 2018   195
R. Eftimie, *Hyperbolic and Kinetic Models for Self-organised Biological Aggregations*, Lecture Notes in Mathematics 2232,
https://doi.org/10.1007/978-3-030-02586-1_7

7.1 Numerical Schemes for Hyperbolic Models

As we have seen in the previous chapters, many hyperbolic equations for collective movement of cells/animals/particles can be formulated using the conservation law

$$\mathbf{u}_t + \nabla \cdot F(\mathbf{u}) = S(\mathbf{u}), \tag{7.1}$$

where $\mathbf{u} = (u^1, u^2, \ldots, u^m)$ is the vector for the variables of the system, F describes the conserved flux, and S describes the dynamics of the populations (birth/death, or transition between various sub-populations). The most common approach to solve numerically these types of equations and systems of equations is to use finite differences schemes [3]. To discretise such an equation, one can use (1) *unsplit methods*, where a finite difference formula is derived to propagate the whole equation to the next time step, or (2) *splitting methods*, where the equation is split into different components (e.g., the advection part, the reaction/source part), and different numerical methods are used to discretise each component separately [3]. Since the unsplit methods depend on the expression of the reaction/source term which interacts with the advection (thus influencing the accuracy and stability of the scheme, and slowing-down the computation of the solution), in the following we focus mainly on the faster splitting methods (although even these methods introduce an error [8]). While there are many applications of this splitting approach to solving mathematical models [8], throughout this chapter we discuss only the splitting of the advection from the reaction components, splitting the x-direction from the y-direction, or splitting the larger domain into smaller domains (i.e., domain decomposition). In particular, in this section our focus is on splitting advection from reaction, where the basic idea is to re-write the conservation law (7.1) as

$$\mathbf{u}_t = (A(\mathbf{u}) + R(\mathbf{u}))\mathbf{u}, \tag{7.2}$$

where $A(u)$ describes the discretised advection term (i.e., $\nabla \cdot F(\mathbf{u})$; see below) and $R(\mathbf{u})$ describes the reaction term (i.e., $S(\mathbf{u})$). Then the solution of each equation is computed separately (over one time step) [8]:

$$\mathbf{u}_t = A(\mathbf{u}), \quad \mathbf{u}(t = 0) = \mathbf{u}_0 \;\Rightarrow\; \text{intermediate solution: } \mathbf{u}^*, \tag{7.3a}$$

$$\mathbf{u}_t = R(\mathbf{u}), \quad \mathbf{u}(t = 0) = \mathbf{u}^* \;\Rightarrow\; \text{final solution over 1 time step: } \mathbf{u}^{**}. \tag{7.3b}$$

Before discussing various methods for the discretisation of the advection component (7.3a), we emphasise that the reaction term $\mathbf{u}_t = R(\mathbf{u}) = S(\mathbf{u})$ can be discretised using classical numerical schemes for ODEs (such as the Runge-Kutta schemes [9]). If these terms contain also nonlocal interactions, the integrals need to be discretised first using, for example, classical Newton-Cotes formulas (trapezoidal rule or Simpson's rule [10, 11]), or more advanced Gaussian quadrature formulas [11]. (Note that the 1D integrals that appear in Table 5.1 were discretised using an extended Simpson's rule; for details see [11].) We will return to the

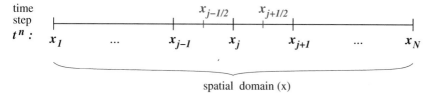

Fig. 7.1 Discretisation of the 1D domain into N points: x_j, $j = 1, \ldots, N$. In some cases, it is useful to consider the intermediate points $x_{j+1/2} = x_j + \Delta x/2$

discretisation of these nonlocal terms in Sect. 7.2, in the context of the discretisation of the collision operator for higher dimensional kinetic equations.

In the following, we start the discussion of numerical finite-difference schemes used to discretise hyperbolic models, by focusing on single advection equations, i.e., $\mathbf{u} = u$, with no source terms, $S(u) = 0$. (Note that systems of equations can be approached in a similar manner.) In Sect. 7.1.1 we discuss the 1D case, in Sect. 7.1.2 we focus on higher resolution numerical schemes, while in Sect. 7.1.3 we discuss the multi-dimensional (e.g., 2D) case (where we return briefly to the discussion about splitting/unsplit methods).

7.1.1 1D Finite Difference and Finite Volume Methods

Finite Difference Schemes To start our discussion on the numerical schemes developed to approximate the solution of a hyperbolic equation, we assume that variable u (which represents the density of particles/animals/cells/etc.) is a function of a single space variable: $u(x)$. The spatial domain of u can be discretised into equal-size intervals using a finite number of points N: $x_j = j\Delta x$, $j = 1, \ldots, N$, with $\Delta x = x_j - x_{j-1} = x_{j+1} - x_j$ (see Fig. 7.1). Similarly, one can discretise into equal-size intervals the time domain on which u is defined: $t^n = n\Delta t$, $n \geq 0$, with $\Delta t = t^n - t^{n-1} = t^{n+1} - t^n$. The value of u at these grid points is $u(x_j, t^n) = u_j^n$.

In the following, we discuss briefly some of the most common finite difference schemes developed for 1D hyperbolic equations with no reaction/source terms (i.e., $S(u) = 0$):

$$\frac{\partial u}{\partial t} + \frac{\partial F(u)}{\partial x} = 0. \tag{7.4}$$

If $F(u)$ is differentiable, then one can replace this nonlinear equation with its quasi-linear form

$$\frac{\partial u}{\partial t} + F'(u)\frac{\partial u}{\partial x} = 0. \tag{7.5}$$

If $F'(u) = a =$constant, we obtain the linear advection equation

$$\frac{\partial u}{\partial t} + a\frac{\partial u}{\partial x} = 0. \tag{7.6}$$

Next, we summarise some numerical schemes can be used to discretise the linear equation (7.6), as well as Eqs. (7.5) and (7.4). For a detailed review of a wider range of schemes see [3]. Moreover, we focus only on explicit schemes (i.e., schemes where calculating u^{n+1} depends on calculating u^n), since implicit numerical schemes are not commonly used for time-dependent hyperbolic problems [3].

- *Upwind scheme.* This first-order scheme depends on the sign of the derivative of F. For the linear advection (7.6) this numerical scheme reads:

$$a > 0 : u_j^{n+1} = u_j^n - \alpha(u_j^n - u_{j-1}^n), \tag{7.7}$$

$$a < 0 : u_j^{n+1} = u_j^n - \alpha(u_{j+1}^n - u_j^n), \tag{7.8}$$

  where $\alpha = a(\Delta t)/(\Delta x)$. The numerical scheme is stable only if $0 \le \alpha \le 1$ for $a > 0$, and only if $-1 \le \alpha \le 0$ for $a < 0$. There is also an implicit version of this upwind scheme, obtained when we discretise the time derivative using the time steps n and $n - 1$. For example, when $a > 0$, this implicit scheme reads

$$u_j^n + \alpha(u_j^n - u_{j-1}^n) = u_j^{n-1}. \tag{7.9}$$

  For the quasi-linear conservation law (7.5), the upwind numerical scheme reads:

$$a_j > 0 : u_j^{n+1} = u_j^n - (\Delta t/\Delta x)(F_j^n - F_{j-1}^n), \tag{7.10}$$

$$a_j < 0 : u_j^{n+1} = u_j^n - (\Delta t/\Delta x)(F_{j+1}^n - F_j^n), \tag{7.11}$$

  with the local Courant number for the conservation law, a_j, being given by

$$a_j = \frac{F_{j+1}^n - F_j^n}{u_{j+1}^n - u_j^n}, \quad \text{if } u_{j+1}^n \ne u_j^n, \text{ and } a_j = F'(u_j^n) \text{ otherwise.} \tag{7.12}$$

  It is well known that the upwind scheme induced diffusive (or dissipation) errors [12], where the discontinuous profile of a solution is being damped numerically; see also Fig. 7.2a.
- *Leapfrog scheme:* For the linear advection equation (7.6), this second-order scheme reads:

$$u_j^{n+1} = u_j^{n-1} - \alpha(u_{j+1}^n - u_{j-1}^n). \tag{7.13}$$

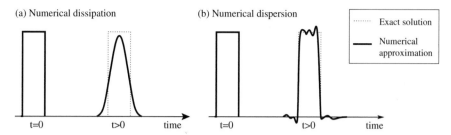

Fig. 7.2 Caricature description of the effect (**a**) dissipation errors, and (**b**) dispersion errors, which could appear following the numerical discretisation of equations. Dissipative errors dampen high wave number components of the solution. Dispersion errors occur when waves associated with different wave numbers travel at different speed. The numerical dissipation is often used to avoid unwanted oscillations in the approximated solution (see our discussion about artificial viscosity in Chap. 2)

This numerical scheme is stable provided that $|a \Delta t / \Delta x| \leq 1$. The nonlinear equation (7.4) has a similar discretisation:

$$u_j^{n+1} = u_j^{n-1} - \alpha(F_{j+1}^n - F_{j-1}^n). \tag{7.14}$$

We note that this numerical scheme induces numerical dispersion errors [12], where the discontinuous profile of a solution develops small oscillations; see also Fig. 7.2b. This aspect is particularly problematic for biological problems, where population densities must be non-negative to ensure biological realism.

- *Lax-Friedrichs scheme.* For the linear advection equation (7.6), the first-order scheme reads:

$$u_j^{n+1} = \frac{1}{2}(u_{j+1}^n + u_{j-1}^n) - \frac{\alpha}{2}(u_{j+1}^n - u_{j-1}^n). \tag{7.15}$$

This scheme is stable if $|a \Delta t / \Delta x| \leq 1$. For the nonlinear equation (7.4) we have

$$u_j^{n+1} = \frac{1}{2}(u_{j+1}^n + u_{j-1}^n) - \frac{\alpha}{2}(F_{j+1}^n - F_{j-1}^n). \tag{7.16}$$

This scheme induces numerical dissipative errors, as well as dispersion errors [13].

- *Lax-Wendroff scheme.* For the linear advection equation (7.6) we have

$$u_j^{n+1} = u_j^n - \frac{\alpha}{2}(u_{j+1}^n - u_{j-1}^n) - \frac{\alpha^2}{2}(u_{j+1}^n - 2u_j^n + u_{j-1}^n). \tag{7.17}$$

This numerical scheme is stable if $|a \Delta t / \Delta x| \leq 1$. We note that also this scheme induces dissipation errors in the solution.

For nonlinear hyperbolic equations (7.4), the Lax-Wendroff method is not unique, a few other methods being developed to generalise this scheme. One such method (a direct generalisation of (7.17)) is:

$$u_j^{n+1} = u_j^n - \frac{\alpha}{2}(F_{j+1}^n - F_{j-1}^n) + \frac{\alpha^2}{2}(u_{j+1}^n - 2u_j^n + u_{j-1}^n). \tag{7.18}$$

Another method that generalises the classical Lax-Wendroff scheme to nonlinear equations is the MacCormack method, discussed next.

- *MacCormack scheme.* For linear advection equations, this scheme takes the form

$$u_j^* = u_j^n - \alpha(u_{j+1}^n - u_j^n), \tag{7.19}$$

$$u_j^{n+1} = \frac{1}{2}(u_j^n + u_j^*) - \alpha(u_j^* - u_{j-1}^*), \tag{7.20}$$

which is the equivalent to the Lax-Wendroff scheme (for the linear case). The corresponding scheme for the nonlinear equation takes the form

$$u_j^* = u_j^n - (\Delta t/\Delta x)\Big(F(u_j^n) - F(u_{j-1}^n)\Big), \tag{7.21}$$

$$u_j^{n+1} = \frac{1}{2}\Big(u_j^n + u_j^*\Big) + (\Delta t/\Delta x)\Big(F(u_{j+1}^*) - F(u_j^*)\Big), \tag{7.22}$$

This predictor-corrector scheme is well suited for discretising nonlinear equations (e.g., Euler equations, Navier-Stokes equations [14]). Unfortunately, it can induce dispersive errors near sharp transitions, leading to oscillatory solutions.

These methods for the discretisation of the advection equation can be easily extended to systems of equations of the form $U_t + AU_x = 0$, with $U = (u_1, \ldots, u_m) : \mathbb{R} \times \mathbb{R} \to \mathbb{R}^m$ and $A \in \mathbb{R}^{m \times m}$ a constant matrix [3]. The stability conditions for the numerical schemes applied to systems of hyperbolic equations are defined in terms of the eigenvalues of matrix A. For example, the upwind method

$$U_j^{n+1} = U_j^n - A(\Delta t/\Delta x)(U_j^n - U_{j-1}^n), \tag{7.23}$$

is stable only if $0 \leq \lambda_l(\Delta t/\Delta x) \leq 1$ for all $l = 1, \ldots, m$ (with λ_l the eigenvalues of matrix A).

Finally, we acknowledge that all these schemes need to be accompanied by an appropriate discretisation of the boundary conditions discussed in Chap. 4. We will return to this aspect in Sect. 7.3.

Finite Volume Schemes The solutions of the hyperbolic systems may contain discontinuities, which leads to difficulties in applying the classical finite difference methods discussed above (since the differential equations are not valid at these discontinuity points). This problem can be overcome by considering finite volume methods based on the integral forms of the conservation laws. Instead of focusing on

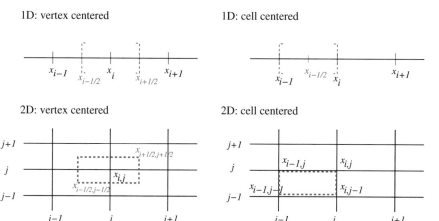

Fig. 7.3 Examples of simple discretisation grids for 1D and 2D domains, for the finite volume methods. The cells can be chosen either vertex-centred (at the mesh points x_i, $i = 1, \ldots, N$ for the 1D domain, or $x_{i,j}$, $i, j = 1, \ldots, N$ for the 2D domain), or cell-centred (at the half mesh points $x_{i\pm1/2}$, $i = 1, \ldots, N$ for the 1D domain, or $x_{i\pm1/2,j\pm1/2}$, $i, j = 1, \ldots, N$ for the 2D domain). Other types of grids (non-orthogonal and even unstructured) could be also chosen to discretise more complex domains [15]

the values u calculated at the grid points j (i.e., $u_j^n = u(x_j, t^n)$), one could focus on the average values of u over each grid interval (or cell) $[x_j, x_{j+1}]$ centred at $x_{j+1/2}$, or over the translated interval $[x_{j-1/2}, x_{j+1/2}]$ centred at x_j (see also Fig. 7.3). For example, one can consider the average value of u at the mid-point of the interval:

$$U_{j+1/2}^n = \frac{1}{x_{j+1} - x_j} \int_{x_j}^{x_{j+1}} u(x, t^n)dx, \qquad (7.24)$$

with $x_{j+1} - x_j = \Delta x$. For the initial conditions ($u(x, 0) = u_0$), one could choose the pointwise values of u_j^0, or the average value $U_j^0 = \frac{1}{x_{j+1}-x_j} \int_{x_j}^{x_{j+1}} u(x, t^0)dx$. The interpretation of u as an averaged value over the interval is consistent with the integral formulation of the conservation law [3].

This integral approach can be used to develop finite-volume versions of the previous numerical schemes (or other discrete schemes). To this end, the original hyperbolic equation is integrated on each cell of the domain. For example, a finite-volume version of the upwind scheme (7.7) is described by:

$$u_{i+1/2}^{n+1} = u_{i+1/2}^n - \alpha(u_{i+1/2} - u_{i-1/2}), \qquad (7.25)$$

with $\alpha = a\Delta t/\Delta x$.

For a non-linear hyperbolic equation, one has

$$u_{i+1/2}^{n+1} = u_{i+1/2}^n - \alpha\left(\bar{F}_{i+1/2}^n - \bar{F}_{i-1/2}^n\right), \qquad (7.26)$$

with the numerical flux $\tilde{F}^n_{i+1/2}$ approximating the physical flux $F(u(x_{i+1/2}))$ over each cell

$$\bar{F}^n_{i+1/2} = \frac{1}{\Delta t} \int_{t^n}^{t^{n+1}} F(u(x_{i+1/2}, t)) dt.$$

Godunov [16] observed that this numerical flux can be given in terms of the Riemann solution at each cell interface $x_{i+1/2}$, and can be computed explicitly as follows

$$\bar{F}^n_{i+1/2} = \begin{cases} \min_{u^n_i \leq u \leq u^n_{i+1}} F(u), & \text{if } u^n_i \leq u^n_{i+1}, \\ \max_{u^n_{i+1} \leq u \leq u^n_i} F(u), & \text{if } u^n_{i+1} \leq u^n_i. \end{cases}$$

This discretisation is referred to as the Godunov method. We note that this formula is similar to the formula for the finite difference upwind scheme (shifted half a cell—to calculate the centre of the cells). This holds true for lower order schemes, but not for higher order schemes that require the reconstruction of polynomials of certain degrees from the cells averages. Moreover, since the higher order schemes usually involve oscillations around the discontinuities, specific numerical methods have been developed over the last four decades. In the next section, we will discuss in more detail some of these methods.

7.1.2   Higher Resolution Schemes for Nonlinear Equations

To capture the sharp solution profiles exhibited by nonlinear hyperbolic equations, without having any misleading oscillations, Harten [17] introduced total variation diminishing (TVD) schemes.

Definition 7.1 A numerical scheme is called a *TVD scheme* (or as Harten [17] called it, *total variation nonincreasing* scheme (TVNI)) if $TV(u^{n+1}) \leq TV(u^n)$, where the total variation for the discrete case is $TV(u^n) = \sum_j |u^n_{j+1} - u^n_j|$.

Harten [17] showed that a monotone numerical scheme is TVD, and a TVD scheme is preserving the monotonicity. However, Godunov's theorem [16] proves that linear schemes that preserve the monotonicity can be at most first order accurate. To overcome this problem Harten [17] introduced a new class of high resolution second order accurate TVD schemes which use flux/slope limiters. There are currently three types of techniques that can be used to develop numerical schemes which are accurate in the presence of shocks or discontinuities: ENO (Essentially Non-Oscillatory) techniques [4, 18, 19], WENO (Weighted Essentially Non-Oscillatory) techniques [4, 19, 20], and MUSCL (Monotone Upstream-Centred Schemes for Conservation Laws) techniques [21]. For a comparison of the ENO and WENO methods see [4, 19]. In the following we will discuss briefly a ENO scheme introduced in [18] and a MUSCL-based scheme introduced in [22].

Harten et al. [18] proposed the following essentially non-oscillatory scheme based on cell averaging to approximate a hyperbolic system of the form (7.1):

$$u_i^{n+1} = u_i^n - \alpha(\bar{F}_{i+1/2} - \bar{F}_{i-1/2}), \quad \text{with} \tag{7.27}$$

$$\bar{F}_{i+1/2} = \sum_{k=0}^{m} \lambda_k F^R(\bar{u}_i(x_{i+1/2}, \beta_k \Delta t), \bar{u}_{i+1}(x_{i+1/2}, \beta_k \Delta t)), \tag{7.28}$$

where $\alpha = (\Delta t)/(\Delta x)$,

$$F^R(u_1, u_2) = \begin{cases} \min_{u_1 \le u \le u_2} F(u), & \text{if } u_1 \le u_2, \\ \max_{u_1 \ge u \ge u_2} F(u), & \text{if } u_1 > u_2. \end{cases} \tag{7.29}$$

In the equations above, the numerical flux $\bar{F}_{i+1/2} = (1/\Delta t) \int_0^{\Delta t} F(u(x_{i+1/2}, t)) dt$ is being discretised at each interface $x_{i+1/2}$ using a numerical quadrature

$$\bar{F}_{i+1/2} \approx \sum_{k=0}^{m} \lambda_k F(u(x_{i+1/2}, \beta_k \Delta t)). \tag{7.30}$$

Finally, $\bar{u}_i$ is the lth-order Taylor approximation of u_i:

$$\bar{u}_i(x, t) = \sum_{l=0}^{\Delta t - l} \sum_{k=0}^{l} \frac{\partial^l u(x_i, 0)}{\partial x^k \partial t^{l-k}} \frac{(x - x_i)^k}{k!} \frac{t^{l-k}}{(l-k)!}, \quad 0 \le k \le l. \tag{7.31}$$

The high resolution schemes derived in [17, 18] are based on the finite volume Godunov's method. However, this method has some drawbacks as it relies on the explicit formula for the solution of the Riemann problem. While this formula can be obtained for scalar conservation laws, it might not be possible to obtain it for more complicated conservation laws (e.g., the MHD equations in physics, or nonlocal and nonlinear transport models in biology). Moreover, for Godunov's scheme one requires only the value of the flux at the interface $(x_{i+1/2})$, and solving the whole Riemann problem to obtain this value is time consuming.

One way of addressing these drawbacks is by approximating the exact solutions of the Riemann problem, and using these approximations to define the numerical flux $\bar{F}_{i+1/2}$. These numerical schemes are called *approximate Riemann solvers*. An example of such a scheme is the Roe scheme [23], where the nonlinear flux $F(u)_x$ is replaced by a linearised version: $F(u)_x = F'(u)u_x \approx A_{i+1/2} u_x$. The coefficient $A_{i+1/2}$ is given by the following Roe average:

$$A_{i+1/2} = \begin{cases} \frac{F(u_{i+1}^n) - F(u_i^n)}{u_{i+1}^n - u_i^n}, & \text{if } u_{i+1}^n \ne u_i^n, \\ F'(u_i^n), & \text{if } u_{i+1}^n = u_i^n. \end{cases}$$

Solving the linearised Riemann problem

$$u_t + A_{i+1/2} u_x = 0, \quad \text{with } u(x, t^n) = \begin{cases} u_i^n, & \text{if } x < x_{i+1/2}, \\ u_{i+1}^n, & \text{if } x > x_{i+1/2}, \end{cases}$$

leads to the following flux

$$F_{i+1/2}^n = F^{Roe}(u_i^n, u_{i+1}^n) = \begin{cases} F(u_i^n), & \text{if } A_{i+1/2} \geq 0, \\ F(u_{i+1}^n), & \text{if } A_{i+1/2} < 0. \end{cases}$$

For a more detailed discussion of approximate Riemann solvers we refer the reader to [21].

Another approach to address the drawbacks of using an upwind Godunov solver for a high resolution numerical scheme is to use a central difference solver, such as the Lax-Friedrichs solver. Nessyahu and Tadmor [22] introduced a family of second-order central difference high resolution schemes based on the Lax-Friedrichs solver combined with a slope limiter approach. The schemes have the following predictor-corrector form:

$$u_i^{n+1/2} = u_i^n - \frac{1}{2} \alpha F_i', \tag{7.32}$$

$$u_{i+1/2}^{n+1} = \frac{1}{2}(u_i^n + u_{i+1}^n) + \frac{1}{8}(u_i' - u_{i+1}') - \alpha\left(F(u_{i+1}^{n+1/2}) - F(u_i^{n+1/2})\right), \tag{7.33}$$

with u_i' an approximate slope at the grid point x_i and F_i' the numerical derivative of the function F_i. To ensure that this scheme is non-oscillatory and TDV, the following condition for derivatives $w_i' = (u_i', F_i')$ needs to be satisfied:

$$0 \leq w_i' \cdot sgn(\Delta u_{i\pm 1/2}) \leq C_w |MinMod\{\Delta w_{i+1/2}, \Delta w_{i-1/2}\}|, \tag{7.34}$$

where the *MinMod* limiter is given by

$$MinMod\{a, b\} = \frac{1}{2}\left(sgn(a) + sgn(b)\right) \cdot Min(|a|, |b|).$$

In the above inequality, we defined $\Delta w_{i+1/2} = w_{i+1} - w_i$. To satisfy these conditions, the numerical derivatives can be chosen for example to be

$$u_i' = MinMod\{\Delta u_{i+1/2}, \Delta u_{i-1/2}\}, \tag{7.35}$$

$$F_i' = MinMod\{\Delta F_{i+1/2}, \Delta F_{i-1/2}\}. \tag{7.36}$$

Note that in addition to this *MinMod* limiter, there are many other possible choices for the slope/flux limiters; e.g., monotonized central [24], superbee [25], Sweby

[26], or van Leer [27] limiters. For a more in-depth discussion of these types of non-oscillatory central differencing schemes we refer the reader to [22, 28].

7.1.3 Multi-Dimensional Finite Difference and Finite Volume Schemes

The 2D version of a system of conservation laws is described by

$$\frac{\partial u}{\partial t} + \frac{\partial f(u)}{\partial x} + \frac{\partial g(u)}{\partial y} = 0, \tag{7.37}$$

where $u = u(x, y, t) \in \mathbb{R}^n$. The methods used to discretise this multi-dimensional (in space) equation range from simple generalisations of previous 1D numerical (finite difference and finite volume) methods to 2D equations, to splitting methods [3].

- *Generalisation of 1D finite difference methods.* Some of the previously discussed 1D methods can be easily generalised in 2D. For example, the 1D finite difference Lax-Friedrichs scheme (7.15) can take the following form in 2D:

$$\begin{aligned} u_{i,j}^{n+1} &= \frac{1}{4}\left(u_{i+1,j}^n + u_{i-1,j}^n + u_{i,j+1}^n + u_{i,j-1}^n\right) \\ &- \frac{\Delta t}{2}\left(\frac{f_{i+1,j}^n - f_{i-1,j}^n}{\Delta x} + \frac{g_{i,j+1}^n - g_{i,j-1}^n}{\Delta y}\right), \end{aligned} \tag{7.38}$$

  where $f_{i,j}^n = f(u_{i,j}^n)$ and $g_{i,j}^n = g(u_{i,j}^n)$.
- *Generalisation of 1D high-resolution finite volume methods.* Jiang and Tadmor [28] generalised the 1D high-resolution scheme introduced in [22] to a 2D domain by considering cells $C_{i+1/2,k+1/2}$ centred around the generic point $(x_{i+1/2}, y_{k+1/2})$. Defining a normalised cell average as

$$\bar{u}_{i+1/2,k+1/2}^n = \frac{1}{|C_{i+1/2,k+1/2}|} \int_{C_{i+1/2,k+1/2}} u(x, y, t)dxdy, \tag{7.39}$$

and the space steps in the x and y directions as $\alpha_x = \Delta t/\Delta x$ and $\alpha_y = \Delta t/\Delta y$, one can write the predictor-corrector steps for the 2D scheme as follows:

$$u_{ik}^{n+1/2} = \bar{u}_{ik}^n - \frac{\alpha_x}{2} f(u)_{ik}' - \frac{\alpha_y}{2} g(u)_{ik}', \tag{7.40a}$$

$$\bar{u}_{i+1/2,k+1/2}^{n+1} = \frac{1}{4}\left(\bar{u}_{ik}^n + \bar{u}_{i+1,k}^n + \bar{u}_{i,k+1}^n + \bar{u}_{i+1,k+1}^n\right)$$

$$+ \frac{1}{16}\left(u'_{ik} - u'_{i+1,k}\right) - \frac{\alpha_x}{2}\left(f(u^{n+1/2}_{i+1,k}) - f(u^{n+1/2}_{i,k})\right)$$

$$+ \frac{1}{16}\left(u'_{i,k+1} - u'_{i+1,k+1}\right) - \frac{\alpha_x}{2}\left(f(u^{n+1/2}_{i+1,k+1}) - f(u^{n+1/2}_{i,k+1})\right)$$

$$+ \frac{1}{16}\left(u`_{i,k} - u`_{i,k+1}\right) - \frac{\alpha_y}{2}\left(g(u^{n+1/2}_{i,k+1}) - g(u_{i,k})\right)$$

$$+ \frac{1}{16}\left(u`_{i+1,k} - u`_{i+1,k+1}\right) - \frac{\alpha_y}{2}\left(g(u^{n+1/2}_{i+1,k+1}) - g(u^{n+1/2}_{i+1,k})\right). \tag{7.40b}$$

In the equations above, $f(u)'_{ik}$ and $g(u)'_{ik}$ are the 1D discrete slopes in the x and y directions, respectively. Hence, starting with a cell average $\bar{u}^n_{ik}$, the predictor step computes the midpoint values $u^{n+1/2}_{ik}$ while the corrector step computes the new cell averages $\bar{u}^{n+1}_{ik}$.

- *Splitting methods.* Equation (7.37) can be solved by splitting it into two 1D problems along each spatial dimension: the first 1D problem can be written as

$$\frac{\partial u^*}{\partial t} + \frac{\partial f(u^*)}{\partial x} = 0, \quad \text{with } u^*(x, y, 0) = u_0(x, y), \tag{7.41}$$

which has the solution $u^*(x, y, t)$, and the second 1D problem

$$\frac{\partial u^{**}}{\partial t} + \frac{\partial g(u^{**})}{\partial y} = 0, \quad \text{with } u^{**}(x, y, 0) = u^*(x, y, t), \tag{7.42}$$

which has the solution $u^{**}(x, y, t)$. Each of these 1D problems can be solved using previously-discussed numerical schemes. For a more detailed discussion of these splitting methods, see [3].

We note here that even the 1D TVD schemes discussed above can be generalised to 2D,

$$\frac{\partial u_{i,j}}{\partial t} + \frac{1}{\Delta x}\left(\bar{F}_{i+1/2,j} - \bar{F}_{i-1/2,j}\right) + \frac{1}{\Delta y}\left(\bar{F}_{i,j+1/2} - \bar{F}_{i,j-1/2}\right) = 0, \tag{7.43}$$

but the TVD is not in more than one dimension.

While the methods used to reduce the 2D equations (and even the 3D equations) to 1D equations are easy to implement numerically, they have some disadvantages as the 2D (or 3D) effects could impact the behaviour of the solution locally by introducing a directional bias in the coordinate directions [3]. A relevant example is given by the possibility of having shock waves at angles that are between the coordinate directions [3]. Therefore, current approaches in the development of multi-dimensional numerical methods focus on schemes that incorporate more information about the multidimensional behaviour of the physical/biological problem (such as the direction of wave propagation) [3, 29].

Remark 7.1 An important aspect in the numerical discretisation of hyperbolic models for collective movement is the implementation of boundary conditions. On one hand such boundary conditions (BCs) need to describe the biological realism of the problem (e.g., movement inside a box-like domain as described by Neumann BCs, or movement on a ring-like domain as described by periodic BCs), while on the other hand they need to preserve the conservative character of the equations (where is the case). For example, the upwind/downwind method (7.7) combined with Neumann boundary conditions does not preserve the conservation of total density for the hyperbolic system. We will return to the discussion of the boundary conditions in Sect. 7.3.

Remark 7.2 Hyperbolic equations can be easily discretised using the *method of lines*, which consists in first applying a discretisation scheme (e.g., a finite difference scheme) in space only. This approach transforms the initial equation into a system of ODEs of the form:

$$\frac{\partial U}{\partial t} = \mathscr{L}U, \tag{7.44}$$

where $U(t) = (\ldots, u_i(t), \ldots)$ are the values of the variable u at the grid points i and time t, and the operator $\mathscr{L}$ contains the discretisation of the transport term (e.g., $\nabla \cdot F(U)$ in Eq. (7.1)) at the grid points i, as well as the discretisation of any nonlinear reaction term that can appear in the right-hand-side of the hyperbolic equations (e.g., $S(U)$ in Eq. (7.1)). The solution of this system can be propagated in time with the help of any numerical method for ODEs. For example, if one uses an explicit Euler method [11] for the time-propagation, and an upwind method for the space discretisation, it leads to an explicit upwind scheme of the form (7.7). On the other hand, if one uses an implicit Euler method [11] for the time propagation, and an upwind method for the space discretisation, it leads to an implicit upwind scheme of the form (7.9).

7.2 Numerical Schemes for Higher Dimension Kinetic Models

The literature on numerical methods for kinetic equations (and in particular the kinetic Boltzmann equations) is vast and diverse, with numerical schemes being developed since 1960s. Due to the intense activity in this research field, it is impossible to do a comprehensive review of these methods. Therefore, in the following we discuss briefly a few the approaches used to approximate the solutions of kinetic equations.

The numerical approaches for kinetic equations can be classified as: probabilistic Monte Carlo methods, deterministic finite difference methods, semi-Lagrangian methods, discrete velocity methods, and spectral methods [30–39]. While the deterministic methods are more accurate, they are also computationally expensive (due to the choice of integration points that preserve the conservation properties of the collision operator; see Eqs. (6.14)–(6.17) and the discussion therein). In contrast, the probabilistic methods have lower computational requirements. In fact, throughout this section we aim to emphasise the difficulties of solving numerically kinetic equations: from the high dimensionality of the equations (depending on variables $(x, t, v) \in \mathbb{R}^d \times \mathbb{R}^+ \times \mathbb{R}^d$, for some $d \geq 1$), to maintaining the conservation properties of the equations, the increased computational cost when discretising the reaction operator, or even the presence of multiple scales [1].

As mentioned before, one of the most common approaches to solving (hyperbolic and) kinetic equations is based on operator splitting, which solves separately the collision/interaction step and the transport step (see the discussion in Sect. 7.1.3). While the transport operator can be approximated with the help of finite difference or finite volume schemes [40], the collision operator can be approximated using finite difference schemes, spectral methods involving Fourier transforms [5], or Monte Carlo methods [1]. In the following we review briefly the last two approaches.

7.2.1 Spectral Methods

To exemplify this class of numerical methods, we focus on the approach in Filbet and Russo [5], who used the following operator splitting to approximate the solution of the Boltzmann equation (6.14):

- a finite volume approach to solve the transport step (over the small time interval $\Delta t = [t^n, t^{n+1}]$):

$$\frac{\partial u^*}{\partial t} + v \cdot \nabla_x u^* = 0, \tag{7.45a}$$

$$u^*(x, 0, v) = u^n(x, v), \quad \text{with } v \in [-\pi, \pi]^2, \ x \in \mathbb{R}^2, \ t \in \mathbb{R}^+, \tag{7.45b}$$

where $u^n(x, v)$ describes the solution of Boltzmann equation at the previous time step t^n, while $u^*(x, t, v)$ describes the solution of the transport equation alone (which uses u^n as initial condition). In 1D, denoting by u_i^n the density average over the interval $[x_{i-1/2}, x_{i+1/2}]$ (where $x_{i+1/2}, i \in I$ are a finite set of mesh points over the computational domain),

$$f_i^n = \frac{1}{\Delta x} \int_{x_{i-1/2}}^{x_{i+1/2}} u(x, t^n) dx, \quad \text{with } \Delta x = x_{i+1/2} - x_{i-1/2},$$

leads to the following approximation of the distribution function (with slope correctors to avoid spurious oscillations):

$$u_h(x, t^n) = u_i^n$$

$$+ \frac{\epsilon_i^+}{6\Delta x^2} \Big[2(x - x_i)(x - x_{i-3/2}) + (x - x_{i-1/2})(x - x_{i+1/2}) \Big]$$
$$\times (u_{i+1}^n - u_i^n)$$

$$- \frac{\epsilon_i^-}{6\Delta x^2} \Big[2(x - x_i)(x - x_{i+3/2}) + (x - x_{i-1/2})(x - x_{i+1/2}) \Big]$$
$$\times (u_i^n - u_{i-1}^n),$$

with

$$\epsilon_i^\pm = \begin{cases} \min(1; \frac{2u_i^n}{u_{i\pm1}^n - u_i^n}), & \text{if } u_{i\pm1}^n - u_i^n > 0, \\ \min(1; -\frac{2(u_\infty - u_i^n)}{u_{i\pm1}^n - u_i^n}), & \text{if } u_{i\pm1}^n - u_i^n < 0, \end{cases}$$

and $u_\infty = \max_{j \in I}\{u_j^n\}$ a local maximum.

- a Fourier spectral approach to evolve the collision operator:

$$\frac{\partial u^{**}}{\partial t} = Q(u^{**}, u^{**}), \tag{7.46a}$$

$$u^{**}(x, 0, v) = u^*(x, \Delta t, v), \tag{7.46b}$$

where $u^{**}(x, t, v)$ denotes the solution of the above ODE, which has as initial condition the previous solution $u^*(x, t, v)$ of the transport equation. Re-write the collision operator Q as a difference between two operators describing the gain (Q^+) and loss (L) of particles/individuals/cells with velocity v:

$$Q(u, u)(v) = Q^+(u, u) - L[u]u, \tag{7.47}$$

with

$$Q^+(u, u) = \int_{\mathbb{R}^2} \int_{S^1} B(|v - v_*|, \theta)u(v')u(v'_*)d\omega dv_*, \tag{7.48}$$

$$L[u] = \int_{\mathbb{R}^2} \int_{S^1} B(|v - v_*|, \theta)u(v_*)d\omega dv_*, \tag{7.49}$$

where v and v_* are the post-collision velocities of two particles that have pre-collision velocities v' and v'_*. Moreover, θ is the angle between $v - v_*$ and $v' - v'_*$ (i.e., the deflection angle). Finally, the collision kernel B describes the details of the particle-particle interactions (see also Chap. 6).

The distribution function u restricted to the velocity space $[-\pi, \pi]$ is approximated by the truncated Fourier series

$$u_N(v) = \sum_{k=-N}^{N} \hat{u}_k e^{ikv}, \quad \text{with } \hat{u}_k = \frac{1}{(2\pi)^2} \int_{[-\pi,\pi]^2} u(v) e^{-ikv} dv. \quad (7.50)$$

The coefficients $\hat{u}_k$ can be calculated by requiring that the u_N functions satisfy

$$\int_{[-\pi,\pi]^2} \left(\frac{\partial u_N}{\partial t} + u_N L(u_N) - Q^+(u_N, u_N) \right) e^{-ikv} dv = 0. \quad (7.51)$$

Substituting (7.50) into (7.48) and (7.49) leads to the following approximations for the collision operators (7.48)–(7.49):

$$Q^+(u_N, u_N) = \sum_{l=-N}^{N} \sum_{m=-N}^{N} \hat{u}_l \hat{u}_m \hat{B}(l, m) e^{i(l+m)v}, \quad (7.52)$$

$$u_N L(u_N) = \sum_{l=-N}^{N} \sum_{m=-N}^{N} \hat{u}_l \hat{u}_m \hat{B}(m, m) e^{i(l+m)v}, \quad (7.53)$$

where the kernel modes $\hat{B}(l, m)$ are described by

$$\hat{B}(l, m) = \int_{\mathbb{R}^2} \int_{S^1} B(|v - v_*|, \theta) e^{-i(v-v_*)\frac{l+m}{2} - i|v - v_*|\omega \frac{m-l}{2}} d\omega dg.$$

Further, using (7.51), it leads to the following differential equations for the Fourier coefficients $\hat{u}_k$:

$$\frac{\partial \hat{u}_k}{\partial t} = \sum_{m=k-N}^{N} \hat{u}_{k-m} \hat{u}_m \left(\hat{B}(k - m, m) - \hat{B}(m, m) \right), \quad (7.54)$$

with the initial condition

$$\hat{u}_k(0) = \frac{1}{(2\pi)^2} \int_{[-\pi,\pi]^2} u_0(v) e^{-ikv} dv. \quad (7.55)$$

Calculating the solution for these differential equations (7.54) is computationally much faster than calculating the solution of (7.46), which usually contains a high-dimensional integral in the velocity/orientation space that needs to be computed at every point in the physical space [1].

Following the computation of the transport and collision steps, the solution of the Boltzmann equation at time t^{n+1} is approximated by $u^{n+1}(x, v) = u^{**}(\Delta t, x, v)$.

We conclude this discussion by noting that in the context of collective movement in biology, a spectral method was recently used in [38] to approximate the solution of a kinetic Vicsek model for swarming, with nonlocal interactions describing particle alignment. The model was shown to exhibit features such as vortex formation and travelling waves.

7.2.2 Monte Carlo Methods

When collisions between particles occur at a very fast rate, there is a large ratio of time scales between the macroscopic and microscopic dynamics. This renders the numerical investigation of the kinetic model—with the help of explicit numerical schemes—computationally very expensive (since the model becomes numerically "stiff") [30]. In this case, a more efficient approach is offered by the use of Monte Carlo algorithms, which are a class of algorithms that use random sampling to approximate numerically the solutions. These algorithms have started to be developed in the 1960s by Bird [41, 42], and have been used to approximate the solution of the following Boltzmann equation:

$$\frac{\partial u}{\partial t} + v \cdot \nabla_x u = \frac{1}{\epsilon} Q(u, u), \quad u(x, 0, v) = u_0(x, v), \tag{7.56}$$

where the collision operator is $Q(u, u)(v) = Q^+(u, u) - \mu u$ (i.e., the operator for the probability of changing the velocity v is defined by a constant: $L[u] = \mu$). Parameter ϵ is called the Knudsen number, which is the ratio between the collision mean free path of the particles and the characteristic length of variation of macroscopic variables [43]. We assume here that $\epsilon \ll 1$.

A classical probabilistic approach to approximate the solution of the Boltzmann equation is described by the Direct Simulation Monte Carlo (DSMC) schemes [30, 41, 42, 44, 45]. For these schemes, the non-spatial Boltzmann equation (7.56) is discretised with respect to time using, for example, a forward Euler scheme [30]:

$$u^{n+1} = \left(1 - \frac{\mu \Delta t}{\epsilon}\right) u^n + \frac{\mu \Delta t}{\epsilon} \frac{Q^+(u, u)}{\mu}. \tag{7.57}$$

This equation can be given a probabilistic interpretation: a particle with velocity v avoids collision with other particles at a probability $(1 - \mu \Delta t / \epsilon)$, and collides with other particles at a probability $\mu \Delta t / \epsilon$ and according to the rule offered by $Q^+(u, u)$ [30].

The following DSMC conservative algorithm was introduced by Babovsky [45] to describe the selection of independent pairs of particles with specific velocities. To start, consider N particles. Then, the expected number of particles that collide in a time step Δt is $N \mu \Delta t / \epsilon$, and the expected number of collision pairs is $N \mu \Delta t / (2\epsilon)$ [43].

- Step 1. The initial velocity of particles v_i^0 (with $i = 1, \dots, N$) is computed by sampling them from the initial density $u_0(v)$.
- Step 2. Given particles' velocities v_i^n (with $i = 1, \dots, N$), the velocities at the next time step, v_i^{n+1}, are computed as follows:

  - define $N_c = Iround(\mu N \Delta t / 2\epsilon)$, where $Iround(z)$ gives the rounding of a positive real number z:

$$Iround(z) = \begin{cases} [z], & \text{with probability } [z] + 1 - z, \\ [z] + 1 & \text{with probability } z - [z], \end{cases}$$

    with $[z]$ denoting the integer part of z.
  - select N_c pairs of (i, j) particles, uniformly among all possible pairs.
  - for the selected pairs (i, j) perform the collision between the particles (according the collision law $Q(u(v), u(v))$), and calculate the post-collisional velocities v_i' and v_j'.
  - define the velocities at the next time step: $v_i^{n+1} = v_i'$, $v_j^{n+1} = v_j'$.
  - define $v_k^{n+1} = v_k^n$ for all particles that have not been selected (i.e., $N - 2N_c$ particles velocities are not changed).

- This Step 2 is repeated up to a chosen total time $t = T$.

Note that the post-collisional velocities are calculated from the pre-collisional velocities as follows:

$$v_i' = \frac{v_i + v_j}{2} + \frac{|v_i - v_j|}{2}\omega, \quad v_j' = \frac{v_i + v_j}{2} - \frac{|v_i - v_j|}{2}\omega,$$

with the 2D value of ω [30]

$$\omega = (\cos\theta, \sin\theta), \quad \text{with } \theta = 2\pi\psi_2,$$

or the 3D value of ω [43]

$$\omega = (\cos\phi \sin\theta, \sin\phi \sin\theta, \cos\theta), \quad \text{with } \theta = \arccos(2\psi_1 - 1), \quad \phi = 2\pi\psi_2.$$

and ψ_1, ψ_2 two uniformly distributed random variables in $[0, 1]$. As emphasised in [43], this approach is equivalent to sampling the post collisional velocities based on the rule $Q(u, u)/\mu$.

In addition to this classical DSMC scheme introduced in [45], there are many other schemes developed to improve the efficiency of calculating numerically the solutions of nonlinear kinetic equations. For example, a similar DSMC scheme was introduced in the 1960s by Bird [41, 42]. While in the Babovsky's algorithm [45] the particles collide only once per time step, in the Bird's algorithm [42] multiple collisions are allowed per time step, and this difference impacts the time accuracy of the methods (with the Bird's scheme being considered of infinite order in time [30]). More recently, modified DSMC methods have been developed to reduce the statistical error through increasing the sample size [46–48]. These different Monte Carlo methods could be adapted to be applied to the various kinetic models for cells and animal behaviours discussed in Chap. 6, depending on the assumptions of the models.

To conclude this discussion, we note that these Monte Carlo methods break down for $\epsilon \to 0$ (i.e., in the hydrodynamic regime). To address this problem, new numerical schemes have been developed: Time Relaxed Monte Carlo (TRMC) methods [43], which are asymptotic preserving methods. In the following, we will summarise briefly different asymptotic preserving numerical methods developed in the context of both 1D and 2D/3D models.

7.2.3 Multi-Scale Stiff Kinetic Problems: Asymptotic Preserving Methods and Domain Decomposition Methods

Mathematical models described by hyperbolic and kinetic equations sometimes include parameters that vary widely over the parameter space, and thus the standard numerical solvers might not work well due to high computational costs. One could think about using implicit schemes, which allow for larger time steps and faster computation times, but since the collision operator is nonlinear and nonlocal, there are problems with inverting this operator. To address this numerical problem, there are different methods, such as domain decomposition methods and asymptotic preserving methods [49–51]. The domain decomposition techniques are used to solve the kinetic equations only on some small regions of the domain, where the dynamics of the model departs from the thermodynamic equilibrium. For the rest of the domain one uses classical numerical schemes to approximate the macroscopic (hydrodynamic) equations. In contrast, the asymptotic preserving methods use the same numerical scheme for both the perturbation problem and its limit problem. These techniques have been used to explore the collective behaviour of particles in the transition regime between mesoscopic and macroscopic dynamics. In the following we briefly illustrate both approaches. For the asymptotic preserving methods we first discuss a numerical scheme introduced in [52] in the context of nonlocal 1D models for collective animal behaviours, and then discuss a time relaxation scheme introduced in [53] in the context of general kinetic Boltzmann equations. However, for a more detailed review of asymptotic preserving schemes in the context of other types of hyperbolic and kinetic models with applications to physics, we direct the reader to the study in [54]. For the domain decomposition techniques we focus on a scheme proposed in [55].

- **Asymptotic preserving methods.** Before we give examples of asymptotic preserving numerical schemes, it is useful to first define the concept. To this end, we adopt the same definition as in [1]:

  Definition 7.2 Consider a kinetic equation given by (7.56). A consistent and stable numerical discretisation method that uses a time step size Δt is called *asymptotic preserving* if, for fixed Δt, in the limit $\epsilon \to 0$ it becomes a consistent and stable numerical method for the limiting system.

  We say that a numerical method is *stable* if the numerical errors grow slowly over many time steps [8]. Moreover, a method is *consistent* if the numerical solution is closer and closer to the exact solution as the time step becomes smaller [8].

  As mentioned above, the basic framework for solving kinetic equations employs a time-splitting approach. Therefore, in the following we briefly discuss two such time-splitting schemes used for 1D and 2D/3D kinetic models.

  - **1D model.** A numerical asymptotic preserving method has been used in [52] to investigate the preservation of numerical patterns exhibited by the following nonlocal kinetic two-speed model (with $v = \pm\gamma$; see also Eq. (5.14))

$$\frac{\partial u^+}{\partial t} + \gamma \frac{\partial u^+}{\partial x} = -u^+ \lambda^+ [u^+, u^-] + u^- \lambda^- [u^+, u^-], \tag{7.58a}$$

$$\frac{\partial u^-}{\partial t} - \gamma \frac{\partial u^-}{\partial x} = u^+ \lambda^+ [u^+, u^-] - u^- \lambda^- [u^+, u^-], \tag{7.58b}$$

  following the parabolic scaling of space and time variables: $x = \tilde{x}/\epsilon$ and $t = \tilde{t}/\epsilon^2$. For notational simplicity, in the following we ignore the "$\sim$" from these time and space variables. The limiting parabolic equation (obtained for $\epsilon \to 0$) is

$$\frac{\partial u}{\partial t} = D_0 \frac{\partial^2 u}{\partial x^2} - B_0 \frac{\partial}{\partial x}\big(u(f^-[u] - f^+[u])\big), \tag{7.59}$$

  where, $u = u^+ + u^-$, $\lambda^\pm = \lambda_1 + \lambda_2 f(y^\pm[u])$ and $f^\pm[u] := f(y^\pm[u])$.

  The stiff reaction part in (7.58), can be re-written in terms of the total density (u) and the rescaled flux ($J = (u^+ - u^-)/(\epsilon)$) [52],

$$\frac{\partial u}{\partial t} = 0$$

$$\frac{\partial J}{\partial t} = \frac{1}{\epsilon^2} u \lambda_2 (f[y^-] - f[y^+]) + (1 - \frac{1}{\epsilon^2})\gamma \frac{\partial u}{\partial x}$$

$$\qquad - \frac{1}{\epsilon^2} J\big(2\lambda_1 + \epsilon\lambda_2(f[y^+] + f[y^-])\big),$$

  and then discretised using an implicit Euler scheme, where the total density is evaluated at full grid points $x_i = i\Delta x$, and the flux is evaluated at the half grid points $x_{i+1/2} = (i + 1/2)\Delta x$. The discrete time variable is defined as $t_n = n\Delta t$. This leads to an explicit expression for the flux J^* at the intermediate time step:

$$J^*_{i+1/2} = \frac{\epsilon^2 J^n_{i+1/2} + \gamma \frac{\Delta t}{\Delta x}(\epsilon^2 - 1)(u^n_{i+1} - u^n_i)}{\epsilon^2 + 2\lambda_1 \Delta t + \epsilon\lambda_2 \Delta t (f^+[u^n] + f^-[u^n])_{i+1/2}}$$

$$+ \frac{\lambda_2 \Delta t \left((f^-[r^n] - f^+[u^n])^+_{i+1/2} u^n_i + (f^-[u^n] - f^+[u^n])^-_{i+1/2} u^n_{i+1} \right)}{\epsilon^2 + 2\lambda_1 \Delta t + \epsilon\lambda_3 \Delta t (f^+[u^n] + f^-[u^n])_{i+1/2}}.$$

The transport part,

$$\frac{\partial u}{\partial t} + \gamma \frac{\partial J}{\partial x} = 0,$$

$$\frac{\partial J}{\partial t} + \gamma \frac{\partial u}{\partial x} = 0,$$

was discretised in [52] using a classical upwind scheme. Because of the evaluation of the flux at the half-grid points, the discretisation of the transport part was chosen independently of the sign of the drift:

$$\frac{1}{\Delta t}(u^{n+1}_i - u^*_i) + \frac{1}{\Delta x}(J^*_{i+1/2} - J^*_{i-1/2}) = 0,$$

$$\frac{1}{\Delta t}(J^{n+1}_{i+1/2} - J^*_{i+1/2}) + \frac{1}{\Delta x}(u^*_{i+1} - u^*_i) = 0.$$

Finally, taking the limit $\epsilon \to 0$ in the formula for $J^*_{i+1/2}$, and substituting the result in the discretised equation for the transport part, gives following discretisation for the limiting parabolic equation (7.59):

$$\frac{u^{n+1}_i - u^n_i}{\Delta t} = \frac{D_0}{(\Delta x)^2}(\partial^c_{xx}u^n)_i$$

$$- \frac{B_0}{\Delta x}\left(u^n_i(f^-[u^n] - f^+[u^n])^+_{i+1/2} \right.$$

$$\left. - u^n_{i-1}(f^-[u^n] - f^+[u^n])^+_{i-1/2} \right)$$

$$- \frac{B_0}{\Delta x}\left(u^n_{i+1}(f^-[u^n] - f^+[u^n])^-_{i+1/2} \right.$$

$$\left. - u^n_i(f^-[u^n] - f^+[u^n])^-_{i-1/2} \right).$$

- **2D/3D models.** As discussed before, the Boltzmann equations that include a relaxation process (as a result of fast collision between particles) take the form

$$\frac{\partial u}{\partial t} + v \cdot \nabla_x u = \frac{1}{\epsilon} Q(u, u), \quad u(\mathbf{x}, t = 0, v) = u_0(\mathbf{x}, v). \tag{7.60}$$

The limit $\epsilon \to 0$ corresponds to the hydrodynamic approximation of the kinetic model, where the local mass density (ρ), momentum (ρw) and temperature (T) converge to the solution of compressible Euler equations [53, 56]:

$$\frac{\partial \rho}{\partial t} + \nabla \cdot (\rho f) = 0, \tag{7.61a}$$

$$\frac{\partial (\rho w)}{\partial t} + \nabla \cdot (\rho w \otimes w) + \nabla(\rho T) = 0, \tag{7.61b}$$

$$\frac{\partial E}{\partial t} + \nabla \cdot (Ew + \rho T w) = 0, \quad \text{with } E = \frac{3}{2}\rho T + \frac{1}{2}\rho w^2, \tag{7.61c}$$

where "$\otimes$" describes the tensor product between two vectors, i.e., for any two vectors $a = (a_1, a_2)$ and $b = (b_1, b_2)$, their tensor product is

$$a \otimes b = \begin{pmatrix} a_1 b_1 & a_1 b_2 \\ a_2 b_1 & a_2 b_2 \end{pmatrix}. \tag{7.62}$$

The macroscopic variables are described as follows (in terms of the kinetic variables u and v):

$$\rho(\mathbf{x}, t) = \int_{\mathbb{R}^3} u(\mathbf{x}, t, v) dv, \tag{7.63a}$$

$$\rho w(\mathbf{x}, t) = \int_{\mathbb{R}^3} vu(\mathbf{x}, t, v) dv, \tag{7.63b}$$

$$T(\mathbf{x}, t) = \frac{1}{3\rho} \int_{\mathbb{R}^3} [v - w(\mathbf{x}, t)]^2 u(\mathbf{x}, t, v) dv. \tag{7.63c}$$

In this hydrodynamic case, the probabilistic numerical methods that use random particles need to consider a very large number of particles, which slows down the computation of the solution.

From a physical point of view, the collisions cause the velocity distribution to approach the local Maxwellian faster before the transport part has any effect [53]. Therefore, a numerical approximation of the solution should bring the density near the local Maxwellian. Gabetta et al. [53] considered this aspect and proposed a class of numerical scheme for the relaxation process: the so-called Time Relaxed (TR) methods. These schemes start by representing the solution in a power series for the mean values of successive iterations of the bilinear operator $Q(u, u)$. To this end, the solution of the discretised collision problem (on the time interval $[t_n, t_{n+1}]$)

$$\frac{\partial u^{n+1}}{\partial t} = \frac{1}{\epsilon} Q(u^{n+1}, u^{n+1}), \tag{7.64}$$

with $u^{n+1}(\mathbf{x}, t_n, v) = u^*(x, t_n, v)$ and u^* the solution of the transport problem $\frac{\partial u}{\partial t} + v \cdot \nabla_{\mathbf{x}} u = 0$ (calculated during the first step of the time-splitting approach), can be written in the form of a power series. To obtain a approximation that is valid in the $\epsilon \ll 1$ regime, Gabetta et al. [53] rescaled the time variable $t' = (1 - e^{-\mu t/\epsilon})$, and $U(t', v) = u(t, v)e^{\mu t/\epsilon}$, which lead to the following formal expansion for the density of particles:

$$u(t) = e^{-\mu t/\epsilon} \sum_{k=0}^{\infty} (1 - e^{-\mu t/\epsilon})^k u_k, \tag{7.65}$$

with functions u_k given by the following recurrent formula:

$$u^{k+1}(v) = \frac{1}{k+1} \sum_{h=0}^{k} \frac{1}{\mu} Q(u^h, u^{k-h}), \quad k = 0, 1, \ldots \tag{7.66}$$

Using this representation, Gabetta et al. [53] constructed the following class of numerical schemes:

$$u^{n+1}(v) = e^{-\mu \Delta t/\epsilon} \sum_{k=0}^{m} (1 - e^{-\mu \Delta t/\epsilon})^k u_k^n(v) + (1 - e^{-\mu \Delta t/\epsilon})^{m+1} M(v),$$

$$\tag{7.67}$$

where M is the asymptotic stationary solution of the kinetic equation (i.e., the local Maxwellian associated with u; see also Chap. 6). It has been shown that these schemes are of order m in time, and ensure the conservation of mass, momentum and energy for the Boltzmann equation [53]. Moreover, the schemes guarantee the correct hydrodynamic limit. Finally, these schemes could be written using the following general formula [50]:

$$u^{n+1} = \sum_{k=0}^{m} A_k u_k + A_{m+1} M, \tag{7.68}$$

where coefficients u_k are given by (7.66), and the weights $A_k = A_k(t')$ are non-negative functions that satisfy the following three properties:

consistency:

$$\lim_{t'\to 0} A_1(t')/t' = 1, \quad \lim_{t'\to 0} A_k(t')/t' = 0, \quad k = 2,\dots,m+1;$$

conservation:

$$\sum_{k=0}^{m+1} A_k = 1, \quad t' \in [0, 1];$$

asymptotic preserving:

$$\lim_{t'\to 1} A_k(t') = 0, \quad k = 0,\dots m.$$

An example of such weight functions is given in [50]:

$$A_k = (1 - t')(t')^k, \quad k = 0,\dots,m, \quad A_{k+1} = (t')^{m+1}.$$

- **Domain decomposition methods.** A natural approach of approximating the solutions of multi-scale kinetic equations is to sub-divide the domain into regions corresponding to the hydrodynamic and kinetic descriptions of the models, and solve the corresponding models on each sub-domain using appropriate solvers. However, since in general the two regions are not known, the boundary between them might have to be computed at every time step as part of the solution [1].

  There are many numerical methods derived to solve efficiently the solution of kinetic equations on different subdomains: from moving interface methods [55], to moment guided Monte Carlo methods [57], low-variance deviational Monte Carlo methods [58] or hybrid multiscale methods [59, 60]. In the following we discuss briefly the moving interface method proposed in [55, 61], and refer the reader to [1] for a review of other methods.

  Consider the following 1D kinetic model (where the collision term is given by the simplified Bhatnagar-Gross-Krook (BKG) operator) [55],

$$\frac{\partial u}{\partial t} + v\frac{\partial u}{\partial x} = \frac{1}{\epsilon}\big(M(u) - u\big), \quad \text{with } u(x, v, 0) = u_0(x, v), \tag{7.69}$$

where $x, v \in \mathbb{R}, t \in \mathbb{R}^+$, ϵ is a relaxation time, and $M(u)$ describes the local Maxwellian distribution (6.17) (i.e., the collisions are modelled by a relaxation towards the equilibrium $M(u)$). In the limit $\epsilon \to 0$ the density u converges to the Maxwellian distribution, and one can obtain the (macroscopic) Euler equations (see Eq. (7.61)).

To couple the macroscopic Euler equations with the mesoscopic equation (7.69), Degond et al. [55] assumed that the domain (i.e., the real line) can be written as $\mathbb{R}^1 = \Omega_1 \cup \Omega_2 \cup \Omega_3$, where the three disjoint subdomains were chosen as follows: on Ω_1 the model behaves kinetically, on Ω_2 the model behaves macroscopically (i.e., the flow is close to the Maxwellian equilibrium), and Ω_3 is chosen as the buffer zone between Ω_1 and Ω_2. Degond et al. [55] defined a time-dependent function $h(x, t)$ which encoded the topology and geometry of the three subdomains:

$$h(x, t) = \begin{cases} 1, & x \in \Omega_1, \\ 0, & x \in \Omega_2, \qquad \text{with } h(x, 0) = h_0(x). \\ 0 \leq h(x, t) \leq 1, & x \in \Omega_3, \end{cases} \qquad (7.70)$$

For example, if $\Omega_1 = (-\infty, a)$, $\Omega_2 = (b, \infty)$ and $\Omega_3 = [a, b]$, the function $h(x, t)$ can be chosen piecewise linear: $h(x, t) = (x - b)/(a - b)$, for $x \in [a, b]$. With the help of this function $h(x, t)$, Degond et al. [55] defined two distribution functions,

$$u_R = hu, \quad u_L = (1 - h)u,$$

whose evolution is given by

$$\frac{\partial u_R}{\partial t} = \frac{\partial (hu)}{\partial t} = u \frac{\partial h}{\partial t} + h \frac{\partial u}{\partial t},$$

$$\frac{\partial u_L}{\partial t} = \frac{\partial (1 - h)u}{\partial t} = -u \frac{\partial h}{\partial t} + (1 - h) \frac{\partial u}{\partial t}.$$

Substituting $\partial u / \partial t$ from Eq. (7.69) into the above equations leads to the following coupled equations for u_R and u_L (where $u = u_L + u_R$):

$$\frac{\partial u_R}{\partial t} + hv \left(\frac{\partial u_R}{\partial x} + \frac{\partial u_L}{\partial x} \right) = \frac{h}{\epsilon} (M(u) - u) + u \frac{\partial h}{\partial t}, \qquad (7.71a)$$

$$\frac{\partial u_L}{\partial t} + (1 - h)v \left(\frac{\partial u_L}{\partial x} + \frac{\partial u_R}{\partial x} \right) = \frac{(1 - h)}{\epsilon} (M(u) - u) - u \frac{\partial h}{\partial t}, \qquad (7.71b)$$

with initial conditions

$$u_R(x, v, 0) = h_0(x)u_0(x, v), \quad u_L(x, v, 0) = (1 - h_0(x))u_0(x, v).$$

In region Ω_2 one has $h = 0$, and the distribution $u = u_L$ is close to its associated local Maxwellian $M(u) = M(u_L)$. Replacing u_L by $M(u_L)$ in Eq. (7.71b) and collecting the hydrodynamic moments (mass, momentum, energy) leads to the following modified Euler equations (for $x \leq b$) [55]:

$$\frac{\partial \rho_L}{\partial t} + (1 - h) \frac{\partial}{\partial x} (\rho_L w_L) = -(1 - h) \frac{\partial}{\partial x} \left(\int_{\mathbb{R}} v u_R dv \right) - \rho \frac{\partial h}{\partial t},$$

$$(7.72a)$$

$$\frac{\rho_L w_L}{\partial t} + (1 - h)\frac{\partial}{\partial x}(\rho_L w_L^2 + p_L) = -(1 - h)\frac{\partial}{\partial x}\left(\int_{\mathbb{R}} v^2 u_R dv\right) - \rho w \frac{\partial h}{\partial t},$$
$$\tag{7.72b}$$

$$\frac{\partial E_L}{\partial t} + (1 - h)\frac{\partial}{\partial x}\left((E_L + p_L)w_L\right) = -(1 - h)\frac{\partial}{\partial x}\left(\int_{\mathbb{R}} v\frac{|v|^2}{2} u_R dv\right) - E\frac{\partial h}{\partial t}.$$
$$\tag{7.72c}$$

Here, $\rho = \int_{\mathbb{R}} u \, dv$ is the density, $w = \int_{\mathbb{R}} vu \, dv$ is the mean velocity, and $T = (1/2\rho)\int_{\mathbb{R}} |v - w|^2 u \, dv$ is the temperature in the local Maxwellian distribution $M[\rho, w, T](v) = (\rho/(2\pi T)^{3/2})\exp(-|w - v|^2/(2T))$. Moreover, $p_L = \rho_L T_L$ and the energy $E_L = (1/2)\int_{\mathbb{R}} |v|^2 u_L \, dv = \rho_L(3T_L + w_L^2)/2$. The initial conditions for this system are

$$(\rho_L, w_L, T_L)(x, 0) = (1 - h_0(x))(\rho, w, T)(x, 0).$$

In region Ω_1, the solution is given by u_R that satisfies

$$\frac{\partial u_R}{\partial t} + hv\frac{\partial u_R}{\partial x} + hv\frac{\partial}{\partial x}M[w_L, T_L] = \frac{h}{\epsilon}\left(M(u) - u\right) + u\frac{\partial h}{\partial t}. \tag{7.73}$$

Finally, in region Ω_3, the solution is given by $M(u_L) + u_R$. Once these solutions are specified on each of the sub-domains, one can use classical numerical methods (e.g., finite volume schemes) to discretise Eqs. (7.72) and (7.73). Note that this method can be applied also in higher dimensions (i.e., $x, v \in \mathbb{R}^3$ [55]).

We conclude this discussion by emphasising that different domain decomposition methods have been applied to investigate multiscale problems that contain isolated structures inside the domains, such as shocks, contacts and corners of rarefaction fans [62]. For more details on the various domain decomposition methods, we refer the reader to [49, 60, 63].

While the time-splitting approaches discussed throughout this chapter are widely used when approximating kinetic equations, they usually suffer from order reduction [1]. The last few years have seen the development of a new class of numerical schemes without time splitting (e.g., exponential Runge-Kutta methods), which allow for the derivation of uniformly accurate higher order methods [1]. The difficulty with implementing the schemes without time splitting is related to the time-dependence of the local Maxwellian, which does not allow a direct application of the methods previously developed for the collision part (without time dependance).

Since the purpose of the results of this chapter is not to describe in detail the various asymptotic preserving numerical methods for kinetic and hyperbolic equations derived over the last few decades, but rather to give the reader an overview of the various approaches that could be considered when approximating numerically the solution of the models discussed in Chaps. 3–6, we conclude our discussion on these numerical methods by referring the reader to the more detailed studies and reviews in [1, 7, 32, 43, 50, 54, 64, 65] and the references therein.

7.3 Boundary Conditions

While choosing the most efficient numerical scheme to discretise the hyperbolic and kinetic models discussed throughout this monograph is an important aspect, equally important is to deal with the boundary of the domain (as animal/cell/bacterial dynamics usually occurs on bounded domains). Even when considering an initial value problem on an infinite domain (e.g., $\mathbb{R}^n$, with $n = 1, 2, 3$), one needs to truncate this domain for computational purposes. Boundary conditions also need to be carefully chosen for the domain decomposition methods, where the different regions (for the kinetic/mesoscopic and macroscopic dynamics) are coupled together [66, 67]. Generally, the boundary conditions depend on the geometry of the domain, but for models describing the collective movement of animals the domains have usually classical shapes (e.g., rectangular or circular). Moreover, these boundary conditions need to preserve the positivity of the solution, and for conservation equations they also need to preserve the total mass.

We need to emphasise that while for parabolic equations one can prescribe a condition for population density at each boundary, for the hyperbolic equations one can prescribe conditions for the population density only at those boundaries where the characteristics are pointing inwards [68, 69]. (Note that solutions are already known at the boundaries where the characteristics are pointing outwards, since these solutions have been transported by the characteristic lines from inside the domain. If we impose extra conditions at these boundaries, then the solutions arriving at the boundaries might not match the imposed conditions.) Hence, for the kinetic/hyperbolic transport models discussed throughout this study, the boundary conditions will be imposed only for population densities moving away from the boundaries into the domain (i.e., densities with ingoing velocities).

As mentioned briefly in Chap. 4, relevant boundary conditions (BCs) for models describing biological aggregations are: (a) periodic BCs, to describe arena-like domains (on which many experiments are performed; e.g., [70, 71]); (b) reflecting (Neumann) BCs, to describe walled domains; or (c) zero (Dirichlet) BCs, to describe domains surrounded by regions which do not support the survival of the population. Note that these BCs can be applied also to Boltzmann-like equations for the collective movement of animals/cells/bacteria (although the classical BCs for Boltzmann equations assume that a fraction $(1 - \alpha)$ of particles are reflected elastically at the wall, while the remaining fraction α leaves the wall in a Maxwellian distribution [1, 72, 73]). In Fig. 7.4 we describes these different types of BCs applied to (a) the hyperbolic 1D systems (4.3) and (5.14) [68, 69], and (b) the kinetic 2D systems (6.14) [68]. Note that the boundary conditions in Fig. 7.4b, are a simplified

(a) **1D hyperbolic domain: D = [0,L]**

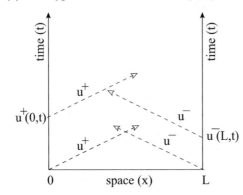

Homogenous Dirichlet BCs:
$$u^+(0,t)=0, \quad u^-(L,t)=0;$$

Homogenous Neumann BCs:
$$u^+(0,t)=u^-(0,t), \quad u^-(L,t)=u^+(L,t);$$

Periodic BCs:
$$u^+(0,t)=u^+(L,t), \quad u^-(L,t)=u^-(0,t);$$

(b) **2D kinetic domain: D = $[0,L_{x1}] \times [0,L_{x2}]$ at some arbitrary time "t"**

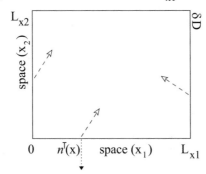

Homogenous Dirichlet BCs:
$$u(x,t,v)=0, \quad \text{for} \quad x \in \delta D, x=(x_1,x_2)$$
$$\text{and} \quad n^T(x)v<0$$

Homogenous Neumann BCs:
$$u(x,t,\bar{v})=u(x,t,v), \quad \text{for} \quad x \in \delta D$$
$$\text{and} \quad n^T(x)\bar{v} = -n^T(x)v$$

Fig. 7.4 (**a**) Examples of three types of boundary conditions (BCs) for the 1D hyperbolic systems (4.3) and (5.14), as described in [68, 69]. (**b**) Examples of two types of boundary conditions for the 2D kinetic systems (6.14), as described in [68]. Here, $n(\mathbf{x})$ is the outer normal at $\mathbf{x}$, and $n^T \mathbf{v} = n(\mathbf{x}) \cdot \mathbf{v}$ is the inner product between the outer normal and the velocity. Moreover, δD denotes the boundary of the domain

version of the more complex Boltzmann boundary conditions in [1, 74, 75]:

$$|n(x) \cdot v|u(x,t,v) = \int_{n(x)\cdot\bar{v}>0} R(v,\bar{v})(n(x) \cdot \bar{v})u(x,t,\bar{v})d\bar{v},$$

$$\text{for every } n(x) \cdot v < 0, \qquad (7.74)$$

where $R(v,\bar{v})$ describes the probability that a particle, which hits the boundary surface with velocity $\bar{v}$, will move away from it with velocity v. Moreover, this probability distribution kernel satisfies the following conditions (which guarantee the positivity and mass conservation at the boundaries):

$$R(v,\bar{v}) \geq 0, \quad \int_{n(x)\cdot v<0} R(v,\bar{v}) = 1. \qquad (7.75)$$

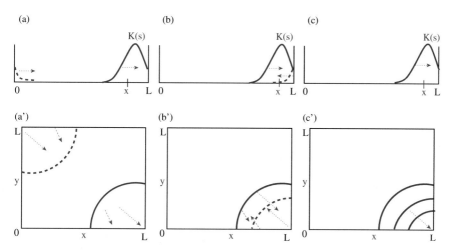

Fig. 7.5 Description of three types of boundary conditions (BCs) and their effects on the kernels that describe the nonlocal terms in both 1D and 2D. (**a**),(**a'**) Periodic BCs; (**b**),(**b'**) Reflective BCs; (**c**),(**c'**) Zero BCs

For the Dirichlet boundary conditions, one could also choose a more general form (when the distribution function of the particles/cells/animals entering the domain D is known) [1]:

$$u(\mathbf{x}, t, v) = g(v, t), \quad \text{for } \mathbf{x} \in \delta D, \quad \text{and} \quad n(x) \cdot v < 0. \tag{7.76}$$

For a larger variety of boundary conditions for Boltzmann equation, we refer the reader to [1, 76].

The presence of nonlocal interaction terms in the hyperbolic and kinetic models requires a special treatment at the boundary of the finite domain. Figure 7.5 depicts the three possible ways of dealing with the integrals (in 1D and 2D) for three different BCs: (a),(a') periodic BCs (where the integrals are wrapped around the domain); (b),(b') reflective BCs (where the integrals are reflected back into the domain); (c),(c') zero BCs (where the integrals are cut off at the boundary (since no population is assumed to survive outside the domain).

To conclude this section, we mention that in the context of asymptotic preserving numerical schemes for kinetic equations (discussed above in Sect. 7.2.3), there is the question of how to choose the boundary conditions so that the numerical solution offers a good approximation for both the mesoscopic and macroscopic regimes, and the different kinetic boundary layers (corresponding to the different order equations obtained during the Hilbert expansion of the density in powers of ϵ). Some studies used boundary conditions from the macroscopic model [77]. However, these conditions might not always provide a good approximation for the parameter regimes where the model is kinetic (i.e., far from the limit $\epsilon \to 0$). To address this issue, Lemou and Méhats [78] proposed a mesoscale-macroscale decomposition of the kinetic problem, where the boundary conditions were generated by the kinetic model but also offered a good approximation of the macroscopic model in the limit $\epsilon \to 0$.

References

1. G. Dimarco, L. Pareschi, Acta Numer. **23**, 369 (2014)
2. G. Colonna, in *Plasma Modelling. Methods and Applications*, ed. by G. Colonna, A. D'Angola (IOP, London, 2016), pp. 1–23
3. R. LeVeque, *Numerical Methods for Conservation Laws* (Birkhäuser, Basel, 1992)
4. C.W. Shu, in *High-Order Methods for Computational Physics*, vol. 9, ed. by T. Barth, H. Deconinck (Springer, Berlin, 1999), pp. 439–582
5. F. Filbet, G. Russo, in *Modelling and Computational Methods for Kinetic Equations*, ed. by P. Degod, L. Pareschi, G. Russo (Birkhäuser, Boston, 2004), pp. 117–145
6. F. Filbet, T. Rey, SIAM J. Sci. Comput. **37**(3), A1218 (2015)
7. L. Pareschi, G. Russo, G. Toscani, *Modelling and Numerics of Kinetic Dissipative Systems* (Nova Science Publ., New York, 2006)
8. S. MacNamara, G. Strang, in *Splitting Methods in Communication, Imaging, Science and Engineering*, ed. by R. Glowinski, S. Osher, W. Yin (Springer, Cham, 2017), pp. 95–114
9. J. Butcher, *Numerical Methods for Ordinary Differential Equations* (Wiley, Hoboken, 2008)
10. W. Hackbusch, *Integral Equations: Theory and Numerical Treatment*. International Series of Numerical Mathematics (Birkhäuser, Basel, 2012)
11. W. Press, S. Teukolsky, W. Vetterling, B. Flannery, *Numerical Recipes in C. The Art of Scientific Computing,* 3rd edn. (Cambridge University Press, Cambridge, 2007)
12. R. LeVeque, *Finite Volume Methods for Hyperbolic Problems* (Cambridge University Press, Cambridge, 2002)
13. C. Chu, *Advances in Applied Mechanics*, vol. 18 (Academic, New York, 1979), pp. 285–331
14. J. Furst, K. Kozel, Math. Bohem. **126**(2), 379 (2001)
15. N. Taniguchi, T. Kobayashi, Comput. Fluids **19**(3–4), 287 (1991)
16. S. Godunov, Math. Sbornik **47**, 271 (1959)
17. A. Harten, J. Comput. Phys. **49**(2), 357 (1983)
18. A. Harten, B. Engquist, S. Osher, S. Chakravarty, J. Comput. Phys. **71**(2), 231 (1987)
19. C.W. Shu, in *Advanced Numerical Approximation of Nonlinear Hyperbolic Equations*, ed. by B. Cockburn, C. Johnson, C.W. Shu, E. Tadmor. Lecture Notes in Mathematics, vol. 1697 (Springer, Berlin, 1998), pp. 325–432
20. C. Shu, SIAM Rev. **51**(1), 82 (2009)
21. E. Toro, *Riemann Solvers and Numerical Methods for Fluid Dynamics* (Springer, Berlin, 2009)
22. H. Nessyahu, E. Tadmor, J. Comput. Phys. **87**, 408 (1990)
23. P. Roe, J. Comput. Phys. **43**, 357 (1981)
24. B.V. Leer, J. Comput. Phys. **23**(3), 263 (1977)
25. P. Roe, Annu. Rev. Fluid Mech. **18**, 337 (1986)
26. P. Sweby, SIAM J. Numer. Anal. **21**(5), 995 (1984)
27. B.V. Leer, J. Comput. Phys. **14**(4), 361 (1974)
28. G.S. Jiang, E. Tadmor, SIAM J. Sci. Comput. **19**(6), 1892 (1998)
29. E. Godlewski, P. Raviart, *Numerical Approximation of Hyperbolic Systems of Conservation Laws* (Springer, New York, 1996)
30. L. Pareschi, G. Russo, ESAIM Proc. **10**, 35–75 (2001)
31. B. Lapeyre, E. Pardoux, R. Sentis, *Introduction to Monte-Carlo Methods for Transport and Diffusion Equations* (Oxford University Press, Oxford, 2003)

32. S. Rajasanow, W. Wagner, *Stochastic Numerics for the Boltzmann Equation* (Springer, Berlin, 2005)
33. L. Pareschi, G. Toscani, *Interacting Multiagent Systems: Kinetic Equations and Monte Carlo Methods* (Oxford University Press, Oxford, 2014)
34. L. Pareschi, G. Toscani, C. Villani, Numer. Math. **93**, 527 (2003)
35. L. Pareschi, B. Perthame, Transp. Theory Stat. Phys. **25**(3–5), 369 (1996)
36. L. Pareschi, G. Russo, Transp. Theory Stat. Phys. **29**(3–5), 431 (2000)
37. L. Pareschi, G. Russo, SIAM J. Numer. Anal. **37**, 1217 (2000)
38. I. Gamba, J. Haack, S. Motsch, J. Comput. Phys. **297**, 32 (2015)
39. P. Degond, L. Pareschi, G. Russo (eds.), *Modelling and Computational Methods for Kinetic Equations* (Birkhäuser, Boston, 2004)
40. F. Rogier, J. Schneider, Transp. Theory Stat. Phys. **23**(1–3), 313 (1994)
41. G. Bird, *Molecular Gas Dynamics* (Oxford University Press, London, 1976)
42. G. Bird, *Molecular Gas Dynamics and Direct Simulation of Gas Flows* (Clarendon Press, Oxford, 1994)
43. L. Pareschi, G. Russo, SIAM J. Sci. Comput. **23**(4), 1253 (2001)
44. K. Nanbu, in *Proceedings of the 15th International Symposium on Rarefied Gas Dynamics*, ed. by V. Boffi, C. Cercignani (1986), pp. 369–383
45. H. Babovsky, Math. Methods Appl. Sci. **8**, 223 (1986)
46. I. Boyd, J. Stark, J. Comput. Phys. **80**(2), 374 (1989)
47. L. Pan, G. Liu, B. Khoo, B. Song, J. Micromech. Microeng. **10**(1), 21 (2000)
48. L. Chao, S. Kwak, S. Ansumali, Int. J. Mod. Phys. **25**, 1340023 (2014)
49. G. Dimarco, L. Pareschi, in *Hyperbolic Problems: Theory, Numerics, Applications*, ed. by S. Benzoni-Gavage, D. Serre (Springer, Berlin, 2008)
50. L. Pareschi, G. Russo, Transp. Theory Stat. Phys. **29**(3–5), 415 (2000)
51. P. Degond, Panoramas et Syntheses **39–40**, 1 (2013)
52. J. Carrillo, R. Eftimie, F. Hoffmann, Kinetic Relat. Model. **8**(3), 413 (2015)
53. E. Gabetta, L. Pareschi, G. Toscani, SIAM. J. Numer. Anal. **34**(6), 2168 (1997)
54. S. Jin, Riv. Mat. Univ. Parma **3**, 177 (2012)
55. P. Degond, G. Dimarco, L. Mieussens, J. Comput. Phys. **227**(2), 1176 (2007)
56. C. Cercignani, *The Boltzmann Equation and Its Applications* (Springer, New York, 1987)
57. G.D.P. Degond, L. Pareschi, Int. J. Numer. Methods Fluids **67**(2), 189 (2011)
58. G. Radtke, N. Hadjiconstantinou, Phys. Rev. E **79**, 056711 (2009)
59. L. Pareschi, R. Caflisch, IMA J. Appl. Math. **135**, 57 (2004)
60. L. Pareschi, ESAIM Proc. **15**, 87 (2005)
61. P. Degond, S. Jin, L. Mieussens, J. Comput. Phys. **209**, 665 (2005)
62. S. Chen, E. Weinan, Y. Liu, C.W. Shu, J. Comput. Phys. **225**(2), 1314 (2007)
63. G. Radtke, J.P. Péraud, N. Hadjiconstantinou, Philos. Trans. R. Soc. A **23**, 030606 (2013)
64. W. Ren, H. Liu, S. Jin, J. Comput. Phys. **276**, 380 (2014)
65. B. Zhang, H. Liu, S. Jin, J. Comput. Phys. **305**, 575 (2016)
66. J. Bourgat, P. LeTallec, B. Berthame, Y. Qiu, Contemp. Math. **157**, 377 (1994)
67. S. Tiwari, J. Comput. Phys. **144**(2), 710 (1998)
68. K. Hadeler, Math. Comput. Model. **31**(4–5), 75 (2000)
69. T. Hillen, Can. Appl. Math. Q. **18**(1), 1 (2010)
70. J. Buhl, D.J.T. Sumpter, I.D. Couzin, J.J. Hale, E. Despland, E.R. Miller, S.J. Simpson, Science **312**, 1402 (2006)
71. A. Portz, A. Seyfried, in *Pedestrian and Evacuation Dynamics*, ed. by R. Peacock, E. Kuligowski, J. Averill (Springer, Boston, 2011), pp. 577–586
72. F. Filbert, Multiscale Model. Simul. **10**(3), 792 (2012)
73. J.P. Péraud, C. Landon, N. Hadjiconstantinou, Annu. Rev. Heat Tranf. **17**, 205 (2014)
74. C. Cercignani, *Theory and Application of the Boltzmann Equation* (Scottish Academic Press, Edinburgh, 1975)
75. S. Ansumali, I. Karlin, Phys. Rev. E **66**(2), 026311 (2002)

76. C.D. Wilson, R.K. Agarwal, F.G. Tcheremissine, Evaluation of various types of wall boundary conditions for the Boltzmann equation. AIP Conf. Proc. **1333**, 146–151 (2011)
77. S. Jin, L. Pareschi, G. Toscani, SIAM J. Numer. Anal. **38**, 312 (2000)
78. M. Lemou, F. Méhats, SIAM J. Sci. Comput. **34**(6), B734 (2012)

Chapter 8
A Few Notions of Stability and Bifurcation Theory

While numerical approaches are a very important step in investigating the patterns exhibited by the hyperbolic and kinetic models discussed in the previous chapters, they could be slow and might not offer a full understanding of the models' dynamics due to the very large parameter space associated with some models. Moreover, the analytical approaches discussed in Chap. 2 could offer an understanding of the parameter space where different types of solutions could occur (e.g., finite vs. density-blow up solutions, shocks vs. rarefaction waves, etc.). However, they cannot offer much insight into the conditions for the formation of patterns, as well as the transitions between different patterns.

Stability theory could identify the parameter conditions under which a pattern could form, and eventually could become unstable giving rise to a different pattern. While linear and weakly-linear stability analyses of homogeneous steady states are relatively easy tasks, stability analysis of spatially heterogeneous solutions is complicated by the complexity of the hyperbolic and kinetic models discussed throughout this monograph, and in particular the nonlinear and nonlocal structure of some of these models. Also difficult is the fully nonlinear stability analysis, which is often specific to the system being investigated [1]. As already mentioned, the nonlocality of the models presented in this monograph complicates the analysis even more—which explains the lack of studies focused on the nonlinear analysis of nonlocal (hyperbolic and kinetic) models for collective dynamics in biological aggregations.

A deeper understanding of the formation of various spatial and spatio-temporal patterns is offered by the bifurcation theory, which can distil the mathematical and biological mechanisms not only behind the formation of patterns, but also behind the transitions between different spatial and spatio-temporal patterns. In the following, we will review some basic notions of linear stability analysis for pattern formation in partial differential equations, as well as basic notions of symmetry theory and bifurcation theory. These will help the reader understand better the approaches taken by some of the studies reviewed in Chaps. 3–6. For more detailed discussions of

© Springer Nature Switzerland AG 2018

R. Eftimie, *Hyperbolic and Kinetic Models for Self-organised Biological Aggregations*, Lecture Notes in Mathematics 2232,
https://doi.org/10.1007/978-3-030-02586-1_8

these topics in stability and bifurcation theory, we refer the reader to the books by Hoyle [2], Golubitsky and Stewart [3], Chossat and Lauterbach [4], Haragus and Iooss [5], Kuznetsov [6], and Strogatz [7].

8.1 Basic Notions of Linear Stability Analysis

The first step in the investigation of pattern formation, is the identification of steady states (spatially homogeneous and, if possible, spatially heterogeneous) and their stability—since unstable states are usually associated with pattern formation and transitions between different patterns. The linear stability technique involves the identification of the eigenvalues of the linearised equation/system at the equilibrium (steady state) points, with the goal of understanding the quantitative behaviour of the solution near these points

We start the discussion of linear stability analysis by focusing first on ODEs, and then on PDEs. Since the majority of studies in the mathematical literature exemplify the linear stability analysis by focusing on parabolic reaction-diffusion equations [8], here we decided to change a bit the approach and to focus on nonlocal hyperbolic systems. This is particularly relevant in the context of the models discussed throughout the previous chapters.

8.1.1 Linear Stability Analysis for ODE Models

Consider the following ODE model

$$\frac{du}{dt} = f(u), \quad \text{with } u, f \in \mathbb{R}^n. \tag{8.1}$$

The dynamics of this system is controlled, in the long term, by the *steady states* (or *fixed points* or *equilibrium points*) of the system. A *steady state* of system (8.1) is a time-independent solution $u(t) = u^*$ that satisfies $f(u^*) = 0$.

To investigate the linear stability of these steady states u^*, we consider small temporal perturbations: $u(t) = u^* + ae^{\lambda t}$. After substituting these expressions back into (8.1) and linearising the nonlinear terms $f(u)$ about the steady states, we obtain that the linear stability of these states is controlled by the eigenvalues of the Jacobian matrix J:

$$J(u^*) = D_u f(u^*) = \begin{pmatrix} \frac{\partial f_1}{\partial u_1} & \cdots & \frac{\partial f_1}{\partial u_n} \\ \cdots & \cdots & \cdots \\ \frac{\partial f_n}{\partial u_n} & \cdots & \frac{\partial f_n}{\partial u_n} \end{pmatrix}_{u=u^*}$$

$$\text{for } f = (f_1, \ldots, f_n), \ u = (u_1, \ldots, u_n). \tag{8.2}$$

If all eigenvalues λ of the characteristic equation $\det(J(u) - \lambda I) = 0$ have negative real parts, we say that the steady state is linearly stable. If there are eigenvalues with positive real parts we say that the steady state is unstable.

For a system in $\mathbb{R}^2$ one can classify the fixed points in terms of the determinant $\det(J) = \lambda_1 \lambda_2$ and trace $Tr(J) = \lambda_1 + \lambda_2$ of the Jacobian matrix (where λ_1 and λ_2 are the two eigenvalues). For example, if $\det(J) < 0$, the eigenvalues are real and of opposite signs and the fixed point is a saddle point. If $\det(J) > 0$ the eigenvalues are either real and of opposite signs (and thus they are nodes) or complex conjugates (and thus they are spirals or centres). If $\det(J) = 0$, at least one eigenvalue is zero. The stability of the nodes and spirals is given by $Tr(J)$: the fixed points are stable for $Tr(J) < 0$, and unstable for $Tr(J) > 0$. When $Tr(J) = 0$ the eigenvalues are purely imaginary, and the fixed points are centres. For more details on this fixed point classification, see [7, 9].

Definition 8.1 A fixed point u^* of (8.1) is called *hyperbolic* if and only if the Jacobian matrix $J(u^*)$ does not have any eigenvalues with zero real parts, i.e., $Re(\lambda_i) \neq 0$ for $i = 1, 2, \ldots, n$.

The stability of hyperbolic fixed points is not affected by small perturbations caused by nonlinear small terms (the local phase portrait near a hyperbolic fixed point being topologically equivalent to the phase portrait of the linearised system—see the Hartman-Grobman theorem in [9, 10]). In regard to pattern formation, the important cases are those where the eigenvalues have zero real parts ($Re(\lambda) = 0$), i.e., the fixed points are *non-hyperbolic*. The changes in the stability of fixed points, which suggest the possibility of a bifurcation, can only happen at non-hyperbolic fixed points. To conclude this section, we note that the qualitative behaviour of system (8.1) in the neighborhood of a nonhyperbolic fixed point u^* is determined by its behaviour on the centre manifold near u^*. Moreover, since the dimension of the centre manifold is usually smaller than the dimension of the full system (8.1), it becomes easier to investigate the qualitative behaviour of the system near a nonhyperbolic fixed point. We will return to the discussion of the centre manifold theory in Sect. 8.6. For a more comprehensive study on the stability of fixed points of ODE systems, we refer the reader to [9].

8.1.2 Linear Stability Analysis for PDE Models

Let us focus now on partial differential equations (PDEs), and assume that the models described in the previous chapters can be written in a general form as

$$\frac{\partial u(\mathbf{x}, t)}{\partial t} = \mathscr{L}[u(\mathbf{x}, t)] + \mathscr{N}[u(\mathbf{x}, t)], \quad (\mathbf{x}, t) \in \mathbb{R}^d \times \mathbb{R}^1_+, \tag{8.3}$$

where $\mathscr{L}[u]$ is a linear operator and $\mathscr{N}[u]$ is a nonlinear operator (containing higher order terms $O(u^k)$, $k \geq 2$, both local and nonlocal). Note that we have used $[\cdot]$

instead of (·) to emphasise that these linear and nonlinear terms can depend also nonlocally on $u(\mathbf{x}, t)$.

Definition 8.2 A *spatially homogeneous steady state* of (8.3) is a solution $u(\mathbf{x}, t) = u^* =$ constant which satisfies

$$0 = \mathscr{L}[u^*] + \mathscr{N}[u^*]. \tag{8.4}$$

A *spatially heterogeneous steady state* of (8.3) is a solution $u(\mathbf{x}, t) = u^{**}(\mathbf{x})$ which satisfies

$$0 = \mathscr{L}[u^{**}(\mathbf{x})] + \mathscr{N}[u^{**}(\mathbf{x})]. \tag{8.5}$$

We characterise a steady state as being stable or unstable if small perturbations of this steady state decay or grow. Since a large part of this monograph focuses on nonlocal hyperbolic systems, to exemplify the linear stability technique, we focus on a generic 1D nonlocal hyperbolic system

$$\frac{\partial u^+}{\partial t} + \gamma \frac{\partial u^+}{\partial x} = -\lambda^+[u^+, u^-]u^+ + \lambda^-[u^+, u^-]u^-, \tag{8.6a}$$

$$\frac{\partial u^-}{\partial t} - \gamma \frac{\partial u^-}{\partial x} = \lambda^+[u^+, u^-]u^+ - \lambda^-[u^+, u^-]u^-, \tag{8.6b}$$

on a finite domain $[0, L]$ with periodic boundary conditions, and investigate the linear stability of a generic spatially homogeneous steady state $(u^+(t, x), u^-(t, x)) = (u_+^*, u_-^*)$. For example, the class of nonlocal hyperbolic systems (8.6) with the five communication mechanisms M1–M5 (see Table 5.1) introduced in [11] can exhibit one, three or five steady states; see Fig. 8.1a.

In the following we focus on the linear stability of a generic spatially homogeneous steady state (u_+^*, u_-^*)—any one of the states depicted in Fig. 8.1. (For a detailed discussion of the linear stability approach in reaction diffusion systems, i.e., Turing mechanisms, see [8].) We note that the stability of a spatially heterogeneous steady state $(u_+^{**}(x), u_-^{**}(x))$ follows the same approach, but the calculations are more challenging; see, for example, the studies in [12, 13] in the context of local parabolic equations; to our knowledge, studies on the stability of heterogeneous states exhibited by nonlocal hyperbolic equations/systems are very scarce, due to the challenge posed by dealing with the nonlocal terms. Also challenging is the application of nonlinear stability methods, which can offer more information about the formation of patterns, compared to the classical linear methods [1].

We start the linear stability analysis of a steady state (u_+^*, u_-^*) of system (8.6) by considering small-amplitude perturbations of the steady state: $u^+(x, t) = u_+^* + v^+(x, t)$ and $u^-(x, t) = u_-^* + v^-(x, t)$ with $v^\pm(x, t) \propto a_\pm e^{\sigma t + ikx}$ and $|a_\pm| \ll 1$. Here $\sigma \in \mathbb{C}$ is an eigenvalue that gives the temporal growth/decay of the small perturbations (if $Re(\sigma) > 0$ or $Re(\sigma) < 0$, respectively), and k is the wavenumber (which is a measure of the wavelike pattern, being proportional to the reciprocal of

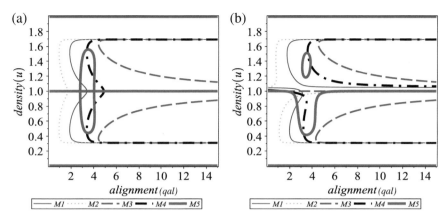

Fig. 8.1 Plot of steady state u_*^+ corresponding to $(u_*^+, u_*^-)=(u_*^+, A - u_*^+)$, as a function of the magnitude of alignment parameter q_{al}, for the five communication models introduced in [11]. In the expression of the steady states, we have $A = \frac{1}{L} \int_0^L (u^+(x) + u^-(x))dx$ the total population density on a finite domain $[0, L]$. We compare the cases of (**a**) *symmetric perception* of neighbours, versus (**b**) *asymmetric perception* of neighbours, as described by Eq. (5.20) and Fig. 5.6. In (**a**) we have $p_+ = p_- = 1.0$ (corresponding to symmetric perception), while in (**b**) we have $p_+ = 1.05$ and $p_- = 0.95$ (corresponding to asymmetric perception)

the wavelength ω of the pattern: $k = 2\pi/\omega$). For finite domains, there is a discrete set of possible wavenumbers $k_n = 2n\pi/L$, where L is the domain size and n is an integer. Substituting the perturbed solutions $u^\pm(x, t) = u_*^\pm + v^\pm(x, t)$ into the linearised hyperbolic system leads to the following equations

$$\frac{\partial v^+}{\partial t} + \gamma \frac{\partial v^+}{\partial x} = -\lambda^+[u_*^+, u_*^-]v^+ + \lambda^-[u_*^+, u_*^-]v^-$$
$$- u_*^+ \lambda_u^+ (K * v^+) + u_*^- \lambda_u^- (K * v^-),$$

$$\frac{\partial v^-}{\partial t} - \gamma \frac{\partial v^-}{\partial x} = \lambda^+[u_*^+, u_*^-]v^+ - \lambda^-[u_*^+, u_*^-]v^-$$
$$+ u_*^+ \lambda_u^+ (K * v^+) - u_*^- \lambda_u^- (K * v^-).$$

Here, $\lambda_u^\pm$ are the derivatives of $\lambda^\pm$ with respect to $u = (u^+, u^-)$, which appear in the Taylor expansion of $\lambda^\pm$ about the steady states (u_*^+, u_*^-). Re-writing these equations in terms of $a_\pm e^{\sigma t + ikx}$, we obtain (after simplifying the exponentials $e^{\sigma t + ikx}$)

$$a_+ \left(\sigma + \gamma ik + \lambda^+[u_*^+, u_*^-] + u_*^+ \lambda_u^+ \hat{K}(k) \right)$$
$$+ a_- \left(-\lambda^-[u_*^+, u_*^-] - u_*^- \lambda_u^- \hat{K}(k) \right) = 0,$$

$$a_+ \left(-\lambda^+[u_*^+, u_*^-] - u_*^+ \lambda_u^+ \hat{K}(k) \right)$$
$$+ a_- \left(\sigma - \gamma ik + \lambda^-[u_*^+, u_*^-] + u_*^- \lambda_u^- \hat{K}(k) \right) = 0,$$

where $\hat{K}(k)$ is the Fourier transform of the interaction kernel $K(s)$:

$$\hat{K}(k) = \int K(s)e^{iks}ds. \tag{8.7}$$

To find a non-trivial solution for this algebraic system, we impose that the determinant is zero, and obtain the characteristic equation that connects the growth rate σ of the perturbations with the wavenumber k:

$$\sigma^2 + \sigma A(k) + B(k) = 0, \tag{8.8}$$

where $A(k)$ and $B(k)$ are nonlinear terms that depend on the parameters of the system and on the steady states. The expression $\sigma = \sigma(k)$ is called a *dispersion relation*.

If $Re(\sigma(k)) > 0$ for some $k = k_n$, we say that the homogeneous steady state (u_*^+, u_*^-) is unstable to spatial perturbations. Otherwise, if $Re(\sigma(k)) < 0$ for all k, we say that the steady state is linearly stable. Note that since we assumed a finite domain (to be able to compare the analytical stability results with the numerical results, as in [11, 14]), the possible unstable wavenumbers k_n and the corresponding spatial wavelengths of allowable patterns could depend on the boundary conditions. The most unstable wavenumber k_n (i.e., the wavenumber for which $\sigma(k_n)$ has the largest positive value) gives—at least for small time where the linear stability analysis is valid—the number of "peaks" (i.e., aggregations) that form in the domain. In Fig. 8.2 we show a caricature description of (a) a typical example of dispersion relation for which the wavenumber k_2 is unstable, and (b) the corresponding two-peak pattern that emerges (at least for small time). (Compare Fig. 8.2a with Fig. 4.7 which showed a non-standard dispersion relation for a class of local hyperbolic systems introduced in [15].) If the eigenvalues $\sigma(k_n)$ have only real parts then the spatial pattern emerges as a result of real

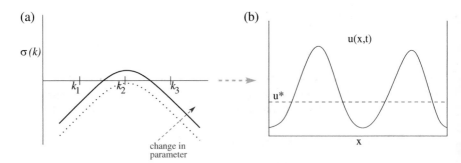

Fig. 8.2 Caricature description of (**a**) a typical example of dispersion relation for which the wavenumber k_2 becomes unstable (as we vary a certain model parameter), and (**b**) the corresponding two-peak pattern $u(x, t)$ that emerges (at least for small time). The dashed line shows the spatially homogeneous solution

(steady state) bifurcations, and the aggregations that form are motionless (e.g., stationary pulses; see also Fig. 1.9a). If, on the other hand, the eigenvalues $\sigma(k_n)$ have complex parts (i.e., $Im(\sigma(k_n)) \neq 0$) then the spatial pattern emerges as a result of complex (Hopf) bifurcations, and the aggregations that form are moving through space (e.g., travelling pulses; see also Fig. 1.9b). We will return to the discussion of real and complex bifurcations in the next section. It is possible that multiple wavenumbers become unstable at the same time; see Fig. 5.10. The spatially heterogeneous solution that emerges, is the sum of the unstable modes: $u(x, t) = \sum_{n1}^{n2} C_n e^{\sigma(k_n)t} cos(k_n x)$. The mode-mode interactions could give rise to more complex spatial and spatio-temporal patterns, as discussed in Chap. 5.

To conclude this brief discussion on linear stability analysis, we emphasise that even if the small perturbations $v^{\pm}$ grow exponentially with time, they are eventually bounded by the nonlinear terms in the reaction-advection equations. If the solution of the PDE is bounded in time, a spatially heterogeneous solution will emerge. For a more detailed discussion of linear stability analysis on pattern formation in partial differential equations in biology (including the stability of steady states for 2D models), we refer the reader to the seminal book by Murray [8].

8.2 Basic Notions of Bifurcation Theory

To be able to understand the changes in the patterns exhibited by various (finite and infinite dimensional) dynamical systems, one needs to have some basic notions of *bifurcation theory*, i.e., the mathematical theory that studies changes in the qualitative or topological structure of a family of differential equations. The term "*bifurcation*" was first introduced by Henri Poincaré in [16]. A bifurcation occurs when a small change in a parameter value (i.e., the bifurcation parameter) leads to a qualitative change in the behaviour of a system. Since in the mathematical literature there are several very good textbooks on bifurcation theory [3, 6, 7, 17, 18], the aim of this chapter is not to give a detailed exposition of the topic, but rather to give the reader enough information to follow the discussion in the previous chapters regarding the mechanisms behind the formation of various patterns. In the following, we will assume that the reader has basic notions of dynamical systems (both finite dimensional and infinite dimensional) and functional analysis; see also the books by Strogatz [7], Robinson [19] and Evans [20].

Even if this monograph focuses on PDEs, we decided to start this brief review of basic notions of bifurcation theory by focusing first on classical bifurcations for ODEs (one dimensional and two dimensional). Our reason for this choice is based on (1) the importance of these classical bifurcations for understanding the long-term dynamics of spatially homogeneous populations (i.e., populations with individuals evenly distributed over the domain), and (2) the importance of these bifurcations in the reduction of infinite-dimensional (PDE) systems to finite-dimensional systems (via Central Manifold reduction, Lyapunov-Schmidt reduction, or weakly nonlinear analysis). In regard to point (1), we note that the class of 1D nonlocal hyperbolic

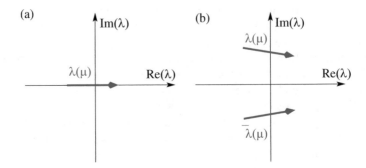

Fig. 8.3 Caricature description of the eigenvalues that generate real and complex bifurcations, as we vary a generic parameter μ. (**a**) Real eigenvalues $\lambda(\mu)$; (**b**) Complex eigenvalues ($\lambda(\mu)$ and $\bar{\lambda}(\mu)$). In general, the complex eigenvalues cross the imaginary axis with nonzero slopes

models introduced in [11] was shown to display spatially homogeneous solutions $(u^+, u^-) = (u_*^+, u_*^-)$ with $u_*^+ \neq u_*^-$ (corresponding to more individuals facing one direction than the other direction), and these solutions arise via saddle-node and (subcritical) pitchfork bifurcations [11, 14]; see also Fig. 8.1. In regard to point (2), we will see in Sect. 8.5 that we could understand the dynamics of a nonlocal hyperbolic system near a bifurcation point via the dynamics of an ODE for the amplitude of the perturbations as given by Eq. (8.38).

The *codimension ("codim") of a bifurcation* is given by the number of parameters that need to be varied to reach the locus of the bifurcation. Throughout Chap. 5 we referred to *codimension-1 bifurcations* (where one parameter μ was varied) and *codimension-2 bifurcations* (where two parameters, μ_1 and μ_2, were varied at the same time). In the following we will review briefly four codimension-1 classical bifurcations from fixed points: saddle-node bifurcations, transcritical bifurcations, pitchfork bifurcations and Hopf bifurcations. The first three types of bifurcations are stationary (or steady state), i.e., they correspond to a real eigenvalue $\lambda(\mu)$ passing through zero (see Fig. 8.3a). The fourth bifurcation is oscillatory, with the real part of the complex eigenvalues passing through zero, while the imaginary part is nonzero (see Fig. 8.3b). Since the majority of bifurcations identified in the literature of hyperbolic and kinetic models for self-organised behaviours are local, here we focus mainly on these local bifurcations. However, towards the end of this section we will also mention briefly some examples of nonlocal bifurcations (e.g., homoclinic loops) exhibited by the nonlocal hyperbolic models (5.14).

The structure of the bifurcations is encoded in their normal forms (i.e., simplified equations that determine the dynamics of the system/bifurcation), and all systems that exhibit a bifurcation are locally topologically equivalent to the normal form of the bifurcation. Thus, in the following we describe briefly the normal forms corresponding to four classical codimension-1 local bifurcations. To this end we start with the following differential equation in $\mathbb{R}$:

$$\frac{du}{dt} = f(u, \mu), \tag{8.9}$$

where u is a real-valued function of time t ($u \in \mathbb{R}^+$) and μ is a real bifurcation parameter ($\mu \in \mathbb{R}$). Assume that the vector field $f(u, \mu)$ satisfies the following two conditions:

$$f(0, 0) = 0, \quad \frac{\partial f(0, 0)}{\partial u} = 0. \tag{8.10}$$

The first condition says that $u = 0$ is an equilibrium point when $\mu = 0$, while the second condition is necessary for the appearance of a local bifurcation at $\mu = 0$. (If $\partial f(0, 0)/\partial u \neq 0$, the implicit function theorem says that $f(u, \mu) = 0$ has a unique solution $u = u(\mu)$ in the neighbourhood of 0, and thus $u = 0$ is the only solution for $\mu = 0$ or sufficiently small μ, leading to the impossibility of having a bifurcation for small values of μ [5]).

- **Saddle-node bifurcations.** Assume that in addition to conditions (8.10), the vector field $f(u, \mu)$ satisfies also two other conditions:

$$\frac{\partial f}{\partial \mu}(0, 0) = 1 \neq 0, \quad \frac{\partial^2 f}{\partial u^2}(0, 0) = c \neq 0. \tag{8.11}$$

Following a Taylor expansion of $f(u, \mu)$ near $(0, 0)$, we obtain the following truncated equation

$$\frac{du}{dt} = \mu + cu^2. \tag{8.12}$$

This normal form Eq. (8.12) approximates the dynamics of the full model (8.9). Equation (8.12) has the following fixed points: $u = 0$ for $\mu = 0$, and $u = \pm\sqrt{-\mu/c}$ for $\mu/c < 0$. The stability of the fixed points is determined by the sign of derivative $\frac{\partial f}{\partial u}$. For $c > 0$ the non-trivial fixed points exist only when $\mu < 0$, and $+2c\sqrt{-\mu/c}$ is unstable while $-2c\sqrt{-\mu/c}$ is stable (see Fig. 8.4a). For $c < 0$ the non-trivial fixed points exist only when $\mu > 0$, and $+2c\sqrt{-\mu/c}$ is stable while $-2c\sqrt{-\mu/c}$ is unstable (see Fig. 8.4a).
- **Transcritical bifurcations.** Assume that in addition to conditions (8.10), the vector field $f(u, \mu)$ satisfies also three other conditions:

$$\frac{\partial f}{\partial \mu}(0, 0) = 0, \quad \frac{\partial^2 f}{\partial u \partial \mu}(0, 0) = 1 \neq 0, \quad \frac{\partial^2 f}{\partial u^2}(0, 0) = 2c \neq 0. \tag{8.13}$$

Expanding $f(u, \mu)$ in Taylor series about $(0, 0)$, and incorporating the above conditions leads to the following truncated normal form equation for a transcritical

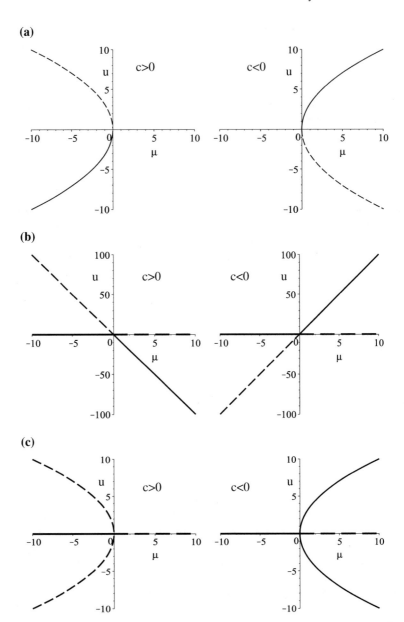

Fig. 8.4 Bifurcation diagrams in the (μ, u) plane of the normal form equations corresponding to:
(**a**) Saddle-node bifurcations; (**b**) Transcritical bifurcations; (**c**) Pitchfork bifurcations. The solid continuous curves describe stable states, while the dashed curves describe unstable states

bifurcation:

$$\frac{du}{dt} = \mu u + c u^2, \quad u, \mu, c \in \mathbb{R}. \tag{8.14}$$

The fixed points of this equation are $u = 0$ and $u = -\mu/c$. Simple linear stability analysis shows that $u = 0$ is stable for $\mu < 0$ and unstable for $\mu > 0$, while $u = -\mu/c$ is stable for $\mu > 0$ and unstable for $\mu < 0$ (see Fig. 8.4b).

- **Pitchfork bifurcations.** In many cases, the models have some sort of symmetry. The simplest symmetry is the reflection symmetry $u \to -u$. We need to emphasise that this symmetry is not biologically realistic in this form, since one cannot have a population with negative density $-u$. However, in some biological systems one could have a slightly different version of this symmetry: $u \to U - u$, with U a maximum populations size; see the steady states in Fig. 8.1a, which are symmetric with respect to $u = A/2$, where $A = (1/L) \int_0^L [u^+(x) + u^-(x)] dx$ denoted the total population density. Assume now that the vector field $f(u, \mu)$ is odd with respect to u, satisfies conditions (8.10) and also the following conditions:

$$\frac{\partial f}{\partial \mu}(0, 0) = 0, \quad \frac{\partial^2 f}{\partial \mu \partial u}(0, 0) = 1 \neq 0, \quad \frac{\partial^3 f}{\partial u^3}(0, 0) = 6c \neq 0. \tag{8.15}$$

Following a Taylor expansion of $f(u, \mu)$ near $(0, 0)$, we obtain the following truncated normal form equation

$$\frac{du}{dt} = \mu u + c u^3. \tag{8.16}$$

This equation has the following fixed points: $u = 0$ for any μ, and $u = \pm\sqrt{-\mu/c}$ for $\mu/c < 0$. These non-trivial points exist for $\mu > 0$ when $c < 0$ and for $\mu < 0$ when $c > 0$. The trivial point is stable for $\mu < 0$ and unstable for $\mu > 0$. The nontrivial point $\pm\sqrt{-\mu/c}$ is unstable for $c > 0$ and $\mu < 0$, and stable for $c < 0$ and $\mu > 0$ (see Fig. 8.4c). The appearance of stable branches for $\mu > 0$ shown in the right panel of Fig. 8.4c occurs through a supercritical bifurcation, while the appearance of unstable branches for $\mu < 0$ in the left panel of Fig. 8.4c occurs through a subcritical bifurcation.

Remark 8.1 The pitchfork bifurcations that give rise to the spatially-homogeneous steady states graphed in Fig. 8.1a are the result of the symmetries of the nonlocal hyperbolic system (8.6) with the five communication mechanisms described in Table 5.1. For symmetric communication mechanisms (i.e., $p_+ = p_- = 1.0$), the bifurcations shown in Fig. 8.1a are perfect. However, as we perturb the perception mechanisms (i.e. $p_+ = 1.05$, $p_- = 0.95$), thus assuming asymmetric communication, we obtain imperfect bifurcation diagrams as a result of symmetry breaking, as shown in Fig. 8.1b.

Remark 8.2 The three bifurcations of fixed points discussed above in the one-dimensional case (i.e., saddle-node, transcritical and pitchfork bifurcations) can be easily generalised to two and higher dimensions. For example, for the normal form equations in 2D (with variables u and v) we can assume that the dynamics in the u-direction is given by the normal forms discussed above, and the dynamics in the v-direction is exponentially damped [7]:

$$\frac{du}{dt} = \mu + cu^2, \quad \frac{dv}{dt} = -v,$$

$$\frac{du}{dt} = \mu u + cu^2, \quad \frac{dv}{dt} = -v,$$

$$\frac{du}{dt} = \mu u + cu^3, \quad \frac{dv}{dt} = -v.$$

For a more detailed discussion regarding the generalisation of these bifurcations to higher dimensions, and the fact that the addition of higher dimensions does not influence the bifurcations (which still occur along a one-dimensional space), see [7].

- **Hopf bifurcations.** Consider now the following differential equation in $\mathbb{R}^2$:

$$\frac{d\mathbf{u}}{dt} = \mathbf{f}(\mathbf{u}, \mu), \quad \text{with } \mathbf{u} = (u, v) \in \mathbb{R}^2, \ \mu \in \mathbb{R}. \tag{8.17}$$

Assume that the vector field $\mathbf{f} \in \mathbb{R}^2$ satisfies $\mathbf{f}(\mathbf{0}, 0) = 0$ (i.e., $\mathbf{u} = \mathbf{0}$ at $\mu = 0$). The presence of a bifurcation is determined by the linearisation of $\mathbf{f}(\mathbf{u}, \mu)$ at $(\mathbf{0}, 0)$, as given by the Jacobian matrix $J = D_{\mathbf{u}}\mathbf{f}(\mathbf{0}, 0)$. Moreover, assume that the Jacobian matrix has the following canonical form

$$J = \begin{pmatrix} \alpha(\mu) & \beta(\mu) \\ -\beta(\mu) & \alpha(\mu) \end{pmatrix}, \tag{8.18}$$

and at $\mu = 0$ we have $\alpha(0) = 0$, $\alpha'(0) \neq 0$ and $\beta(0) = \omega \neq 0$ (so that in the neighbourhood of $\mu = 0$, we have $det(J) \neq 0$). The linearised equations (8.17) are

$$\frac{du}{dt} = \alpha(\mu)u + \beta(\mu)u + O(u^2, v^2, uv), \tag{8.19a}$$

$$\frac{dv}{dt} = -\beta(\mu)u + \alpha(\mu)v + O(u^2, v^2, uv). \tag{8.19b}$$

Let us introduce a new variable $z = u + iv$, which allows us to re-write Eq. (8.19) as

$$\frac{dz}{dt} = \big(\alpha(\mu) - i\beta(\mu)\big)z + O(|z|^2), \quad \text{as } |z| \to 0. \tag{8.20}$$

Making a transformation of the form $\psi = z + S(z, \bar{z}, \mu)$, with $S \approx O(|z|^2)$ leads to a normal form equation

$$\frac{d\psi}{dt} = \big(\alpha(\mu) - i\beta(\mu)\big)\psi + A(\mu)|\psi|^2\psi + O(|\psi|^4), \tag{8.21}$$

with $A(\mu) = a(\mu) + ib(\mu)$ a complex term.

To understand the dynamics of this normal form equation, it is better to introduce the polar coordinates $\psi = re^{i\theta}$ (with $r > 0$ and $0 \le \theta \le 2\pi$), which transforms Eq. (8.21) into the following system:

$$\frac{dr}{dt} = \alpha(\mu)r + a(\mu)r^3, \tag{8.22a}$$

$$\frac{d\theta}{dt} = -\beta(\mu) + b(\mu)r^2. \tag{8.22b}$$

Note that Eq. (8.22a) is the normal form for a pitchfork bifurcation in r. This suggests that a Hopf bifurcation is a pitchfork bifurcation in r direction, with a rotation in θ direction. From the phase equation (8.22b) we obtain $\theta = \theta_0 + \omega(\mu)t$, with $\omega(\mu) = -\beta(\mu) - b(\mu)\alpha(\mu)/a(\mu) \to -\beta(0)$ as $\mu \to 0$ (since we assumed above that $\alpha(0) = 0$).

The fixed points of the amplitude equation (8.22a) are $r = 0$ and $r = \sqrt{-\alpha(\mu)/a(\mu)}$. We assume that $a(\mu) \ne 0$ in the neighbourhood of $\mu = 0$. This non-trivial solution branch corresponds to a periodic solution with period $2\pi/|\omega(\mu)| \to 2\pi/|\omega(0)|$ as $\mu \to 0$. The stability of these two solutions depends on the signs of $\alpha(\mu)$ and $a(\mu)$. While we keep these two functions general enough to not discuss their signs, we graph in Fig. 8.5 the two possible Hopf bifurcations: (a) a supercritical bifurcation, and (b) a subcritical bifurcation.

The Hopf bifurcation represents one way through which limit cycles are created or destroyed. However, limit cycles can be destroyed when two different cycles (a

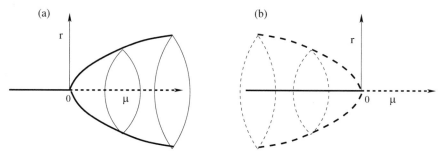

Fig. 8.5 Bifurcation diagram for: (**a**) a supercritical Hopf bifurcation; (**b**) a subcritical Hopf bifurcation. Continuous solid curves indicate stable branches, while dashed curves indicate unstable branches

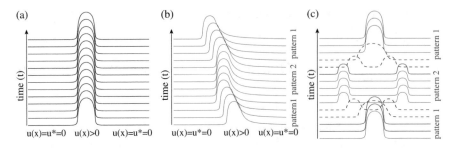

Fig. 8.6 Caricature description of (**a**) a homoclinic orbit (in the form of a stationary pulse connecting two zero steady states); (**b**) a heteroclinic cycle connecting a travelling pulse (pattern 1—in red) and a stationary pulse (pattern 2—in blue); (**c**) a heteroclinic cycle connecting two stationary states (pattern 1—in red; and pattern 2—in blue)

stable and an unstable cycle) coalesce via a *saddle-node bifurcation of cycles* [7]. This is a *global bifurcation* since it does not occur anymore near a fixed point. Note that global bifurcations occur when larger invariant sets, such as periodic orbits, collide with each other or with other equilibria. In the following we discuss briefly two types of global bifurcations that have been shown to be exhibited by the nonlocal hyperbolic systems (8.7).

- **Homoclinic bifurcations** occur when a limit cycle moves closer to a saddle point until it coalesces with it becoming a homoclinic loop. In the context of partial differential equations, a homoclinic loop describes a travelling pulse or a stationary pulse, i.e., a nonzero solution $u(z) = u(x - ct)$ which connects (as $z \to \pm\infty$) the stable and unstable manifolds of a spatially homogeneous steady state (usually $u^* = 0$; see Fig. 5.7a, b) and propagates with speed $c > 0$ (for travelling pulses) or $c = 0$ (for stationary pulses). See Fig. 8.6a for a caricature description of a homoclinic loop.
- **Heteroclinic bifurcations** occur when a cycle connects different unstable (spatially homogeneous or heterogeneous) states, via their stable and unstable manifolds. These bifurcations are more difficult to be identified for the hyperbolic and kinetic models discussed in the previous chapters, due to the large parameter space that needs to be investigated. The nonlocal hyperbolic models (8.7) can exhibit heteroclinic cycles that connect two stationary states (R. Eftimie—unpublished results), which are the result of Steady-state/Steady-state heteroclinic bifurcations; see Fig. 8.6c for a caricature description of a heteroclinic cycle connecting two different stationary states. Figure 8.6b shows a caricature description of a heteroclinic cycle connecting a stationary state and a travelling pulse (which is somehow similar—at macroscopic level—to the semi-zigzag dynamics shown in Fig. 5.7f, but for one aggregation peak). We need to emphasise here that the analytical study of heteroclinic bifurcations in hyperbolic and kinetic models is still an open problem at this moment.

Remark 8.3 As emphasised by Knobloch and Aulbach [21], the various bifurcating objects (which are formed of small bounded solutions, such as steady states or orbits) always lie on centre manifolds. Therefore, from a bifurcation point of view, it is enough to study the flow of a system on the centre manifold. We will return to the existence of these centre manifolds (for both ODEs and PDEs) in Sect. 8.6.

8.3 Symmetry of Hyperbolic and Kinetic Equations

Symmetry and symmetry breaking phenomena are very common in self-organised biological communities, as shown by various experiments [22–26]. From a mathematical point of view, many of the models for self-organised patterns in biological aggregations that we discussed in the previous chapters have some sort of symmetry (either as a result of the equations themselves, or as a result of the domain and the boundary conditions—e.g., periodic conditions). The presence of symmetries is welcomed since it allows us to reduce the size of the system (which reduces the cost of solving the equations). However, very few studies on kinetic and hyperbolic models for biological aggregations recognised the importance of these symmetries on cell/bacterial/animal pattern formation [27–30]. Since the symmetries of systems of differential equations (ODEs and PDEs) are usually discussed in terms of a group of transformations of variables that preserve the structure of the equations and their solutions, in the following we give a brief introduction to the most common notions of group theory that are used to understand the symmetries of a model. To this end, we follow the approaches in [2, 3].

Definition 8.3 A *group* Γ is a set $\{\gamma_1, \gamma_2, \gamma_3, \ldots\}$ together with an operation "·" (which maps $\Gamma \times \Gamma \to \Gamma$) that satisfies the group axioms:

- The group is closed under the group operation: for any $\gamma_1, \gamma_2 \in \Gamma$, then $\gamma_3 = \gamma_1 \cdot \gamma_2 \in \Gamma$;
- Associativity axiom: for any $\gamma_1, \gamma_2, \gamma_3 \in \Gamma$, then $(\gamma_1 \cdot \gamma_2) \cdot \gamma_3 = \gamma_1 \cdot (\gamma_2 \cdot \gamma_3)$;
- Identity axiom: there exists an element $e \in \Gamma$ such that $\gamma \cdot e = e \cdot \gamma = \gamma$, for any $\gamma \in \Gamma$;
- Inverse axiom: for any $\gamma \in \Gamma$, there exists an element $\gamma^{-1} \in \Gamma$ such that $\gamma \cdot \gamma^{-1} = \gamma^{-1} \cdot \gamma = e$.

Let us now summarise some of the most common groups that are important in pattern formation [3]:

- **Lie** group: a finite-dimensional smooth manifold together with a group structure, such that the group operations are smooth maps;
- $\mathbf{D}_n$: the dihedral group of order $2n$, generated by rotations and reflections in the plane that preserve a regular polygon with n sides. For example, $\mathbf{D}_2$ is the symmetry group of a rectangle, and is isomorphic with the direct product $\mathbf{Z}_2 \times \mathbf{Z}_2$;
- $\mathbf{Z}_n$: the cyclic group of order n, generated only by rotations;

- **S^1**: the circle group of unit complex numbers. The group characterises the periodic solutions;
- **O**(n): the orthogonal group in $\mathbb{R}^n$, which consists of $n \times n$ orthogonal matrices (i.e., real matrices A with the property that their transposes are equal to their inverses: $A^\top = A^{-1}$). The group is isomorphic to the group of all rotations and reflections in $\mathbb{R}^n$ that keep the origin fixed;
- **SO**(n): the special orthogonal group consisting of $n \times n$ orthogonal matrices with determinant 1. It is a subgroup of **O**(n), and is sometimes called the *rotation group*, since in $\mathbb{R}^2$ and $\mathbb{R}^3$ its elements are the rotations around a point ($n = 2$) and around a line ($n = 3$);
- **SO**(2): the special orthogonal group in $\mathbb{R}^2$, which consists of rotations

$$R_\theta = \begin{pmatrix} \cos(\theta) & -\sin(\theta) \\ \sin(\theta) & \cos(\theta) \end{pmatrix} \tag{8.23}$$

in the plane. This group is isomorphic with **S**^{1}, since if we write a complex number $e^{i\theta} = \cos(\theta) + i\sin(\theta)$ as a 2×2 real matrix

$$e^{i\theta} \leftrightarrow \begin{pmatrix} \cos(\theta) & -\sin(\theta) \\ \sin(\theta) & \cos(\theta) \end{pmatrix}, \tag{8.24}$$

then the unit complex number corresponds to the 2×2 orthogonal matrix with unit determinant.
- **T**^{n} = **S**^{1} × ... × **S**^{1}: the n-torus;
- **E**(2): the Euclidean group of the plane, generated by rotations, reflections and translations.

To describe how the elements of a group act on some space in a way that preserves the structure of that space, we introduce the notion of *group action* [3]:

Definition 8.4 Consider Γ a Lie group and V a vector space. The *action* of Γ on V is a homomorphism $\rho : \Gamma \to GL(V)$ (with $GL(V)$ the general linear group of invertible matrices on V). We denote the group action $\rho(\gamma)(v) = \gamma \cdot v$.

Definition 8.5 A dynamical system that has an appropriate symmetry is called an *equivariant dynamical system*. In this case, the bifurcation theory is called *equivariant bifurcation theory*.

Consider the following generic dynamical system that depends on a parameter $\mu \in \mathbb{R}$:

$$\frac{du}{dt} = f(u, \mu), \quad \text{with } u \in \mathbb{R}^n, \ f : \mathbb{R}^n \times \mathbb{R} \to \mathbb{R}^n. \tag{8.25}$$

Definition 8.6 We say that system (8.25) is *equivariant* with respect to a group Γ if $f(\gamma \cdot u, c) = \gamma \cdot f(u, c)$, for all $\gamma \in \Gamma$. Here "·" denotes the group action; see [17].

Note that a group element $\gamma \in \Gamma$ is a symmetry of system (8.25) if given a solution $u(t)$, then $\gamma \cdot u(t)$ is also a solution of (8.25). We define the *group orbit* of a solution u as $\Gamma u = \{\gamma \cdot u, \text{ for } \gamma \in \Gamma\}$ (i.e., the group orbit is the set of solutions connected by the group action). Therefore, if one knows a solution of a differential equation, the whole group orbit of this solution will be solutions, too.

One can classify the solutions with respect to their symmetry groups by computing the isotropy subgroups [3, 27]:

Definition 8.7 Consider the action "·" of a group Γ on a vector space V. The *isotropy subgroup* of a point $v \in V$ is defined as

$$\Sigma_v := \{\gamma \in \Gamma | \gamma \cdot v = v\}. \tag{8.26}$$

In other words, the isotropy subgroup of v is the set of all elements (symmetries) that leaves v invariant.

Let us focus now on the notion of *conjugacy*:

Definition 8.8 We say that two group elements $a_1, a_2 \in \Gamma$ are *conjugate* (or in the same conjugacy class) if there exist a group element $\gamma \in \Gamma$ such that $a_1 = \gamma \cdot a_2 \cdot \gamma^{-1}$.

One can further show that solutions u and $\gamma \cdot u$ of (8.25) have *conjugate isotropy subgroups*: $\Sigma_{\gamma \cdot u} = \gamma \cdot \Sigma_u \cdot \gamma^{-1}$. This result is important since it allows us to classify solutions in terms of the conjugacy classes of their isotropy subgroups. More precisely, the isotropy subgroups of all points on an orbit of the action of a group Γ belong to the same conjugacy class. Because the points on the same group orbit have similar existence and stability characteristics, we usually assume (in a loose sense) that the isotropy subgroups are similar [2]. When classifying the solutions of a system of differential equations, we can simplify our analysis by ignoring those solutions corresponding to similar isotropy subgroups (see for example [27] for the classification of steady states solutions for the amplitude equations that resulted from a weakly nonlinear analysis of a Hopf/Hopf bifurcation with $O(2)$ symmetry).

For the dynamical system (8.25), to find an equilibrium solution u with isotropy subgroup Σ_u, we can restrict our search to the fixed point subspace of this isotropy subgroup [27]:

Definition 8.9 Consider an isotropy subgroup $\Sigma_v \in \Gamma$. The *fixed point subspace* of Σ_v is defined as

$$Fix(\Sigma_v) := \{v \in V | \sigma \cdot v = v, \text{ for all } \sigma \in \Sigma\}. \tag{8.27}$$

We conclude this list of definitions necessary for understanding the symmetries of differential equations, by discussing subspaces that are invariant under the action of a group Γ (since these are the spaces that support bifurcations) [3]:

Definition 8.10 Consider a subspace $V \in \mathbb{R}^n$. We say that V is Γ-*invariant* if $\gamma \cdot V = V$ for any $\gamma \in V$.

Definition 8.11 If the subspace $V \in \mathbb{R}^n$ is such that it has only two Γ invariant subspaces, namely V and $\{0\}$, we say that V is Γ-*irreducible*. We say that the action of Γ is *absolutely irreducible* if the only linear maps that commute with the action of Γ on V are the scalar multiples of the identity: $\{aI, a \in \mathbb{R}\}$.

The above notion of absolute irreducibility is important to the Equivariant Branching Lemma, which predicts the existence of branches of symmetry-breaking solutions near bifurcations:

Theorem 8.1 (Equivariant Branching Lemma [3]) *Consider* $\Gamma \subset \mathbf{O}(n)$ *a compact Lie group acting absolutely irreducible on* $\mathbb{R}^n$. *Consider the* Γ*-equivariant bifurcation problem*

$$\frac{du}{dt} = f(u, \mu), \tag{8.28}$$

with $f : \mathbb{R}^n \times \mathbb{R} \to \mathbb{R}^n$ *satisfying the following conditions:* $f(0, \mu) = 0$, $D_u f(0, \mu) = c(\mu)I$ *(where* I=*identity operator), with* $c(0) = 0$ *(bifurcation condition) and* $c'(0) \neq 0$ *(eigenvalue crossing condition). If* Σ *is an isotropy subgroup of* Γ *with* $\dim Fix(\Sigma) = 1$, *then there exist a unique smooth branch of solutions to* $f(u, \mu) = 0$, *with symmetry given by the isotropy group* Σ.

Note that, depending on some conditions for the bifurcation equation $f(u, \mu) = 0$ in $Fix(\Sigma)$, and on whether $\Sigma = \Gamma$ or $\Sigma < \Gamma$, one can distinguish between saddle-node bifurcation, transcritical bifurcation or pitchfork bifurcation (see Theorem 2.3.2 in [4]).

The solution branches that bifurcate from the fixed point $u = 0$ are called *primary branches* (see Fig. 5.11b, c). It is possible to have other solutions that bifurcate from these primary branches (further away from the original fixed point), and they give rise to *secondary branches* (see Fig. 5.11b, c). These secondary branches can lead to an exchange in the stability of solutions.

Returning now to the nonlocal hyperbolic and kinetic equations discussed in Chaps. 4–6, we note that the majority of those models exhibit $\mathbf{O}(2)$ or $\mathbf{SO}(2)$ symmetries:

- translations: $T_\theta \cdot u(x, t) = u(x - \theta, t)$, with $\theta \in [0, L)$;
- reflections (with respect to the domain boundary): $\kappa \cdot (u^+(x, t), u^-(x, t)) = (u^-(L - x, t), u^+(L - x, t))$;

Since the boundary conditions used for the majority of 1D nonlocal hyperbolic and kinetic models discussed in this study [11, 14, 27, 28] are periodic, the translation operator T_θ (= rotation operator on a 1D line) generates a group isomorphic to $\mathbf{SO}(2)$. Moreover, one can check that $T_\theta \circ \kappa = \kappa \circ T_\theta^{-1}$, and thus the translation and reflection operators generate a group isomorphic to $\mathbf{O}(2)$ [30]. It was also shown in various studies [27–30] that the 1D nonlocal hyperbolic system (5.14) with communication mechanisms M1–M5 (see also Fig. 5.5 and Table 5.1) is $\mathbf{O}(2)$-invariant: if $u(x, t) = (u^+(x, t), u^-(x, t))$ is a solution of (5.14), then $T_\theta \cdot u(x, t)$ and $\kappa \cdot u(x, t)$ are also solutions of (5.14).

The symmetry structure of nonlocal 1D hyperbolic models (5.14), together with the types of bifurcations exhibited by these models (i.e., real or complex bifurcations), was used in [27, 28] to classify rigorously the patterns emerging near codimension-2 Hopf/Hopf and Hopf/Steady-state bifurcations, via the isotropy subgroups generated by the $\mathbf{O}(2)$ action on the elements of the hyperbolic system. As an example, we show in Table 8.1 the isotropy subgroups of the $\mathbf{O}(2) \times \mathbb{T}^2$ action, together with the corresponding types of solutions. For more details, we refer the reader to the studies by Buono and Eftimie [27, 28].

Regarding the 2D kinetic models discussed in this study, we note that there are different studies in the literature which investigate the symmetry and invariance properties of various Vlasov-type and Boltzmann-type equations (mainly in 2D, but a few also in 1D) [31–37]. Some of these studies have shown that the collision integral operator for the Boltzmann equation is $\mathbf{SO}(2)$-invariant [31]. Although none of these studies focused on the application of Boltzmann-like equations to describe the collective movement of cells/bacteria/animals, we expect that many of the models discussed in Chap. 6 are also $\mathbf{SO}(2)$-invariant. Finally, since in Chap. 6 we mentioned the Fokker-Planck equations that were derived from Boltzmann-type models via grazing collision limits, it is worth noting that over the last three decades various mathematical studies in the literature have investigated the symmetries of such Fokker-Planck equations [38–41]. A few studies also focused on the bifurcations around homogeneous and heterogeneous states in Vlasov and Vlasov-Fokker-Plank systems used to describe different physics problems [42]. However, it is expected that biological applications of such systems (see some of the biologically-inspired kinetic models described in Chap. 6) could lead to more complex bifurcations.

8.4 Compact Operators and the Fredholm Alternative

Since the Lyapunov-Schmidt reduction (not discussed in this monograph, but reviewed in [30]) and the weakly-nonlinear analysis (discussed in Sect. 8.5) approaches used to reduce the infinite-dimensional nonlinear PDE systems to finite-dimensional ODE systems to study bifurcation dynamics, are based on Fredholm operators and the Fredholm alternative, in the following we present a few definitions related to these two topics. Consider thus two Banach spaces, X and Y (see also Table 2.1).

Definition 8.12 A *linear operator* (or a *linear transformation*) $T : X \to Y$ is *bounded* if there is a constant M such that

$$||Tu||_Y \leq M||u||_X, \quad \text{for all } u \in X.$$

Table 8.1 Summary of patterns emerging near a Hopf/Hopf bifurcation ($k_3 : k_4$), and the isotropy subgroups for the $\mathbf{O}(2) \times \mathbb{T}^2$ action, as given in [27]

Isotropy subgroups	Solutions	Eigenfunctions
$\Sigma_0 = \mathbf{O}(2) \times \mathbb{T}^2$ (full group)	Spatially homogen.	0
$\Sigma_1 = S(0,0,1) \times S(1,3,0)$	Rotating (travelling) waves	$\alpha_1(T)\mathbf{v_1}e^{i\omega_1 t + ik_3 x}$
$\Sigma_2 = S(0,1,0) \times S(1,0,4)$	Rotating (travelling) waves	$\beta_2(T)\mathbf{w_2}e^{i\omega_2 t + ik_4 x}$
$\Sigma_3 = S(0,0,1) \times \mathbb{Z}_2(\kappa) \times \mathbb{Z}\left(\frac{L}{6}, \frac{L}{2}, 0\right)$	Standing waves (ripples)	$\alpha_1(T)(\mathbf{v_1}e^{i\omega_1 t + ik_3 x} + \mathbf{v_2}e^{i\omega_1 t - ik_3 x})$
$\Sigma_4 = S(0,1,0) \times \mathbb{Z}_2(\kappa) \times \mathbb{Z}\left(\frac{L}{8}, 0, \frac{L}{2}\right)$	Standing waves (ripples)	$\beta_1(T)(\mathbf{w_1}e^{i\omega_2 t + ik_4 x} + \mathbf{w_2}e^{i\omega_2 t - ik_4 x})$
$\Sigma_5 = S(0,0,1) \times \mathbb{Z}\left(\frac{L}{6}, \frac{L}{2}, 0\right)$	Modulated standing waves	$\alpha_1(T)\mathbf{v_1}e^{i\omega_1 t + ik_3 x} + \alpha_2(T)\mathbf{v_2}e^{i\omega_1 t - ik_3 x}$
$\Sigma_6 = S(0,1,0) \times \mathbb{Z}\left(\frac{L}{8}, 0, \frac{L}{2}\right)$	Modulated standing waves	$\beta_1(T)\mathbf{w_1}e^{i\omega_2 t + ik_4 x} + \beta_2(T)\mathbf{w_2}e^{i\omega_2 t - ik_4 x}$
$\Sigma_7 = S(1,3,4)$	Modulated rotating waves	$\alpha_2(T)\mathbf{v_2}e^{i\omega_1 t - ik_3 x} + \beta_2(T)\mathbf{w_2}e^{i\omega_2 t - ik_4 x}$
$\Sigma_8 = S(1,3,-4)$	Modulated rotating waves	$\alpha_2(T)\mathbf{v_2}e^{i\omega_1 t - ik_3 x} + \beta_1(T)\mathbf{w_1}e^{i\omega_2 t + ik_4 x}$
$\Sigma_9 = \mathbb{Z}_2 \times \mathbb{Z}\left(\frac{L}{2}, \frac{3L}{2}, \frac{4L}{2}\right)$	Modulated rotating waves	$\alpha_1(T)(\mathbf{v_1}e^{i\omega_1 t + ik_3 x} + \mathbf{v_2}e^{i\omega_2 t - ik_3 x}) + \beta_1(T)(\mathbf{w_1}e^{i\omega_2 t + ik_4 x} + \mathbf{w_2}e^{i\omega_2 t - ik_4 x})$
$\Sigma_{10} = \mathbb{Z}_\kappa = \mathbb{Z}\left(0, 0, \frac{L}{2}\right) \times \mathbb{Z}\left(\frac{L}{2}, \frac{3L}{2}, \frac{4L}{2}\right)$	Modulated rotating waves	$\alpha_1(T)(\mathbf{v_1}e^{i\omega_1 t + ik_3 x} + \mathbf{v_2}e^{i\omega_2 t - ik_3 x}) + \beta_1(T)(\mathbf{w_1}e^{i\omega_2 t + ik_4 x} - \mathbf{w_2}e^{i\omega_2 t - ik_4 x})$

The patterns are defined in the second column, and the approximated solutions (see Eq. (8.46)) are shown in the third column. L describes the domain length

Definition 8.13 A *bounded linear operator* $T : X \to Y$ is *compact* if for each bounded sequence $u_i \subset X$, there exists a subsequence $\{u_{i_k}\}$ such that $\{Tu_{ik}\}$ is convergent.

If $T : X \to Y$ is a bounded linear operator, we define its range $(Ran(T))$ by

$$Ran(T) := \{y \in Y | \exists\, x \in X, \ s.t. \ Tx = y\},$$

and its kernel $(ker(T))$ by

$$ker(T) = \{x \in X | Tx = 0\}.$$

Definition 8.14 A *Fredholm operator* is a bounded linear operator $T : X \to Y$ with finite-dimensional kernel $ker(T)$ and cokernel $coker(T) = Y/Ran(T)$, and closed range $Ran(T)$. We denote by $\mathscr{F}(X, Y)$ the space of all Fredholm operators between X and Y.

Definition 8.15 The *index* of a Fredholm operator is defined as

$$index(T) = \dim\, ker(T) - \dim\, coker(T)$$

Note that the index of an operator is a measure of how invertible an operator is. In particular, if T is an invertible operator then $index(T) = 0$. The index of a Fredholm operator has some properties:

- If T is a Fredholm operator and K is a compact operator, then $T + K$ is a Fredholm operator and $index(T + K) = index(T)$.
- If T and S are Fredholm operators, the TS is Fredholm and $index(TS) = index(T) + index(S)$.
- If T is a Fredholm operator, the adjoint T^* is also Fredholm, and $index(T^*) = -index(T)$.

Theorem 8.2 (Fredholm Alternative) *Consider a compact operator* $T : X \to X$, *and* $\lambda \in \mathbb{C}$ *non-zero. Only one of the following statements hold true:*

(i) Equation $Tu = \lambda u$ *has a non-trivial solution* $u \in X$;
(ii) The operator $T - \lambda$ *has a bounded inverse* $(T - \lambda)^{-1}$ *on* X.

The second statement is equivalent to the fact that the non-homogeneous equation $Tu = \lambda u - f$ has a unique solution for each $f \in X$. One can prove the Fredholm alternative using the index theory of Fredholm operators (by showing that $index(T - \lambda) = 0$, which implies that $T - \lambda$ is surjective whenever there is no eigenvalue).

Moreover, the Fredholm alternative can be restated in terms of Fredholm indices [43]: if K is a compact operator and $\lambda \neq 0$, then $\lambda I - K$ is Fredholm, and $index(\lambda I - K) = 0$

It should be mentioned that the Fredholm alternative can be used to establish spectral results for compact operators. Note that the spectrum of an operator T is defined as [43]:

$$\sigma(T) = \{\lambda \in \mathbb{C} | \lambda I - T \text{ is not invertible}\}.$$

Buono and Eftimie [30] showed that for the nonlocal hyperbolic systems (5.14), the operator $T = \frac{d}{dt} - \mathscr{L}$ (with $\mathscr{L}$ the linearised operator at a steady state $u_*(x)$) is a Fredholm operator of index zero. It was also proven that the spectrum of $\mathscr{L}$ is made up of a finite number of eigenvalues with finite multiplicity [30]. This result ensures that the Centre Manifold Theorem (see Sect. 8.6) holds for this class of nonlocal hyperbolic systems (5.14).

8.5   Analytical Approaches for the Investigation of Patterns: Weakly Nonlinear Analysis

The method of weakly nonlinear analysis generalises the linear stability analysis performed near a bifurcation point, by including also nonlinear terms (via an asymptotic expansion). As discussed above, the linear stability analysis is valid only for small time and infinitesimal perturbations, and cannot capture the long-time effect of the nonlinear terms which dominate the growth of the unstable modes. To overcome this impediment, the weakly nonlinear analysis uses two separate time scales: a fast time scale described by the original time variable t (which gives the time region where the solution starts to develop), and a slow time scale ($T = \epsilon^m t$, for some $m > 0$) on which the effects of the nonlinear terms become important. (Note that close to the bifurcation point, the amplitude of the patterns evolves on a slow temporal scale.) It is assumed that as $\epsilon \to 0$, the two time variables (t and T) are independent. The weakly nonlinear analysis then reduces the dynamics of the full system to the temporal evolution (on the slow time scale) of the amplitude of the perturbations of the steady state, and these differential equations (either ODEs or PDEs) are faster to solve than the full nonlinear systems.

Although the weakly nonlinear analysis can be performed in the neighborhood of codim-1 [44] and codim-2 points [27, 28], in the following we focus on the simpler case of codimension-1 bifurcations (with real eigenvalues) and describe the main steps of this approach. To this end, we consider the 1D nonlocal hyperbolic system (5.14) introduced in [14, 44], which can exhibit codimension-1 steady-state bifurcations as we vary, for example, the magnitude of attractive interactions q_a. Denote by q_a^* the critical value of q_a for which the dispersion relation satisfies $\sigma(q_a^*, k_c) = 0$ (where $k = k_c > 0$ is the critical wavenumber; e.g., $k_c = k_2$ in Fig. 8.2a for the dispersion relation described by the dotted curve). A solution of (5.14) near the bifurcation point is given by (see [44])

$$u^{\pm}(x, t) \propto e^{\sigma t + i k_c x} + \text{c.c.}, \tag{8.29}$$

where "c.c" stands for "complex conjugate". Perturbing the parameter q_a in the neighborhood of its critical value, $q_a = q_a^* + \nu\epsilon^2$ (with $0 < \epsilon \ll 1$ and $\nu = \pm 1$ a parameter that will give the direction of the bifurcating solution branches), substituting this expression into the dispersion relation $\sigma(q_a, k_c)$ and expanding it into Taylor series about q_a^* leads to the following eigenfunction for the solution:

$$e^{\sigma(q_a,k_c)t+ik_cx} \approx e^{ik_cx+\sigma(q_a^*,k_c)+\frac{d\sigma(q_a^*,k_c)}{dq_a}} = e^{ik_cx+\frac{d\sigma(q_a^*,k_c)}{dq_a}} = \alpha(\epsilon^2t)e^{ik_cx}. \tag{8.30}$$

Since the amplitude of the solution depends on the slow time ϵ^2t, the authors in [44] introduced a slow-time variable $T = \epsilon^2t$. The left-moving and right-moving densities were re-written as $u^\pm(x,t) = \tilde{u}^\pm(x,t,\epsilon,T)$. After dropping the tilde for simplicity and assuming a formal expansion of $u^\pm$ in powers of ϵ,

$$u^+(x,t,\epsilon,T) = u_*^+ + \epsilon u_1^+ + \epsilon^2 u_2^+ + \epsilon^3 u_3^+ + O(\epsilon^4),$$

$$u^-(x,t,\epsilon,T) = u_*^- + \epsilon u_1^- + \epsilon^2 u_2^- + \epsilon^3 u_3^- + O(\epsilon^4),$$

these expressions can be substituted into the nonlinear system (5.14). The nonlinear turning rates $\lambda^\pm[u^+, u^-]$ are then expanded in Taylor series about the steady states $u_*^\pm$. Overall, the nonlinear hyperbolic system (5.14) can be re-written as

$$0 = N\left(\sum_{j\geq1}\epsilon^j u_j\right) \approx \sum_{j\geq1}(\mathscr{L}(u_j) + \mathscr{N}_j(u_{j-k}) + \mathscr{E}_j), \quad k \geq 1. \tag{8.31}$$

Here, $\mathscr{L}(u_j)$ describes the linear part of the system (5.14), $\mathscr{N}_j(u_{j-k})$ contains nonlinear terms formed of $u_{j-1}^\pm$, $u_{j-2}^\pm$, etc. (which were calculated at previous $O(\epsilon^{j-k})$ steps, where $k \geq 1$), and $\mathscr{E}_j$ contains the slow time derivatives $\partial_T u_{j-2}^\pm$ (for $j \geq 3$) and the terms multiplied by ν. While the linear operator $\mathscr{L}$ is the same at each $O(\epsilon^j)$, the nonlinear operators $\mathscr{N}_j$ and $\mathscr{E}_j$ are calculated at each j-step. For the nonlocal system (5.14) described in [44], the linear operator $\mathscr{L}$ is given by

$$\mathscr{L}(u) = \begin{pmatrix} \gamma\partial_x + L_1 + M_5 K \star \cdot & -L_1 + M_5 K \star \cdot \\ -L_1 - M_5 K \star \cdot & -\gamma\partial_x + L_1 - M_5 K \star \cdot \end{pmatrix}\begin{pmatrix} u^+ \\ u^- \end{pmatrix}, \tag{8.32}$$

where L_1 and M_5 are constants depending on the steady states and the various model parameters, while the convolutions "$K \star \cdot$" are defined as a difference between repulsive and attractive nonlocal interactions:

$$K \star u^\pm = q_r\left(\tilde{K}_r \star u^\pm - K_r \star u^\pm\right) - q_a^*\left(\tilde{K}_a \star u^\pm - K_a \star u^\pm\right), \tag{8.33}$$

with $\tilde{K}_{r,a}(s) = K_{r,a}(-s)$ and $K_{r,a} \star u^\pm(x) = \int_{-\infty}^\infty K_{r,a}(s)u^\pm(x-s)ds$.

At $O(\epsilon^1)$ the nonlinear terms are zero ($\mathscr{N}_1 = \mathscr{E}_1 = 0$), and solving the nonlinear system (5.14) reduces to solving the linear system $\mathscr{L}(u) = 0$, which has a nontrivial solution. For this reason, at each $O(\epsilon^j)$, $j \geq 2$, the nonlinear system $\mathscr{L}(u_j) =$

$\mathcal{N}_j + \mathcal{E}_j$ has a solution if and only of $\mathcal{N}_j + \mathcal{E}_j$ satisfies the Fredholm alternative. To check whether this alternative is applied, one needs to investigate first whether the linear operator $\mathcal{L}$ is compact. Consider the Hilbert space [44]

$$Y = \{\mathbf{v}(x, t) \in [0, L] \times [0, \infty)| \lim_{T \to \infty} \frac{1}{T} \int_0^T \int_0^L |\mathbf{v}|^2 dx dt < \infty\} \tag{8.34}$$

with the inner product

$$\langle \mathbf{v}, \mathbf{w} \rangle = \lim_{T \to \infty} \int_0^T \int_0^{L=2\pi/k_c} (v^1 \bar{w}^1 + v^2 \bar{w}^2) dx dt. \tag{8.35}$$

Here $\mathbf{v} = (v^1, v^2)^\top$ and $\mathbf{w} = (w^1, w^2)^\top$, which satisfy periodic boundary conditions. Since $\mathbf{v}$ are bounded on $L^2([0, L] \times [0, T])$ (see [44]), then $\lim_{T \to \infty}(1/T)\|\mathbf{v}\|^2_{L^2([0,L]\times[0,T])}$ is finite. Because the linear operator $\mathcal{L}$ is given also in terms of the differential operator $\partial/\partial x$ (which is not bounded), one needs to interpret this differential operator as a distribution in a Sobolev subspace of Y. Consider the space

$$V_{bc} = \{(v^+, v^-) \in Y | (\partial_x v^+, \partial_x v^-) \in Y, \text{ and } v^\pm(L, t) = v^\pm(0, t))\}, \tag{8.36}$$

with the norm $\|v\|^2_{V_{bc}} = \|(v^+, v^-)\|^2_Y + \|(\partial_x v^+, \partial_x v^-)\|^2_Y$, which is associated with the inner product (8.35). As discussed in [27], the linear operator $\mathcal{L} : V_{bc} \to Y$ is bounded, and following the approach in Kmit and Recke [45] for linear local hyperbolic systems, one can show that $\mathcal{L}$ is a Fredholm operator; see the proof for nonlocal hyperbolic systems in [30].

Since the Fredholm alternative can be applied, the term $\mathcal{N}_j + \mathcal{E}_j$ has to be orthogonal on the bounded solution of the adjoint homogeneous problem $\mathcal{L}^*(\hat{u}) = 0$:

$$\langle \hat{u}, \overline{(\mathcal{N}_j + \mathcal{E}_j)} \rangle = 0. \tag{8.37}$$

Focusing only on those terms in $\mathcal{N}_j + \mathcal{E}_j$ that contain the exponentials $e^{\pm i k_c x}$ (since they give rise to secular solutions that grow unbounded), and substituting these terms into the inner product (8.37) one eventually obtains the following differential equation for the evolution of the amplitude $\alpha(T)$ (truncated here at the third order $\alpha|\alpha|^2$):

$$\frac{d\alpha}{dT} = -\nu\alpha Y - \alpha|\alpha|^2 X, \tag{8.38}$$

where X and Y are constant terms that depend on model parameters. This complex amplitude can be re-written as $\alpha(T) = R(t)e^{i\theta(T)}$, with real terms $R(T) = |\alpha|$ and

$\theta(T)$ that satisfy

$$\frac{dR}{dT} = -\nu R \mathfrak{R}(Y) - R^3 \mathfrak{R}(X), \tag{8.39}$$

$$\frac{d\theta}{dT} = -\nu \mathfrak{I}(Y) - R^3 \mathfrak{I}(X). \tag{8.40}$$

Here $\mathfrak{I}$ and $\mathfrak{R}$ denote the imaginary and real parts of the coefficients X and Y. The equation for the real amplitude $R(T)$ exhibits two steady states: $R = 0$ and $R = \sqrt{-\nu \mathfrak{R}(Y)/\mathfrak{R}(X)}$. The stability of these states can be investigated in a classical way via small perturbations: $R(T) = R^* + R_\delta$, with R^* denoting the steady state and R_δ being a small perturbation which satisfies

$$\frac{dR_\delta}{dT} = R_\delta \left(-\nu \mathfrak{R}(Y) - 2R^{*2} \mathfrak{R}X \right). \tag{8.41}$$

It is easy to observe that the zero state $R^* = 0$ is stable for $\nu \mathfrak{R}(Y) > 0$ and unstable otherwise. In contrast, the nonzero state $R^* = \sqrt{-\nu \mathfrak{R}(Y)/\mathfrak{R}(X)}$ is stable for $\nu \mathfrak{R}(Y) < 0$ and unstable otherwise. One can graph the solution of (8.39) (and its stability) for the exact parameter values used during numerical simulations, to obtain bifurcation diagrams for the amplitudes of the solution branches as functions of parameter values (e.g., q_a here). We graph two caricature examples of the bifurcating branches in Fig. 8.7 (see also Fig. 5.9 for bifurcation diagrams based on specific model parameters, as we vary the magnitude of alignment q_{al}).

Remark 8.4 Note in Fig. 8.7a that the 3rd-order truncation of the amplitude equation (8.38) allows only for the detection of the main nontrivial amplitude branch that bifurcates from $\alpha = 0$ at $q_a = q_a^*$. The fact that this branch is unstable, it suggests that there exists also a stable high-amplitude spatially heterogeneous solution towards which the small perturbations of the spatial homogeneous steady state will grow, and which can be detected numerically. Therefore, in the range $q_a \in (q^*, q_a^*]$ two qualitatively different stable states co-exit (together with an unstable state). One could identify the secondary bifurcation point $q_a = q^*$ where the unstable branch $\alpha > 0$ changes stability and becomes stable by considering truncations of (8.38) up to the 5th and even 7th orders.

Remark 8.5 We also note in Fig. 8.7a a *hysteresis phenomenon* characterised by a lack of reversibility in the dynamics of the system: for $q_a > q_a^*$ the zero-amplitude ($\alpha = 0$) solution is unstable and small perturbations of it will grow and give rise to high-amplitude spatially heterogeneous solutions (i.e., the solution jumps fast to the upper red curve). As we decrease q_a below q_a^* the dynamics of the system follows the stable high-amplitude branch, and does not decrease immediately to $\alpha = 0$. The solution jumps back to $\alpha = 0$ only when $q_a = q^*$. The bifurcation at $q_a = q^*$ is a saddle-node bifurcation. Moreover, the high-amplitude state exists only for $q_a > q^*$. It could be possible that for some very large q_a (i.e., q_a further away from the bifurcation point q_a) this high-amplitude state disappears through a

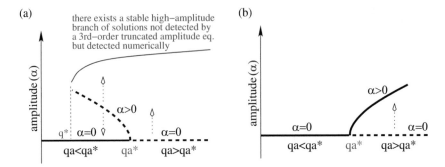

Fig. 8.7 Caricature description of the bifurcation diagram near a codimension-1 bifurcation point, in the (α, q_a) plane. The continuous curves describe the stable states and the dashed curves describe the unstable states. (**a**) Subcritical bifurcation obtained for $\nu < 0$: for $q_a < q_a^*$ small perturbations of the homogeneous steady state (with zero amplitude) decay to zero, while large perturbations (above the nonzero dashed curve) growth even larger towards a high-amplitude spatially heterogeneous solution (which is detected numerically). At the point $q_a = q^*$ (usually located further away from the bifurcation point q_a^*) the unstable branch $\alpha > 0$ becomes unstable through a saddle-node bifurcation. For $q_a > q_a^*$ small perturbations of the homogeneous steady state grow towards a high-amplitude heterogeneous state. (**b**) Supercritical bifurcation obtained for $\nu > 0$: for $q_a > q_a^*$ small perturbations grow and give rise to a small-amplitude spatially heterogeneous solution, while for $q_a < q_a^*$ the small perturbations decay towards zero as the homogeneous steady state is stable

different bifurcation; however, this aspect cannot be investigated through a weakly nonlinear analysis which loses its validity away from the bifurcation points.

Weakly nonlinear analysis has been applied to investigate the solutions emerging in the vicinity of other bifurcation points [27, 28, 44]:

- a Hopf bifurcation point [44], where $\sigma := \pm i\omega$ and the solution can be represented as

$$u^{\pm}(x, t, T) \propto \beta(T)e^{i\omega t + ik_c x} + \text{c.c.} \qquad (8.42)$$

The equation for the variation of amplitude $\beta(T)$ on the slow time scale T is similar to the one for the steady-state bifurcation point (8.38):

$$\frac{d\beta(T)}{dT} = -\beta Y - \beta|\beta|^2 X, \qquad (8.43)$$

with X and Y given in terms of the model parameters. If we now take into consideration also the reflection symmetry of the domain, we can represent the solution as

$$u^{\pm}(x, t, T) \propto \beta_1(T)e^{i\omega t + ik_c x} + \beta_2(T)e^{i\omega t - ik_c x} + \text{c.c.} \qquad (8.44)$$

In this case, the equations for the variation of the amplitudes $\beta_1(T)$ and $\beta_2(T)$ are

$$\frac{d\beta_1(T)}{dT} = -\beta_1 X_1 + \beta_1 |\beta_1|^2 X_2 + \beta_1 |\beta_2|^2 X_3, \tag{8.45a}$$

$$\frac{d\beta_2(T)}{dT} = -\beta_2 Y_1 + \beta_2 |\beta_1|^2 Y_2 + \beta_2 |\beta_2|^2 Y_3. \tag{8.45b}$$

Note that by considering also the reflection symmetry, we obtain a system of coupled normal form equations, which exhibits more complex steady states compared to Eq. (8.43).

- a Hopf/Hopf bifurcation point [27], where the solution can be represented as (considering also the reflection symmetry of the domain):

$$u^{\pm}(x, t) \propto \alpha_1(T) e^{i\omega(k_m)t + ik_m x} + \alpha_2(T) e^{i\omega(k_m)t - ik_m x} + \beta_1(T) e^{i\omega(k_n)t + ik_n x}$$

$$+ \beta_2(T) e^{i\omega(k_n)t - ik_n x} + \text{c.c.}, \tag{8.46}$$

with $k_m \neq k_n$ the two distinct Hopf interacting modes. The equations for the variation of the amplitudes on the slow-time scale T are given as follows:

$$\frac{d\alpha_1(T)}{dT} = -\alpha_1 X_1 + \alpha_1 |\alpha_1|^2 X_2 + \alpha_1 |\alpha_2|^2 X_3 + \alpha_1 |\beta_1|^2 X_4 + \alpha_1 |\beta_2|^2 X_5,$$

$$\frac{d\alpha_2(T)}{dT} = -\alpha_2 Y_1 + \alpha_2 |\alpha_1|^2 Y_2 + \alpha_2 |\alpha_2|^2 Y_3 + \alpha_2 |\beta_1|^2 Y_4 + \alpha_2 |\beta_2|^2 Y_5,$$

$$\frac{d\beta_1(T)}{dT} = -\beta_1 Z_1 + \beta_1 |\alpha_1|^2 Z_2 + \beta_1 |\alpha_1|^2 Z_3 + \beta_1 |\beta_1|^2 Z_4 + \beta_1 |\beta_2|^2 Z_5,$$

$$\frac{d\beta_2(T)}{dT} = -\beta_2 \Psi_1 + \beta_2 |\alpha_1|^2 \Psi_2 + \beta_2 |\alpha_2|^2 \Psi_3 + \beta_2 |\beta_1|^2 \Psi_4 + \beta_2 |\beta_2|^2 \Psi_5.$$

The symmetries of the model lead to similarities between the parameters X_i, Y_i, Z_i and Ψ_i, $i = 1, \ldots, 5$:

$$X_1 = Y_1, \quad Z_1 = \Psi_1, \quad X_2 + X_3 = Y_2 + Y_3, \quad Z_2 + Z_3 = \Psi_2 + \Psi_3,$$

$$X_4 + X_5 = Y_4 + Y_5, \quad Z_4 + Z_5 = \Psi_4 + \Psi_5.$$

Since these similarities in parameter values mean that the steady state solutions have conjugate isotropy subgroups, it allowed the authors in [27] to ignore some of the solutions of the above system of coupled amplitude equations. We summarise in Table 8.1 the various types of solutions that emerge near a Hopf/Hopf bifurcation point (given in terms of the above amplitudes) and their corresponding isotropy subgroups; for details see [27].

- a Hopf/Steady-state bifurcation point [28], where the solution can be represented as (considering also the reflection symmetry):

$$u^{\pm}(x,t) \propto \alpha(T)e^{ik_m x} + \beta_1(T)e^{i\omega(k_n)t+ik_n x} + \beta_2(T)e^{i\omega(k_n)t-ik_n x} + \text{c.c}, \quad (8.47)$$

with k_m the steady-state mode and k_n the Hopf mode. The equations for the variation of the amplitudes on the slow-time scale T are

$$\frac{d\alpha}{dT} = -\alpha X_1 + \alpha|\alpha|^2 X_2 + \alpha|\beta_1|X_3 + \alpha|\beta_2|^2 X_4,$$

$$\frac{d\beta_1}{dT} = -\beta_1 Y_1 + \beta_1|\alpha|^2 Y_2 + \beta_1|\beta_1|^2 Y_3 + \beta_1|\beta_2|^2 Y_4,$$

$$\frac{d\beta_2}{dT} = -\beta_2 Z_1 + \beta_2|\alpha|^2 Z_2 + \beta_2|\beta_2|^2 Z_3 + \beta_2|\beta_1|^2 Z_4.$$

Parameters X_i, Y_i, Z_i, $i = 1, .., 4$, depend on model parameters, and are related through the symmetries of the system [28].

- a steady-state/steady-state bifurcation point [28], where the solution can be represented as

$$u^{\pm}(x,t) \propto \alpha_1(T)e^{ik_m x} + \alpha_2(T)e^{ik_n x} + \text{c.c}, \quad (8.48)$$

with k_m and k_n the two distinct steady-state modes. The equations for the evolution of the amplitudes are

$$\frac{d\alpha_1(T)}{dT} = -\alpha_1 X_1 + \alpha_1|\alpha_1|^2 X_2 + \alpha_1|\alpha_2|^2 X_3,$$

$$\frac{d\alpha_2(T)}{dT} = -\alpha_2 Y_1 + \alpha_2|\alpha_1|^2 Y_2 + \alpha_2|\alpha_2|^2 Y_3.$$

Remark 8.6 We need to discuss briefly the normal form equation (8.38) (and implicitly the coupled systems of normal form equations given above). In general, one assumes that the amplitude α depends not only on the slow time scale T but also on a slow space variable $X = \epsilon^p x$. This leads to a PDE (i.e., a Ginzburg-Landau amplitude equation) for the evolution of $\alpha(X, T)$. However, for the nonlocal hyperbolic system (5.14) with periodic boundary conditions, the zero mode $k = 0$ is not an admissible mode (due to the conservation of the total density, which is not satisfied by eigenfunctions with modes $k = 0$ on finite domains with periodic boundary conditions). In this case, one could assume that $\alpha = \alpha(T)$ and the temporal evolution of the amplitude is reduced to an ODE (8.38) (i.e., a Stuart-Landau amplitude equation).

However, since numerical simulations for these nonlocal hyperbolic models have shown the existence of patterns with space-modulated amplitudes (see the semi-zigzags and travelling breathers in Fig. 5.7f, i) it would be interesting to investigate the evolution of the space-modulated amplitude equation $(\alpha(X, T))$ on infinite domains.

8.6 Centre Manifold Theory

We have seen in Sect. 8.5 that the multiple scales approach for the weakly nonlinear analysis reduces the infinite-dimensional PDE system (5.14) to a finite-dimensional ODE system that can be investigated more easily (and which preserves the stability and bifurcation structure of the original system). Other approaches that can be used to reduce the hyperbolic and kinetic equations/systems to more manageable equations/systems are the Centre Manifold reduction and the Lyapunov-Schmidt reduction. The Centre Manifold reduction focuses on finding a dynamical subsystem invariant under the flow of the full system (and which contains the bifurcation that needs to be investigated), while the Lyapunov-Schmidt reduction focuses on identifying an equation for the equilibria (fixed points or periodic solutions), and then performing a reduction of this equation to a low-dimension set of algebraic equations (also called the bifurcation equations) [46]. A review of these two approaches in the context of nonlocal hyperbolic systems can be found in [30]. Note that the Fredholm property of the linear operator $\mathscr{L}$, which we discussed in Sects. 8.4 and 8.5, is central to the application of the Lyapunov-Schmidt reduction; see [30]. However, in the following we will ignore the Lyapunov-Schmidt reduction, and focus only on the Centre Manifold reduction. For a detailed presentation of the Lyapunov-Schmidt reduction in both finite and infinite dimensions, we refer the reader to [46].

Mirroring the approach in Sect. 8.1 (where we first looked at linear stability results in ODE systems and then in PDE systems), here too we start our discussion on the Centre Manifold Theorem (which ensures the possibility of having a Centre Manifold reduction to simplify the model dynamics around a non-hyperbolic fixed point) by focusing first on the finite-dimensional systems. Then, we consider its generalisation to infinite-dimensional systems. This approach allows us to show how the classical, simple version of the theorem that is listed in almost all books on dynamical systems and bifurcation theory [6, 9], is generalised to infinite dimensions through the addition of some extra assumptions.

- **Finite-dimensional systems.** In the following we give the statement of the Centre Manifold Theorem for ODE systems. We also discuss the form of the extended centre manifold, which includes also a dependence on parameters and thus can be used for bifurcation results. But first let us define the stable, unstable and centre subspaces of a linear system $du/dt = Au$. To this end, assume that matrix A has eigenvalues $\lambda_j = a_j + ib_j$, and corresponding to these eigenvalues there are the generalised eigenvectors $w_j = w_j^1 + iw_j^2$. Then the stable (E^s), unstable (E^u) and centre (E^c) subspaces spanned by the real and imaginary parts of the eigenvectors w_j corresponding to eigenvalues λ_j are:

$$E^s = Span\{w_j^1, w_j^2 | a_j < 0\},$$

$$E^u = Span\{w_j^1, w_j^2 | a_j > 0\},$$

$$E^c = Span\{w_j^1, w_j^2 | a_j = 0\}.$$

Theorem 8.3 (Centre Manifold Theorem [9]) *Consider the nonlinear system*

$$\frac{du}{dt} = f(u), \quad u \in \mathbb{R}^n, \tag{8.49}$$

with $f \in C^r(E)$, $r \geq 1$, and $E \subset \mathbb{R}^n$ which includes the origin. Assume that $f(0) = 0$, and $Df_u(0)$ has p eigenvalues with positive real parts, k eigenvalues with negative real parts, and $m = n - p - k$ eigenvalues with zero real parts. Then there exists a m-dimensional centre manifold $W^c(0)$ tangent to the centre subspace E^c at 0, a k-dimensional stable manifold $W^s(0)$ tangent to the stable subspace E^s at 0, and a p-dimensional unstable manifold $W^u(0)$ tangent to the unstable subspace E^u at 0.

Locally, system (8.49) can be written as

$$\frac{du}{dt} = Cu + F(u, v), \quad \frac{dv}{dt} = Pv + G(u, v), \tag{8.50}$$

with $(u, v) \in \mathbb{R}^m \times \mathbb{R}^{k+p}$, C is a square matrix with m eigenvalues with zero real parts, and P is a square matrix with k eigenvalues with negative real parts and p eigenvalues with positive real parts. Moreover, $F(0) = G(0) = 0$, $DF(0) = DG(0) = 0$, and there is a function $h(u) = v$ (that defines the central manifold) such that the flow on the central manifold is given (locally, for $|u| < \delta$) by

$$\frac{du}{dt} = Cu + F(u, h(u)), \quad \text{for all } u \in \mathbb{R}^m \text{ with } |u| < \delta. \tag{8.51}$$

In the context of bifurcation theory, we need to consider also the effect of a parameter μ. For this reason, we work on *extended central manifolds*, where we generalise Eq. (8.50) through the addition of a trivial equation for the derivative of parameter μ:

$$\frac{du}{dt} = Cu + F(u, v, \mu), \quad \frac{dv}{dt} = Pv + G(u, v, \mu), \quad \frac{d\mu}{dt} = 0. \tag{8.52}$$

We note that the equation for the derivative of μ adds one more dimension to the centre manifold ($= m + 1$), since now we work in the neighbourhood of $(u, v) = (0, 0)$ and $\mu = 0$ (where the bifurcation occurs). Moreover, the equation that parametrises the centre manifold now has the form $v = h(u, \mu)$. Therefore, the equation for the extended central manifold can be written as

$$\frac{du}{dt} = Cu + F(u, h(u, \mu), \mu). \tag{8.53}$$

If u is a scalar and $C = 0$, then the above equation can have three possible descriptions, corresponding to the saddle-node, transcritical and pitchfork bifurcations discussed in Sect. 8.2:

$$\text{Saddle-node:} \quad \frac{du}{dt} = \mu + cu^2,$$

$$\text{Transcritical:} \quad \frac{du}{dt} = \mu u + cu^2,$$

$$\text{Pitchfork:} \quad \frac{du}{dt} = \mu u + cu^3.$$

For a detailed proof of the Centre Manifold Theorem, we refer the reader to [9, 47].

- **Infinite-dimensional systems.** The centre manifold theory for PDEs has been developed in various studies published over the past three decades (see [5, 48–51] and the references therein). In the following we give the assumptions and the statement of the Centre Manifold Theorem for infinite-dimensional systems (following the approach in [5]), and then discuss the applicability of this theorem to the class of nonlocal 1D hyperbolic systems (5.14) discussed in Chap. 5.

  First, let us consider three Banach spaces, X, Y and Z, which satisfy the following continuous embeddings:

$$Z \hookrightarrow Y \hookrightarrow X. \tag{8.54}$$

We define $\mathbb{L}(Z, X)$ to be the Banach space of linear bounded operators $\mathscr{L} : Z \rightarrow X$, with the operator norm

$$||\mathscr{L}||_{\mathbb{L}(Z,X)} = \sup_{||u||_Z=1} (||\mathscr{L}u||_X) \tag{8.55}$$

For some $k \geq 2$ we define $\mathscr{C}^k(Z, X)$ the Banach space of functions $b : Z \rightarrow X$ that are k-times continuously differentiable. The space is equipped with the following norm:

$$||b||_{\mathscr{C}^k} = \max_{j=0,\dots,k} \left(\sup_{y \in Z} ||D^j b(y)||_{\mathbb{L}(Z^j, X)} \right), \tag{8.56}$$

where D denotes the differential operator. Finally, for a constant $\eta > 0$, we define the space $\mathscr{C}_\eta(\mathbb{R}, X)$ of exponentially growing functions with the norm

$$||u||_{\mathscr{C}_\eta} = \sup_{t \in \mathbb{R}} \left(e^{-\eta|t|} ||u(t)||_X \right) < \infty, \quad \text{for } u \in \mathscr{C}^0(\mathbb{R}, X). \tag{8.57}$$

Assume now that a generic PDE can be represented as

$$\frac{du}{dt} = F(u) = \mathscr{L}u + \mathscr{N}(u). \tag{8.58}$$

Here $\mathscr{L}$ and $\mathscr{N}$ are the linear and nonlinear parts of operator F. Next, we list the hypotheses that need to be satisfied by $\mathscr{L}$ and $\mathscr{N}$ for the existence of a central manifold, as given in [5]:

(A) Assume that $\mathscr{L} \in \mathbb{L}(Z, X)$, and for some $k \geq 2$ there exist a neighborhood $\mathscr{V}(0)$ of 0 such that $\mathscr{N} \in \mathscr{C}^k(\mathscr{V}, Y)$ and $\mathscr{N}(0) = 0$ (i.e., $u = 0$ is an equilibrium of (8.58)) and $D\mathscr{N}(0) = 0$ (i.e., $\mathscr{L}$ is the linearisation of the operator $\mathscr{N}$ about 0).

(B) Consider the spectrum σ of the linear operator $\mathscr{L}$, which is defined as $\sigma = \sigma_+ \bigcup \sigma_0 \bigcup \sigma_-$, with

$$\sigma_+ = \{\lambda \in \sigma \,|\, Re(\lambda) > 0\}, \ \sigma_0 = \{\lambda \in \sigma \,|\, Re(\lambda) = 0\},$$

$$\sigma_- = \{\lambda \in \sigma \,|\, Re(\lambda) < 0\}.$$

Assume that there is a positive constant $g > 0$ such that

$$\inf_{\lambda \in \sigma_+} (Re\lambda) > g \ \text{ and } \ \sup_{\lambda \in \sigma_-} (Re\lambda) < -g. \tag{8.59}$$

Moreover, assume that the set σ_0 has a finite number of eigenvalues with finite algebraic multiplicities.

(C) Let P_0 be the projection onto the generalised eigenspaces of σ_0, and define $P_h = I - P_0$. Consider now the linear operator $\mathscr{L}_h$ which is the restriction of $\mathscr{L}$ to $d(\mathscr{L})_h = P_h D(\mathscr{L})$. Then, for any $\eta \in [0, g]$ and any $f \in \mathscr{C}_\eta(\mathbb{R}, Y_h)$, the linear problem

$$\frac{du_h}{dt} = \mathscr{L}_h u_h + f(t) \tag{8.60}$$

has a unique solution $u_h = K_h f \in \mathscr{C}(\mathbb{R}, Z_h)$, with K_h a bounded linear operator from $\mathscr{C}_\eta(\mathbb{R}, Y_h)$ to $\mathscr{C}_h(\mathbb{R}, Z_h))$. Also, there exist a continuous map $\mathscr{C} : [0, g] \to \mathbb{R}$ such that

$$\|K_h\|_{\mathbb{L}\left(\mathscr{C}_\eta(\mathbb{R}, Y_h), \mathscr{C}_h(\mathbb{R}, Z_h)\right)} \leq C(\eta). \tag{8.61}$$

Theorem 8.4 (Centre Manifold Theorem [5]) *Assume that hypotheses (A)–(C) hold. Then there exist a map $\Psi \in \mathscr{C}^k(\mathscr{E}_0, Z_h)$ with $\Psi(0) = 0$, $D\Psi(0) = 0$, and a neighborhood of 0, $\mathscr{O}(0) \in Z$ such that the manifold*

$$\mathscr{M}_0 = \{u_0 + \Psi(u_0), \text{ for } u_0 \in \mathscr{E}_0\} \subset Z \tag{8.62}$$

satisfies the following conditions:

(i) $\mathcal{M}_0$ *is locally invariant;*
(ii) $\mathcal{M}_0$ *contains the set of bounded solutions of (8.58) that stay in $\mathcal{O}$ for all*
    $t \in \mathbb{R}$.

Here, the manifold $\mathcal{M}_0$ is called a *local centre manifold*, while the map Ψ is called the *reduction function*.

If we consider now a solution u of (8.58), with $u \in \mathcal{M}_0$, then we can write $u = u_0 + \Psi(u_0)$, with u_0 satisfying

$$\frac{du_0}{dt} = \mathcal{L}_0 u_0 + P_0 \mathcal{N}(u_0 + \Psi(u_0)). \tag{8.63}$$

Here, $\mathcal{L}_0$ is the restriction of $\mathcal{L}$ to $\mathcal{E}_0 = Range(P_0) = \{P_0 u \in X \,|\, u \in Z\}$.

As for the finite-dimensional case, let us consider a parameter-dependent PDE,

$$\frac{du}{dt} = F(u) = \mathcal{L}u + \mathcal{N}(u, \mu), \tag{8.64}$$

with $\mathcal{N}(u, \mu)$ defined in the neighborhood of $(0, 0) \in Z \times \mathbb{R}^m$. Then we obtain an analogue of the previous Centre Manifold Theorem, with the parameter-dependent local extended centre manifold given by [5]:

$$\mathcal{M}_0(\mu) = \{u_0 + \Psi(u_0, \mu), \text{ for } u_0 \in \mathcal{E}_0\} \subset Z. \tag{8.65}$$

To understand better the difficulties of applying this theorem to hyperbolic systems, let us first give the following spectral property.

Definition 8.16 (Spectral Mapping Property [52]) Consider A an infinitesimal generator of a C_0 semigroup e^{At}. Then A has the *spectral mapping property* if the spectrum of this semigroup, $\sigma(e^{At})$, satisfies:

$$\sigma(e^{At}) \setminus \{0\} = \overline{e^{\sigma(A)t}} \setminus \{0\}, \text{ for } t \geq 0, \tag{8.66}$$

where $\overline{e^{\sigma(A)t}}$ denotes the closure of the set.

It was shown by Renardy [53] that for hyperbolic systems the spectral mapping property does not generally hold (see also the dispersion relation shown in Fig. 4.7, for the local hyperbolic system introduced in [15]). This impacts the validity of the Central Manifold theorems for hyperbolic systems (since the hypothesis (B) given above is violated). Lichtner [52] has proven that this spectral property holds for a class of linear hyperbolic systems. Moreover, different versions of the Central Manifold Theorem were proven for various (local) hyperbolic systems with applications to fluid dynamics or lasers [54–56]. Returning to the class of nonlocal 1D hyperbolic models (5.14) introduced in Chap. 5, it was shown in [30] that if $u^{**}(x)$ is a steady state solution of the nonlocal hyperbolic system (5.14) and $\mathscr{L}$ is the linearised operator at $u^{**}(x)$, then the spectrum of $\mathscr{L}$ contains isolated eigenvalues with finite multiplicity and no accumulation point in $\mathbb{C}$. Moreover, this spectrum has only a finite number of eigenvalues with finite multiplicity on the imaginary axis [30], and thus hypothesis (B) holds true. Using the model symmetries to decompose (5.14) into a family of finite-dimensional systems, it was shown in [30] that also hypothesis (C) was satisfied.

We emphasise that the validity of the spectral property for the various hyperbolic/kinetic models discussed throughout this monograph is still an open problem.

We conclude this section by giving the statement of the Contraction Mapping Theorem (which is used in the construction of the Centre Manifold [47], or in the proof of existence of unique solutions for hyperbolic systems—as mentioned in Chap. 4). To this end, we use the version of the theorem stated in [47].

Theorem 8.5 (Contraction Mapping Theorem [47]) *Consider two Banach spaces X and Y, and a continuous map $F : X \times Y \to Y$ that is a contraction in the second variable:*

$$||F(x, y) - F(x, y')|| \leq k||y - y'||, \quad \forall\, x \in X, \ \forall\, y, y' \in Y, \quad and\ some\ k < 1.$$

The following results hold true:

1. *For every $x \in X$, there exists a unique fixed point $y(x) \in Y$ for the map F:*

$$y(x) = F(x, y(x)).$$

2. *For every $x \in X$, $y \in Y$, the following inequality holds true:*

$$||y - y(x)|| \leq \frac{1}{1 - k}||x - x'||.$$

3. *If the map F is Lipschitz continuous with respect to x,*

$$||F(x, y) - F(x', y)|| \leq L||x - x'||, \quad \forall x, x' \in X, \ \forall y \in Y,$$

then the map $x \to y(x)$ is also Lipschitz continuous with respect to x:

$$||y(x) - y(x')|| \leq \frac{L}{1 - k}||x - x'||.$$

4. *For any convergent sequence $x_n \to \bar{x} \in X$, and any $y_0 \in Y$, the sequence of iterates $y_{n+1} = F(x_n, y_n)$ converges to the fixed point $\bar{y} = y(\bar{x})$.*

For a proof of this theorem see [47].

8.7 Stochastic Bifurcations

As we have seen numerically at the end of Chap. 5, stochastic hyperbolic models could also exhibit various bifurcations when we vary model parameters. These bifurcations could represent transitions between two different deterministic-types of patterns, or transitions between deterministic and random patterns (e.g., from a travelling pulse for low noise, to a chaotic zigzag for medium noise, and a stationary pulse for high noise; see Fig. 5.26a).

Similar to the deterministic case, the stochastic bifurcation theory focuses on qualitative changes in parametrised classes of stochastic dynamical systems [57]. While the bifurcation theory for deterministic PDE systems is well developed [3, 17], the field of stochastic bifurcations for stochastic PDEs is still not fully developed [58], and to our knowledge it was never applied to the very few stochastic hyperbolic/kinetic models derived to investigate pattern formation in animal aggregations. For this reason, we will not detail here the basic concepts of stochastic bifurcation theory, but we refer the reader to the books by Arnold [59] and Blömker [58] (which introduce and develop the concepts of bifurcation theory for stochastic ODEs and PDEs). However, for the completeness of our discussion on bifurcations, in the following we discuss briefly two approaches used to describe bifurcations in the context of random dynamical systems. As noted in [58], these two approaches sometimes can give completely different results.

- A *D-bifurcation* or *dynamical bifurcation* is characterised by changes in the structure of the random attractor (e.g., as shown by the sign changes in the Lyapunov exponents for the random dynamical system);
- A *P-bifurcation* or *phenomenological bifurcation* is characterised by changes in the density function for stationary measures associated with the random dynamical system.

These two concepts can be used to describe the classical types of bifurcations that can appear in a stochastic context (e.g., stochastic pitchfork, stochastic transcritical, saddle node, or Hopf bifurcations) [57, 60, 61].

Similar to the case of deterministic PDEs, one could approximate the stochastic PDEs with amplitude equations for the dominant modes, which could be then used to investigate the impact of noise on the dynamics of the system near points of changes in stability [58]. However, one needs to emphasise that, as for deterministic hyperbolic PDEs, spectral gap properties might impact the possibility of deriving such amplitude equations through random centre manifold reductions.

8.8   Bifurcation and Symmetry Theory in the Context of Hyperbolic/Kinetic Models

To conclude this chapter, we need to review the application of the previous notions of stability and bifurcation theory to the kinetic and hyperbolic models discussed in the previous chapters. In this context, we note that the calculation of the steady states and their stability (as well as the existence of different classical and weak solutions) are relatively common approaches taken when investigating the patterns generated by different models [14, 27, 29, 44, 62–69]. However, the analytical investigation of the branches bifurcating at different points where spatially homogeneous and heterogeneous solutions loose their stability is still an open problem for the vast majority of models discussed throughout this monograph, as well as for many other models in the literature (which were not even mentioned here, due to the limited space and purpose of this study). As we have seen above, one of the main reasons for this lack of results is the applicability of the Centre Manifold Theorem for different classes of hyperbolic systems. While it was shown in [30] that this theorem holds for the nonlocal 1D hyperbolic systems introduced in Chap. 5, its applicability to the majority of all other hyperbolic/kinetic models discussed here is still an open problem.

Similarly, the impact of various symmetries on model dynamics has been mainly investigated for nonlocal hyperbolic models [27–30]. As seen above in our discussion on the symmetries of Boltzmann and Fokker-Planck models, many other models in the literature do exhibit similar $O(2)$ and $SO(2)$ symmetry, which impacts the types of patterns one expects to see [3]. However, this investigation is still an open problem in the context of the models for collective behaviours in biology. Equally an open problem is the understanding of the similar-looking patterns displayed by some deterministic hyperbolic models with symmetry and the corresponding stochastic models without symmetry (see Fig. 5.26).

The impact of this lack of results on the understanding of the bifurcating dynamics of the models summarised in this monograph will become more evident in the next Chapter, as Table 9.1 will show that only the nonlocal hyperbolic systems discussed in Chap. 5 have been observed to exhibit a large variety of spatial and spatio-temporal patterns (some of which were identified through the rigorous investigation of the solution branches bifurcating near codim-1 and codim-2 points). However, it is expected that many other hyperbolic and kinetic classes of models could exhibit equally interesting spatial and spatio-temporal patterns and bifurcations. The identification of these potential patterns can only be done by combining analytical approaches with intensive numerical simulations (which still needs to be performed for the majority of models discussed throughout this monograph).

References

1. W. Holmes, Bull. Math. Biol. **76**(1), 157 (2014)
2. R. Hoyle, *Pattern Formation. An Introduction to Methods* (Cambridge University Press, Cambridge, 2006)
3. M. Golubitsky, I. Stewart, *The Symmetry Perspective: From Equilibrium to Chaos in Phase Space and Physical Space* (Birkhäuser, Basel, 2002)
4. P. Chossat, R. Lauterbach, *Methods in Equivariant Bifurcations and Dynamical Systems* (World Scientific Publishing, Singapore, 2000)
5. M. Haragus, G. Iooss, *Local Bifurcations, Center Manifolds, and Normal Forms in Infinite-Dimensional Dynamical Systems* (Springer, London, 2010)
6. Y. Kuznetsov, *Elements of Applied Bifurcation Theory*, 2nd edn. (Springer, New York, 2000)
7. S. Strogatz, *Nonlinear Dynamics and Chaos* (Westview Press, Boulder, 1994)
8. J.D. Murray, *Mathematical Biology* (Springer, New York, 1989)
9. L. Perko, *Differential Equations and Dynamical Systems* (Springer, New York, 2000)
10. C. Chicone, *Ordinary Differential Equations with Applications* (Springer, New York, 1999)
11. R. Eftimie, G. de Vries, M.A. Lewis, Proc. Natl. Acad. Sci. USA **104**(17), 6974 (2007)
12. T. Kolokolnikov, M. Ward, J. Wei, Discr. Contin. Dyn. Syst. Ser. B **19**(5), 1373 (2014)
13. T. Kolokolnikov, W. Sun, M. Ward, J. Wei, SIAM J. Appl. Dyn. Syst. **5**(2), 313 (2006)
14. R. Eftimie, G. de Vries, M.A. Lewis, F. Lutscher, Bull. Math. Biol. **69**(5), 1537 (2007)
15. F. Lutscher, J. Math. Biol. **45**, 234 (2002)
16. H. Poincaré, Acta Math. **7**, 259 (1885)
17. M. Golubitsky, I. Stewart, D.G. Schaeffer, *Singularities and Groups in Bifurcation Theory. Volume II* (Springer, New York, 1988)
18. R. Seydel, *Practical Bifurcation and Stability Analysis* (Springer, New York, 2009)
19. J.C. Robinson, *Infinite-Dimensional Dynamical Systems* (Cambridge University Press, Cambridge, 2001)
20. L. Evans, *Partial Differential Equations* (American Mathematical Society, Providence, 1997)
21. H. Knobloch, B. Aulbach, in *Equadiff5, Proceedings of the Fifth G. Teubner Verlagsgesellschaft*, ed. by M. Greguš (Teubner, Leipzig, 1982), pp. 179–189
22. E. Altshuler, O. Ramos, Y.N. ez, J. Fernández, A. Batista-Leyva, C. Noda, Am. Nat. **166**(6), 643 (2005)
23. G. Li, D. Huan, B. Roehner, Y. Xu, L. Zeng, Z. Di, Z. Han, PLoS One **9**(12), e114517 (2014)
24. Y.K. Chung, C.C. Lin, PLoS One **12**(3), e0173642 (2017)
25. Q. Ji, C. Xin, S. Tang, J. Huang, Phys. A Stat. Mech. Appl. **492**, 941 (2018)
26. N. Zabzina, A. Dussutour, R. Mann, D. Sumpter, S. Nicolis, PLoS Comput. Biol. **10**(12), e1003960 (2014)
27. P.L. Buono, R. Eftimie, Math. Models Methods Appl. Sci. **24**(2), 327–357 (2014)
28. P.L. Buono, R. Eftimie, SIAM J. Appl. Dyn. Sys. **13**(4), 1542 (2014)
29. P.L. Buono, R. Eftimie, J. Math. Biol. **71**(4), 847 (2014)
30. P.L. Buono, R. Eftimie, *Mathematical Sciences with Multidisciplinary Applications*. Springer Proceedings in Mathematics & Statistics, vol. 157 (Springer, Cham, 2016), pp. 29–59
31. J. Massot, R. Bacis, J. Math. Phys. **17**, 1392 (1976)
32. M. Makai, Transp. Theory Stat. Phys. **15**(3), 249 (1984)
33. A. Bobylev, G. Caraffini, G. Spiga, J. Math. Phys. **37**(6), 2787 (1996)
34. S. Takata, J. Stat. Phys. **136**(4), 751 (2009)
35. Y. Grigoriev, S. Meleshko, N. Ibragimov, V. Kovalev, *Symmetries of Integro-Differential Equations: With Applications in Mechanics and Plasma Physics* (Springer, Dordrecht, 2010)
36. O. Ilyin, Theor. Math. Phys. **186**(2), 183 (2016)
37. A.W.H. Mochaki, J.M. Manale, On Modified Symmetries for the Boltzmann Equation. Proceedings **2**, 7 (2018)
38. I. An, S. Chen, H.Y. Guo, Phys. A Stat. Mech. Appl. **128**(3), 520 (1984)
39. C. Sastri, K. Dunn, J. Math. Phys. **26**, 3042 (1985)

40. P. Rudra, J. Phys. A Math. Gen. **23**(10), 1663 (1990)
41. R. Kozlov, J. Eng. Math. **82**(1), 39 (2013)
42. D. Métivier, Kinetic models, from Kuramoto to Vlasov: bifurcations and experimental analysis of a magneto-optical trap, Université Côte d'Azur (2017) (English)
43. C. Kubrusly, Bull. Belg. Math. Soc. Simon Stevin **15**(1), 153 (2008)
44. R. Eftimie, G. de Vries, M. Lewis, J. Math. Biol. **59**, 37 (2009)
45. I. Kmit, L. Recke, J. Math. Anal. Appl. **335**, 355 (2007)
46. A. Vanderbauwhede, *Lyapunov–Schmidt Method for Dynamical Systems* (Springer, New York, 2011), pp. 937–952
47. A. Bressan, D. Serre, M. Williams, K. Zumbrun, *Hyperbolic Systems of Balance Laws* (Springer, Berlin, 2007)
48. A. Mielke, J. Differ. Equ. **65**, 68 (1986)
49. A. Mielke, Math. Meth. Appl. Sci. **10**, 51 (1988)
50. A. Vanderbauwhede, in *Dynamics in Infinite Dimensional Systems*, ed. by S.N. Chow, J. Hale (Springer, Berlin , 1987), pp. 409–420
51. A. Vanderbauwhede, G. Iooss, in *Dynamics Reported*, vol. 1, ed. by C. Jones, U. Kirchgraber, H. Walter (Springer, Berlin, 1992), pp. 125–163
52. M. Lichtner, Proc. Am. Math. Soc. **136**(6), 2091 (2008)
53. M. Renardy, Z. Angew. Math. Phys. **45**(6), 854 (1994)
54. M. Renardy, Proc. R. Soc. Edin. Sect. A **122**(3–4), 363 (1992)
55. M. Lichtner, M. Radziunas, L. Recke, Math. Methods Appl. Sci. **30**, 931 (2007)
56. W. Liu, M. Oh, in *Infinite Dimensional Dynamical Systems*, ed. by J. Mallet-Paret, J. Wu, H. Zhu (Springer, New York, 2013), pp. 169–183
57. L. Arnold, P. Boxler, *Diffusion Processes and Related Problems in Analysis, Volume II. Progress in Probability*, vol. 27 (Birkhäuser, Boston, 1992), pp. 241–255
58. D. Blömker, *Amplitude Equations for Stochastic Partial Differential Equations* (World Scientific Publishing, Singapore, 2007)
59. L. Arnold, *Random Dynamical Systems* (Springer, Berlin, 1998)
60. H. Crauel, P. Imkeller, M. Steinkamp, *Stochastic Dynamics* (Springer, New York, 1999), pp. 27–47
61. C. Kuehn, Physica D **240**(12), 1020 (2011)
62. K. Lika, T. Hallam, J. Math. Biol. **38**, 346 (1999)
63. C.M. Topaz, A.L. Bertozzi, M.A. Lewis, Bull. Math. Bio. **68**, 1601 (2006)
64. K. Fellner, G. Raoul, Math. Comput. Model. **53**, 1436 (2011)
65. K. Fellner, G. Raoul, Math. Models Methods Appl. Sci. **20**, 2267 (2010)
66. G. Raoul, Differ. Integr. Equ. **25**(5/6), 417 (2012)
67. D. Balagué, J. Carrillo, T. Laurent, G. Raoul, Phys. D Nonlinear Phenom. **260**, 5 (2013)
68. P.H. Chavanis, Phys. A Stat. Mech. Appl. **387**, 5716 (2008)
69. F. Lutscher, A. Stevens, J. Nonlinear Sci. **12**, 619 (2002)

Chapter 9
Discussion and Further Open Problems

9.1 Summary

For the past 30 years, hyperbolic and kinetic models have been used to investigate the growth, movement and self-organisation of cells, animals, and even human pedestrians. One of the main reasons for employing such models—as opposed to the classical parabolic models—is their finite propagation speed that makes them more biologically realistic. Another reason is that these models do seem to exhibit a richer pattern dynamics compared to the parabolic models.

In this study, we reviewed some of the local and nonlocal hyperbolic and kinetics models derived to investigate various biological aggregations and traffic-like movement. We presented models that investigated the movement and aggregation of various bacteria (e.g., *Myxobacteria*, *Escherichia Coli*), cells (e.g., tumour cells), and animal populations (e.g., flocks of birds or herds of ungulates). Moreover, since traffic-like collective movement is such a common behaviour in biology (from intracellular transport, to ant traffic and pedestrian self-organised movement [1–3]), we also discussed some traffic flow models. (Although it may seem unusual to discuss car-traffic models in a monograph focused on biological phenomena, we chose to do so at the beginning of Chap. 3 since the same car-traffic models have been applied to describe pedestrian traffic and collective behaviours, and further generalised to biological traffic, such as cellular, bacterial and ants traffic.)

Our goal was not to provide a very comprehensive review of these hyperbolic and kinetic models. Rather, we wanted to present the complexity of the biological and mathematical problems, and to summarise the patterns exhibited by the models. Moreover, since this study is intended for researchers not familiar with these types of models (and the analytical and numerical approaches derived to investigate them), we took a step-by-step approach to present a clearer view of the motivations and the costs associated with increased model complexity. We started with the simplest one-dimensional models described by advection or advection-reaction equations. Then,

© Springer Nature Switzerland AG 2018

R. Eftimie, *Hyperbolic and Kinetic Models for Self-organised Biological Aggregations*, Lecture Notes in Mathematics 2232,
https://doi.org/10.1007/978-3-030-02586-1_9

we investigated systems of local and non-local hyperbolic models that have constant or density-dependent speeds and turning rates. Finally, we discussed a couple of kinetic models in higher dimensions, and their hydrodynamic limits (and in some cases their parabolic limits). By choosing this structural approach to review the hyperbolic and kinetic models, we were able to highlight the contribution of these models to the investigation of group patterns in various communities of organisms. We also stressed the difficulties that these complex models are confronted with, such as the absence of analytical approaches to investigate some of the resulting spatial and spatio-temporal patterns, and/or the absence of numerical approaches to illustrate the behaviour of the more complex kinetic models.

The mathematical models reviewed in this study were mainly deterministic (even if they were shown to exhibit also chaotic dynamics; see, for example, Fig. 5.23). Even so, the kinetic models contained a stochastic component in the velocity, since the turning events were usually governed by Poisson processes. However, recent studies started to focus on models that incorporate stochasticity in an explicit manner, either additively or multiplicatively, and thus we reviewed a few such models at the end of Chaps. 5 and 6. Another way to incorporate stochasticity into the models was to start with the Langevin equations for the motion of particles (where external noise was added explicitly to particles' velocity), and then derive the corresponding stochastic kinetic and hyperbolic equations [4–6]. Note that this Langevin approach (as well as the stochastic Ornstein-Uhlenbeck processes [7]) lead to distribution functions that satisfied equations which were hyperbolic with respect to the space variable and parabolic with respect to the velocity variable. For this reason, we chose not to describe them here in more detail.

The majority of models discussed here incorporated spatial dynamics. The few exceptions were: (i) the kinetic models for active particles, where the population could also change over an "activity" space; and (ii) the age-structured models, that could describe the formation of human pairs [8], predator-prey dynamics [9], tumour growth [10], or the epidemic spread of diseases [11] in age-structured populations.

9.2   Biological Relevance of Models' Assumptions and Generated Numerical Patterns

The numerical investigation of the patterns exhibited by the kinetic and hyperbolic models reviewed in this study allows for a visual comparison of these theoretical models with the patterns observed in nature, with the final goal of determining whether the assumptions incorporated into the models can explain the reality. In regard to *animal behaviours*, these assumptions usually refer to: (i) the necessity of having repulsive-attractive-alignment interactions versus only alignment interactions, or only attractive-repulsive interactions; (ii) the nature of spatial interactions as determined by the various nonlocal kernels; (iii) the nature of inter-individual communication mechanisms. In regard to *cell behaviours*, these assumptions usually refer to (i') the strength of cell-cell adhesion versus the strength of cell-matrix adhesion; (ii') the appropriate incorporation of specific signalling pathways that control cell macroscopic behaviours (e.g., movement, turning).

The mathematical models reviewed here were shown to exhibit a large variety of spatial and spatio-temporal patterns. Many of these patterns can be connected to empirically observed animal group behaviours: zigzagging flocks of birds, rippling behaviours observed in Myxobacteria colonies, travelling pulses and stationary pulses corresponding to moving (e.g., travelling schools of fish) and resting aggregations, respectively. The expanding and contracting group behaviours that characterise the breather patterns have been observed for example in flocks of birds [12]. Vortices or mills (i.e., rotating groups) are group patterns observed quite frequently in schools of fish (e.g., barracuda, bluefin tuna, or sharks [13, 14]) or groups of ants [15]. The formation of shock waves (i.e., gradient blow-up patterns) is a well known pedestrian behaviour observed during mass events which result in panic stampede [16]. The travelling trains could describe the propagation of density waves through the aggregation (a behaviour observed in schools of herring [17]).

While many of the mathematical patterns can be traced back to empirical observations, there are also some biologically unrealistic patterns, such as the density blow-ups. These patterns are mathematical artefacts caused by the particular assumptions incorporated into the models (e.g., interaction kernels discontinuous at the origin [18], or nonlinear production of an external signal [19]). When such blow-up solutions occur, it is an indication that the model is no longer appropriate to describe the behaviour of the biological system. We note here that these solutions were displayed by both local and nonlocal hyperbolic models (see also Table 9.1 for a summary of the patterns discussed here).

In regard to inter-individual communication, it was shown that some patterns seem to be connected with specific communication mechanisms. For example, mechanism M5—describing interactions with neighbours moving towards the reference individual—seems to generate the observed ripples in Myxobacteria colonies (irrespective of how this mechanism is incorporated into local or nonlocal continuum models [20–22], or into individual-based models [23]). However, the most common pattern observed in every biological aggregation, namely the stationary pulses, is associated with the majority of communication mechanisms (see also Table 5.2). Moreover, when multiple communication mechanisms are used in a combined manner by one group of individuals, it can lead to behaviours (patterns) not predicted by the use of one communication mechanisms, including chaotic behaviours; see the discussion in Sect. 5.6.

A recent review by Bellomo and Dogbé [24] discussed the derivation and use of empirical data to validate models for traffic and crowds dynamics. Available data usually refers to the speed and movement direction of cars and pedestrians [25–27], as well as cells [28], bacteria [29, 30], fish [31], birds [32] or ungulates [33]. However, the incorporation of this data into kinetic and hyperbolic models is still an open research area, with very few studies combining modelling with data analysis.

Another open research area is related to the translation of results obtained with animal crowd models to the understanding of human behaviours [34, 35].

When ethical concerns do not allow to experiment with human subjects, one could focus on non-human subjects (e.g., ants, mice, sheep) to gain some understanding on collective human behaviour (and movement) under specific conditions (e.g., extreme escape from various built environments). However, as recently discussed in [36], there is the need to have a more systematic connection between animal and human experiments, to be able to understand better the context in which reliable inferences can be drawn from experiments with non-human crowds.

9.3 Directions for Future Research

We conclude the discussion of pattern formation in this monograph by summarising some possible directions of future research. To this end, we focus on modelling, numerical and analytical aspects.

Modelling Multiscale models have been developed intensively over the past years in the context of cell dynamics, to connect macroscopic processes related to cell movement and turning behaviours, to microscopic processes that occur inside cells and control cell movement/turning. The majority of kinetic models in the literature consider simplifications of the molecular-level processes (i.e., cell signalling pathways) involved in cell movement and turning. In the future it is expected that more detailed signalling pathways will be incorporated into the multiscale models for collective cell movement (e.g., the Erk/MAPK pathway, the JNK pathway or the p38 signalling pathway that all have roles in cell migration [37]). Moreover, not many multiscale models have been developed in the context of ecological collective movement (among the very few we mentioned for human crowds [38, 39]). It is expected that in the future, research in animal communication, animal psychology and physiology will be combined with mathematical modelling of animal movement, to increase our understanding regarding the collective behaviour of animals [40].

Another aspect related to modelling that will develop further in the next years is the incorporation of stochastic events in these hyperbolic/kinetic models. More and more studies recognise the importance of environmental and demographic stochasticity in animal/cell movement [41–43]. Until now the majority of models for the collective movement of cells/bacteria/animals that incorporated stochasticity have been of discrete type, with stochasticity affecting the individual level (see the IBMs discussed briefly in Chap. 1). However, we expect that the upcoming decades will see a significant increase in the development of stochastic transport models for animal/cell dynamics, where noise will have an impact at the population level. This will lead to the further development of analytical and numerical methods to investigate the patterns generated by these new models.

Numerical Investigation of Patterns One of the most interesting (and most difficult to investigate) aspects of patterns formation focuses on connecting the observed biological patterns to specific mathematical and biological mechanisms. Intensive numerical investigations could provide some understanding of the biological mechanisms behind specific patterns. We mentioned before the connection between the communication mechanism M5 and ripples [20, 21, 44, 45]. Stationary pulses, on the other hand, are one of the most common patterns, being observed in almost every model discussed here (see Tables 5.2 and 9.1, and the majority of models discussed throughout this monograph). Hence, we cannot associate this pattern with a specific biological mechanism. However, since many of the models presented here have not been the subject of very thorough numerical and analytical investigations, it is possible that they could exhibit even more (possible exotic) patterns. The discovery of new spatial and spatio-temporal patterns (especially for multi-dimensional nonlocal kinetic and hyperbolic models) requires the development of fast numerical schemes, to be able to run multiple simulations that would span large parameter spaces. This is particularly relevant for the multi-dimensional (nonlocal) kinetic models. Intensive simulations are also required for the numerical investigation of the bifurcation dynamics of these models, and the tracking of various solution branches that can bifurcate at specific points in the parameter space. In particular, new continuation algorithms need to be developed to take into account the characteristics of these local/nonlocal hyperbolic and kinetic models. One first step was recently made in [46], where the authors described a continuation algorithm that considers the symmetry structure of the nonlocal hyperbolic models presented in Chap. 5; see Eqs. (5.14), (5.18) and (5.19).

Analytical Investigation of Patterns While numerical simulations can offer some insight into the mechanisms behind these patterns, analytical investigations (using, for example, existence results, linear and nonlinear stability, bifurcation and symmetry theory) are necessary to: (i) explain the role of model parameters on the formation (or not) of the patterns, (ii) reduce the size of the parameter space where we look for specific patterns, (iii) rigorously identify and classify all patterns that could be exhibited by a mathematical model, (iv) decide whether the model is biologically realistic (e.g., exhibits finite or blow-up patterns, and if so in which biologically realistic/unrealistic parameter spaces?). As discussed throughout this review, many hyperbolic and kinetic models have not been subjected to detailed analytical investigation of pattern formation (and this could explain the lack of patterns in columns 2, 3, 4 and 6 of Table 9.1). Moreover, apart form classifying these patterns based on their symmetry subgroups (as discussed in Chap. 8), it is unclear how else one could classify them. This classification approach raises another question: how to classify the (similarly-looking) patterns generated by the corresponding stochastic models (see Fig. 5.26), which have lost the initial symmetry. It is likely that further analytical investigations would reveal that these complex hyperbolic and kinetic models might generate new patterns and bifurcations, whose investigation could be very challenging.

Table 9.1 Summary of some of the patterns exhibited by the hyperbolic and kinetic models presented throughout this monograph

Spatio-temporal patterns	Local 1-eq. models (1D)	Non-local 1-eq. models (1D)	hyperbolic systems (1D)	Nonlocal hyperbolic systems (1D)	2D (or 3D) models (mesoscopic, macroscopic)
Stationary pulse		✓	✓	✓	✓
Travelling pulse	✓			✓	✓
Travelling front	✓			✓	
Travelling train				✓	
Ripples (standing waves)			✓	✓	
Feather				✓	
Travelling feather				✓	
Breather				✓	
Travelling breather				✓	
Zigzags (Vortices (2D))				✓	✓
Semi-zigzags (stop and go waves)				✓	
Density blow-up				✓	✓
Shocks	✓	✓			✓
Lanes					✓
Chaos				✓	

The list is not very comprehensive, since we focused mainly on nonlocal (1D) hyperbolic systems for which detailed numerical and analytical investigations have been performed to identify and classify the exhibited patterns

Combining analytical and numerical approaches will lead to the further development of the area of pattern formation in nonlocal kinetic and hyperbolic models. Some of the questions that could be answered in the future are:

- Can the patterns observed in nonlocal models (especially the more exotic ones, such as the feathers, breathers, zigzags) be exhibited also by the local models? If not, why?
- Are the complex (exotic) patterns discussed throughout this monograph specific only to the hyperbolic/kinetic models? In other words, if we develop parabolic or individual-based models that incorporate communication mechanisms similar to the ones discussed in nonlocal hyperbolic systems, can we obtain similar patterns?
- Can the complex 1D patterns exhibited by the 1D nonlocal hyperbolic and kinetic models be generalised to 2D models? If so, what are the mathematical/biological assumptions that need to be incorporated into these 2D models, to generate the appropriate patterns?
- How could one incorporate various 2D communication mechanisms into the existent mesoscopic and macroscopic models for the collective behaviours of cells/bacteria/animals? In this case, could the corresponding 2D patterns be associated with particular communication mechanisms (or combination of mechanisms)?
- How do we connect the 1D and 2D models for collective spatial movement of cells/bacteria/animals to the available data? What kind of data is necessary to be collected to parametrise these models, to allow for quantitative predictions?
- Could the assumption of "pairwise interactions", which is incorporated into the Boltzmann-type kinetic models, impede our understanding of the contribution of other particles/cells to these interactions? This is a valid question since the in vivo dynamics of cells is not always determined by pairwise interactions, but by interactions with a variety of other cells via communication molecules (cytokines, chemokines) produced by these cells. Similarly, it is less likely that animals in group interact with their neighbours via "binary collisions", and is more likely that these interactions involve more than two individuals (if the community comprises multiple individuals). Moreover the derivation of the Boltzmann equation in the limit $N \to \infty$ also requires that collisions involve only uncorrelated particles, in the sense that particles that have collided already will not collide again. This assumption does not seem to carry great biological realism, since animals in a group will likely interact again.
- Can we understand the bifurcation structure of the stochastic PDE models for the collective movement of cells/bacteria/animals? How can we extend the current stochastic bifurcation theory [47–50] (mainly developed for ODEs) to the nonlocal and local transport models discussed in this study (as well as many more other models in the literature)? How can we adapt the (dynamical) D-bifurcation and (phenomenological) P-bifurcation theory to the realities of the chaotic and deterministic patterns generated by the nonlocal hyperbolic models presented briefly in Chap. 5?

To conclude, we remark that the use of hyperbolic and kinetic models to answer biological questions is far from having reached its full potential. On the contrary, they seem to be used more and more to investigate various problems in ecology and medicine. Moreover, in the last few years these models have been applied to new research areas, such as social dynamics [51], economy [52] or human psychology [53]. Furthermore, the authors in [54, 55] suggested that these kinetic models could be the start of a biological mathematical theory for complex systems. In particular, Bellomo and Forni [55] argued that these models can incorporate two of the most important aspects of living matter: the notion of function or purpose for biological organisms, and the multi-scale aspect of biological interactions. While some first steps have been taken in this direction, we note that these are very complex aspects that require further extensive investigations.

References

1. D. Chowdhury, A. Schadschneider, N. Katsuhiro, Phys. Life Rev. **2**(4), 318 (2005)
2. A. Schadschneider, D. Chowdhury, K. Nishinari, *Stochastic Transport in Complex Systems. From Molecules to Vehicles* (Elsevier, Amsterdam, 2011)
3. T. Vicsek, A. Zafeiris, Phys. Rep. **517**(3–4), 71 (2010)
4. K. Hadeler, T. Hillen, F. Lutscher, Math. Models Methods Appl. Sci. **14**(10), 1561 (2004)
5. P.H. Chavanis, Phys. A Stat. Mech. Appl. **387**, 5716 (2008)
6. P.H. Chavanis, Commun. Nonlinear Sci. Numer. Simul. **15**, 60 (2010)
7. P. Degond, S. Motsch, J. Stat. Phys. **131**, 989 (2008)
8. K. Hadeler, Acta Appl. Math. **14**, 91 (1989)
9. S. Busenberg, M. Iannelli, J. Math. Biol. **22**, 145 (1985)
10. M. Gyllenberg, G. Webb, J. Math. Biol. **28**, 671 (1990)
11. H. Inaba, J. Math. Biol. **28**, 411 (1990)
12. H. Pomeroy, F. Heppner, Auk **109**, 256 (1992)
13. I.D. Couzin, J. Krause, R. James, G. Ruxton, N.R. Franks, J. Theor. Biol. **218**, 1 (2002)
14. S. Wilson, Fish. Oceanogr. **13**(4), 283 (2004)
15. T. Schneirla, Am. Mus. Novit. **1253**, 1 (1944)
16. D. Helbing, A. Johansson, H.Z. Al-Abideen, Phys. Rev. E **75**, 046109 (2007)
17. B. Axelsen, T. Anker-Nilssen, P. Fossum, C. Kvamme, L. Nøttestad, Can. J. Zool. **79**, 1586 (2001)
18. R. Fetecau, R. Eftimie, J. Math. Biol. **61**(4), 545 (2010)
19. T. Hillen, A. Stevens, Nonlinear Anal. Real World Appl. **1**, 409 (2000)
20. R. Eftimie, G. de Vries, M.A. Lewis, Proc. Natl. Acad. Sci. USA **104**(17), 6974 (2007)
21. O.A. Igoshin, R. Welch, D. Kaiser, G. Oster, Proc. Natl. Acad. Sci. USA **101**, 4256 (2004)
22. O.A. Igoshin, G. Oster, Math. Biosci. **188**, 221 (2004)
23. U. Börner, A. Deutsch, M. Bär, Phys. Biol. **3**, 138 (2006)
24. N. Bellomo, C. Dogbé, SIAM Rev. **53**, 409 (2011)
25. I. Prigogine, R. Herman, *Kinetic Theory of Vehicular Traffic* (Elsevier, New York, 1971)
26. S. Buchmueller, U. Weidman, Parameters for pedestrians, pedestrian-traffic and walking facilities. Technical Report 132, ETH, Zürich (2006)
27. Y. Liu, C. Sun, Y. Bie, Math. Probl. Eng. **2015**, Article ID 308261, 6pp. (2015)
28. T. Yang, J.S. Park, Y. Choi, W. Choi, T.W. Ko, K. Lee, PLoS One **6**(6), e20255 (2011)
29. C. Qian, C. Wong, S. Swarup, K.H. Chiam, Appl. Environ. Microbiol. **79**(15), 4734 (2013)
30. A. Patterson, A. Gopinath, M. Goulian, P. Arratia, Sci. Rep. **5**, 15761 (2015)

31. V. Mwaffo, S. Butail, M. di Bernardo, M. Porfiri, Zebrafish **12**(3), 250 (2015)
32. M. Ballerini, N. Cabibbo, R. Candelier, A. Cavagna, E. Cisbani, I. Giardina, V. Lecomte, A. Orlandi, G. Parisi, A. Procaccini, M. Viale, V. Zdravkovic, Proc. Natl. Acad. Sci. USA **105**(5), 1232 (2008)
33. J. Killeen, H. Thurfjell, S. Ciuti, D. Paton, M. Musiani, M. Boyce, Mov. Ecol. **2**(1), 15 (2014)
34. N. Shiwakoti, M. Sarvi, M. Burd, Saf. Sci. **66**, 1 (2014)
35. Z. Shahhoseini, M. Sarvi, PLoS One **12**(8), e0182913 (2017)
36. M. Haghani, M. Sarvi, Transp. Res. B **107**, 253 (2018)
37. C. Huang, K. Jacobson, M. Schaller, J. Cell Sci. **117**, 4619 (2004)
38. E. Cristiani, B. Piccoli, A. Tosin, Multiscale Model. Simul. **9**(1), 155 (2011)
39. B. Piccoli, A. Tosin, Arch. Ration. Mech. Anal. **199**, 707 (2011)
40. A. Sieben, J. Schumann, A. Seyfried, PLoS One **12**(6), e0177328 (2017)
41. P. Smouse, S. Focardi, P. Moorcroft, J. Kie, J. Forester, J. Morales, Philos. Trans. R Soc. Lond. B Biol. Sci. **365**(1550), 2201 (2010)
42. M. Bonsall, A. Hastings, J. Anim. Ecol. **73**(6), 1043 (2004)
43. G. Gellner, K. McCann, A. Hastings, Theor. Ecol. **9**(4), 477 (2016)
44. F. Lutscher, A. Stevens, J. Nonlinear Sci. **12**, 619 (2002)
45. U. Börner, A. Deutsch, H. Reichenbach, M. Bär, Phys. Rev. Lett. **89**, 078101 (2002)
46. P.L. Buono, R. Eftimie, M. Kovacic, L. van Veen, in *Active Particles*, vol. 2, ed. by N. Bellomo, P. Degond, E. Tadmor (Birkhäuser, Basel, 2018)
47. L. Arnold, *Random Dynamical Systems* (Springer, Berlin, 1998)
48. L. Arnold, in *IUTAM Symposium on Nonlinearity and Stochastic Structural Dynamics*, ed. by S. Narayanan, R.N. Iyengar (Springer, Dordrecht, 2001), pp. 15–27
49. I. Chueshov, *Monotone Random Systems - Theory and Applications* (Springer, Berlin, 2002)
50. M. Rasmussen, *Attractivity and Bifurcation for Nonautonomous Dynamical Systems* (Springer, Berlin, 2007)
51. M. Bertotti, M. Delitala, Nonlinear Anal. Real World Appl. **9**, 183 (2008)
52. G.A. Marsan, N. Bellomo, M. Egidi, Kinet. Relat. Model. **1**, 249 (2008)
53. B. Carbonaro, C. Giordano, Math. Comput. Model. **41**, 587 (2005)
54. N. Bellomo, M. Delitala, Phys. Life Rev. **5**, 183 (2008)
55. N. Bellomo, G. Forni, Curr. Top. Dev. Biol. **81**, 485 (2008)

Index

active particles, 154, 168
Adams-Bashforth scheme, 160
admissibility condition, 48, 50, 51
admissible solution, 52
age-structured, 74, 97
alignment, 18, 22, 108, 109, 162
amplitude blow-up, 28, 30
amplitude equation, 239, 252
asymmetric perception, 231
asymptotic preserving scheme, 213, 214
attraction, 18, 22, 108, 109, 141
avoidance, 131, 134

bifurcation theory, 24, 227, 233
blow-up, 28, 89, 112, 113, 141, 166, 178, 182, 270
Boltzman, 182
Boltzmann equation, 211, 216
Boltzmann model, 183, 208
Boltzmann operator, 182, 184, 187
boundary conditions, 86, 120, 200, 221, 254
bounded solution, 141
bounded variation, 43
breathers, 13, 28, 29, 120, 270

car traffic, 24, 55
Cauchy problem, 43, 45
cell signalling, 20
central difference scheme, 95, 98
central-upwind scheme, 93
centre manifold, 229, 256
centre manifold reduction, 255

Centre Manifold Theorem, 255, 257
centres, 229
centre subspace, 255
chaotic, 28, 30, 129, 270
characteristic lines, 39, 46, 48
circle group, 242
classical solution, 40
codimension-1, 121, 140, 234
codimension-2, 121, 125, 140, 234
collapsed swarm, 9
communication, 107
communication mechanisms, 20, 108, 118, 129, 131, 138, 162, 169, 230, 237, 267
congested traffic, 64
conjugacy, 243
conjugate isotropy subgroups, 243, 253
conservation laws, 39, 40
contact discontinuity, 47, 63, 67
correlated random walk, 115, 170
Courant number, 198
cross-adhesion, 108
cross-advection, 96
Cucker-Smale model, 176, 183
cyclic group, 241

D-bifurcation, 261, 271
dihedral group, 241
dimensional splitting, 159
Dirac function, 178
Dirac mass, 173
directional perception, 22
dispersion relation, 90, 124, 139, 232, 248

© Springer Nature Switzerland AG 2018
R. Eftimie, *Hyperbolic and Kinetic Models for Self-organised Biological Aggregations*, Lecture Notes in Mathematics 2232,
https://doi.org/10.1007/978-3-030-02586-1

LECTURE NOTES IN MATHEMATICS  Springer

Editors in Chief: J.-M. Morel, B. Teissier;

Editorial Policy

1. Lecture Notes aim to report new developments in all areas of mathematics and their applications – quickly, informally and at a high level. Mathematical texts analysing new developments in modelling and numerical simulation are welcome.

   Manuscripts should be reasonably self-contained and rounded off. Thus they may, and often will, present not only results of the author but also related work by other people. They may be based on specialised lecture courses. Furthermore, the manuscripts should provide sufficient motivation, examples and applications. This clearly distinguishes Lecture Notes from journal articles or technical reports which normally are very concise. Articles intended for a journal but too long to be accepted by most journals, usually do not have this "lecture notes" character. For similar reasons it is unusual for doctoral theses to be accepted for the Lecture Notes series, though habilitation theses may be appropriate.

2. Besides monographs, multi-author manuscripts resulting from SUMMER SCHOOLS or similar INTENSIVE COURSES are welcome, provided their objective was held to present an active mathematical topic to an audience at the beginning or intermediate graduate level (a list of participants should be provided).

   The resulting manuscript should not be just a collection of course notes, but should require advance planning and coordination among the main lecturers. The subject matter should dictate the structure of the book. This structure should be motivated and explained in a scientific introduction, and the notation, references, index and formulation of results should be, if possible, unified by the editors. Each contribution should have an abstract and an introduction referring to the other contributions. In other words, more preparatory work must go into a multi-authored volume than simply assembling a disparate collection of papers, communicated at the event.

3. Manuscripts should be submitted either online at www.editorialmanager.com/lnm to Springer's mathematics editorial in Heidelberg, or electronically to one of the series editors. Authors should be aware that incomplete or insufficiently close-to-final manuscripts almost always result in longer refereeing times and nevertheless unclear referees' recommendations, making further refereeing of a final draft necessary. The strict minimum amount of material that will be considered should include a detailed outline describing the planned contents of each chapter, a bibliography and several sample chapters. Parallel submission of a manuscript to another publisher while under consideration for LNM is not acceptable and can lead to rejection.

4. In general, **monographs** will be sent out to at least 2 external referees for evaluation.

   A final decision to publish can be made only on the basis of the complete manuscript, however a refereeing process leading to a preliminary decision can be based on a pre-final or incomplete manuscript.

   Volume Editors of **multi-author works** are expected to arrange for the refereeing, to the usual scientific standards, of the individual contributions. If the resulting reports can be forwarded to the LNM Editorial Board, this is very helpful. If no reports are forwarded or if other questions remain unclear in respect of homogeneity etc, the series editors may wish to consult external referees for an overall evaluation of the volume.

5. Manuscripts should in general be submitted in English. Final manuscripts should contain at least 100 pages of mathematical text and should always include

   – a table of contents;
   – an informative introduction, with adequate motivation and perhaps some historical remarks: it should be accessible to a reader not intimately familiar with the topic treated;
   – a subject index: as a rule this is genuinely helpful for the reader.
   – For evaluation purposes, manuscripts should be submitted as pdf files.

6. Careful preparation of the manuscripts will help keep production time short besides ensuring satisfactory appearance of the finished book in print and online. After acceptance of the manuscript authors will be asked to prepare the final LaTeX source files (see LaTeX templates online: https://www.springer.com/gb/authors-editors/book-authors-editors/manuscriptpreparation/5636) plus the corresponding pdf- or zipped ps-file. The LaTeX source files are essential for producing the full-text online version of the book, see http://link.springer.com/bookseries/304 for the existing online volumes of LNM). The technical production of a Lecture Notes volume takes approximately 12 weeks. Additional instructions, if necessary, are available on request from lnm@springer.com.

7. Authors receive a total of 30 free copies of their volume and free access to their book on SpringerLink, but no royalties. They are entitled to a discount of 33.3 % on the price of Springer books purchased for their personal use, if ordering directly from Springer.

8. Commitment to publish is made by a *Publishing Agreement*; contributing authors of multiauthor books are requested to sign a *Consent to Publish form*. Springer-Verlag registers the copyright for each volume. Authors are free to reuse material contained in their LNM volumes in later publications: a brief written (or e-mail) request for formal permission is sufficient.

Addresses:
Professor Jean-Michel Morel, CMLA, École Normale Supérieure de Cachan, France
E-mail: moreljeanmichel@gmail.com

Professor Bernard Teissier, Equipe Géométrie et Dynamique,
Institut de Mathématiques de Jussieu – Paris Rive Gauche, Paris, France
E-mail: bernard.teissier@imj-prg.fr

Springer: Ute McCrory, Mathematics, Heidelberg, Germany,
E-mail: lnm@springer.com

Printed in the United States
By Bookmasters